Dynamic Simulations
of
Multibody Systems

Springer
*New York
Berlin
Heidelberg
Barcelona
Hong Kong
London
Milan
Paris
Singapore
Tokyo*

Murilo G. Coutinho

Dynamic Simulations of Multibody Systems

With 148 Illustrations

 Springer

Murilo G. Coutinho
1745 Selby Avenue, #11
Los Angeles, CA 90024
USA
AnimationEngine@aol.com

Library of Congress Cataloging-in-Publication Data
Coutinho, Murilo G. (Murilo Gondim)
   Dynamic simulations of multibody systems / Murilo G. Coutinho.
     p. cm.
   Includes bibliographical references and index.
   ISBN 0-387-95192-X (alk. paper)
    1. Computer simulation.   2. Dynamics, Rigid.   I. Title.
QA76.9.C65 C695 2001
003'.3—dc21   2001018409

Printed on acid-free paper.

© 2001 Springer-Verlag New York, Inc.
All rights reserved. This work may not be translated or copied in whole or in part without the written permission of the publisher (Springer-Verlag New York, Inc., 175 Fifth Avenue, New York, NY 10010, USA), except for brief excerpts in connection with reviews or scholarly analysis. Use in connection with any form of information storage and retrieval, electronic adaptation, computer software, or by similar or dissimilar methodology now known or hereafter developed is forbidden. The use of general descriptive names, trade names, trademarks, etc., in this publication, even if the former are not especially identified, is not to be taken as a sign that such names, as understood by the Trade Marks and Merchandise Marks Act, may accordingly be used freely by anyone.

Production managed by Jenny Wolkowicki; manufacturing supervised by Jeffrey Taub.
Photocomposed copy prepared from the author's LaTeX files.
Printed and bound by Edwards Brothers, Inc., Ann Arbor, MI.
Printed in the United States of America.

9 8 7 6 5 4 3 2 1

ISBN 0-387-95192-X   SPIN 10789999

Springer-Verlag   New York   Berlin   Heidelberg
A member of BertelsmannSpringer Science+Business Media GmbH

To my girls
Izabella and Leticia

Preface

Physically based modeling is increasingly gaining acceptance within the computer graphics and mechanical engineering industries as a way of achieving realistic animations and accurate simulations of complex systems. Such complex systems are usually hard to animate using scripts, and difficult to analyze using conventional mechanics theory, which makes them perfect candidates for physically based modeling and simulation techniques.

The field of physically based modeling is broad. It includes everything from modeling a ball rolling on the floor, to a car engine working, to a hanging shirt being moved by a gust of wind. The theory varies from precise mathematical methods to purpose-specific approximated solutions that are mathematically incorrect, but produce realistic animations for the particular situation being considered. Depending on the case, an approximated solution might serve the purpose, however, there are times when approximations are not admissible, and the use of accurate simulation engines is a requirement. Developing and implementing physically based dynamic-simulation engines that are robust is difficult. The main reason is that it requires a breadth of knowledge in a diverse set of subjects, each of them standing alone as a broad and complex topic.

Instead of attempting to address all types of simulation engines available in the broad area of physically based modeling, this book provides in-depth coverage of the most common simulation engines. These simulation engines restrict the general case of physically based modeling to the particular case wherein the objects interacting are either particles or rigid bodies.

This book is a comprehensive introduction to the techniques needed to produce realistic simulations and animations of particle and rigid-body systems. It focuses on both the theoretical and practical aspects of developing and implementing physically based dynamic-simulation engines that can be used to generate convincing animations of physical events involving particles and rigid bodies, such as a jet flow of water pushing dry flowers away on a patio. It can also be used to produce accurate simulations of mechanical systems, such as a robotic parts feeder where parts are dropped on a conveyor belt and then positioned and aligned as they hit fences strategically placed on the conveyor and used to align the parts at a specific orientation.

Dynamic Simulations of Multibody Systems was written for computer graphics, computer animation, computer-aided mechanical design and modeling software developers who want to learn to incorporate physically based dynamic-simulation features into their own systems. The goal of this book is to make the principles and methods of physically based modeling of particle and rigid-body systems accessible to a broader audience of software developers who are familiar with mainstream computer-graphics techniques, and the associated mathematics.

The book is organized into three main topics: particle systems, rigid-body systems, and articulated rigid-body systems. The first chapter is an overview of how all techniques covered in this book fit together as independent modules constituting a simulation engine. The following chapters and appendices go into more detailed explanations for each technique. The techniques developed can be used to create simulation engines capable of combining particles, rigid bodies and articulated rigid bodies into a single system. Each chapter presents many algorithms and covers them in considerable depth, yet makes their design and analysis accessible to all levels of readers. We have tried to keep explanations elementary without sacrificing depth or mathematical rigor.

The most complex mathematical algorithms are described in detail in the appendices. Our goal here is to focus the reader's attention to the details of the topic being covered, and not be distracted by mathematical issues that can be viewed as "black box" modules having specific functionality (such as a numerical integrator or a rigid-body-mass-properties computation module). Readers should be able to develop their own software implementation of a simulation engine using the techniques covered in-depth in this book, or shorten their software development effort by taking advantage of the several resources available on the Web. Links to several Web sites with commercial and non-commercial software, as well as pointers to all cited references electronically available, can be found at the Web site supporting this book at **http://hometown.aol.com/animationengine**.

Acknowledgments

I wish to thank my wife, Izabella, who supported me from the start in writing this book during my nonexistent spare time, and for not complaining too much about the annoying light emanating from my desk throughout the endless nights of writing. I also wish to thank our daughter Leticia, who was born when I was half way through the book, for being such an incredible baby. I love you both.

As for detailed technical readings of the book, I wish to thank David Remba for commenting on several chapters and appendices. I also wish to thank Thiago Coutinho, Antonio Diegues and Eduardo Campos for preparing an innovative book cover during the pre-production phase.

Lastly, I wish to thank my parents, Murilo and Marilia, for their support, encouragement and friendship throughout my life.

Murilo Coutinho
Los Angeles, CA
April 2001

Contents

Preface   vii

1 Computational Dynamics   1
  1.1 Introduction 1
  1.2 Particle and Rigid-Body Systems 2
  1.3 Dynamic Simulation Engines 3
  1.4 A Computationally Efficient Implementation 7
      1.4.1 Interface with the Rendering Engine 7
      1.4.2 Moving the Objects 10
      1.4.3 Checking for Collisions 11
      1.4.4 Responding to Collisions 12
  1.5 Guide to Readers 14

2 Hierarchical Representation of 3D Polyhedra   17
  2.1 Introduction 17
  2.2 Hierarchical Representation of Objects 19
      2.2.1 Axis-Aligned Bounding Boxes 20
      2.2.2 Oriented Bounding Boxes 22
      2.2.3 Bounding Spheres 25
      2.2.4 Convex Hull 27
  2.3 Hierarchical Representation of the Simulated World 32
      2.3.1 Uniform Grid 33
      2.3.2 Multi-Level Grid 37

xii   Contents

    2.4 Collision Detection Between Different Hierarchical Representations 42
        2.4.1 Computing Box-Box Intersections 42
        2.4.2 Computing Sphere-Sphere Intersections 45
        2.4.3 Computing Triangle-Triangle Intersections 46
        2.4.4 Point-in-Triangle Test 49
        2.4.5 Computing Box-Sphere Intersections 51
        2.4.6 Computing Box-Triangle Intersections 52
        2.4.7 Computing Sphere-Triangle Intersections 53
        2.4.8 Computing Line Segment-Sphere Intersections . . . 54
        2.4.9 Computing Line Segment-Triangle Intersections . . . 55
    2.5 Notes and Comments 57

3 Particle Systems   **61**
    3.1 Introduction 61
    3.2 Particle Dynamics 63
    3.3 Interaction Forces 68
        3.3.1 Gravity 69
        3.3.2 Viscous Drag 69
        3.3.3 Damped Springs 70
        3.3.4 Spatially Dependent Forces 71
        3.3.5 User Interaction 76
    3.4 Collision Detection: Overview 76
        3.4.1 Particle-Particle Collision 78
        3.4.2 Particle-Rigid Body Collision 81
    3.5 Collision-Detection Implementation 86
        3.5.1 Computing Cylinder-Cylinder Intersections 87
        3.5.2 Computing Cylinder-Box Intersections 90
        3.5.3 Computing Cylinder-Sphere Intersections 94
        3.5.4 Computing Cylinder-Triangle Intersections 97
        3.5.5 Point-in-Cylinder Test 105
    3.6 Particle-Particle Collision Response 106
        3.6.1 Computing Impulsive Forces for a Single Collision . 107
        3.6.2 Computing Impulsive Forces for Multiple Simultaneous Collisions 115
        3.6.3 Computing Contact Forces for a Single Contact . . . 120
        3.6.4 Computing Contact Forces for Multiple Contacts . . 129
    3.7 Particle-Rigid Body Collision Response 134
        3.7.1 Computing Impulsive Forces 136
        3.7.2 Computing Contact Forces 137
    3.8 Particle Emitter 138
        3.8.1 User-Definable Parameters 140
    3.9 Specialized Particle Systems 142
        3.9.1 User-Adjustable Parameters 143
        3.9.2 Overview of Cloth Simulation 146

			Contents	xiii

    3.10 Notes and Comments 152

4 Rigid-Body Systems   155
    4.1 Introduction 155
    4.2 Rigid-Body Dynamics 156
    4.3 Interaction Forces 165
        4.3.1 Gravity 165
        4.3.2 Viscous Drag 166
        4.3.3 Damped Springs 167
        4.3.4 User-Interaction Forces 167
    4.4 Collision Detection 168
    4.5 Collision Detection between Non-Convex Bodies 172
    4.6 Collision Detection between Convex Bodies 174
    4.7 The Voronoi Clip Algorithm 174
        4.7.1 Feature b_2 is a Vertex 178
        4.7.2 Feature b_2 is an Edge 179
        4.7.3 Feature b_2 is a Face 193
        4.7.4 Dealing with Interpenetration 193
        4.7.5 Avoiding Local Minima 193
        4.7.6 The GJK Algorithm 195
        4.7.7 Termination Condition 203
    4.8 Rigid Body-Rigid Body Collision Response 204
        4.8.1 Computing Impulsive Forces for a Single Collision . 205
        4.8.2 Computing Impulsive Forces for Multiple Collisions 218
        4.8.3 Computing Contact Forces for a Single Contact . . . 225
        4.8.4 Computing Contact Forces for Multiple Contacts . . 234
    4.9 Particle-Rigid Body Contact Revisited 238
    4.10 Notes and Comments 242

5 Articulated Rigid-Body Systems   245
    5.1 Introduction 245
    5.2 Articulated Rigid-Body Dynamics 247
    5.3 Collision Detection 253
    5.4 Collision Response 256
        5.4.1 Computing Impulsive Forces for Single or Multiple
              External Collisions 258
        5.4.2 Computing Contact Forces for Single or Multiple Ex-
              ternal Contacts 260
    5.5 Notes and Comments 262

A Useful 3D Geometric Constructions   265
    A.1 Introduction 265
    A.2 Projection of a Point on a Line 266
    A.3 Projection of a Point on a Plane 266
    A.4 Intersection of a Line Segment and a Plane 267

xiv   Contents

    A.5 Closest Point between a Line and a Line Segment 268
    A.6 Computing the Collision- or Contact-Local Frame from the Collision- or Contact-Normal Vector 270
    A.7 Representing Cross-Products as Matrix-Vector multiplication 271
    A.8 Suggested Readings 272

B Numerical Solution of Ordinary Differential Equations of Motion   273
    B.1 Introduction 273
    B.2 Euler Method 276
        B.2.1 Explicit Euler 276
        B.2.2 Implicit Euler 278
    B.3 Runge-Kutta Method 280
        B.3.1 Second-Order Runge-Kutta Method 280
        B.3.2 Forth-Order Runge-Kutta Method 283
    B.4 Using Adaptive Time-Step Sizes to Speed Computations . . 285
    B.5 Suggested Readings 287

C Quaternions   289
    C.1 Introduction 289
    C.2 Basic Quaternion Operations 290
        C.2.1 Addition 290
        C.2.2 Dot product 290
        C.2.3 Multiplication 290
        C.2.4 Conjugate 291
        C.2.5 Module 292
        C.2.6 Inverse 292
    C.3 Unit Quaternions 292
    C.4 Rotation-Matrix Representation Using Unit Quaternions . . 293
    C.5 Advantages of Using Unit Quaternions 295
    C.6 Suggested Readings 297

D Rigid-Body Mass Properties   299
    D.1 Introduction 299
    D.2 Mirtich's Algorithm 300
        D.2.1 Volume-Integral to Surface-Integral Reduction . . . 303
        D.2.2 Surface-Integral to Projected-Surface-Integral Reduction 304
        D.2.3 Projected-Surface-Integral to Line-Integral Reduction 308
        D.2.4 Computing the Line Integrals from the Vertex Coordinates 309
    D.3 Suggested Readings 310

E Useful Time Derivatives   313
    E.1 Introduction 313

	E.2	Computing the Time Derivative of a Vector Attached to a Rigid Body	313
	E.3	Computing the Time Derivative of a Contact-Normal Vector	314
		E.3.1 Particle-Particle Contact	314
		E.3.2 Rigid Body-Rigid Body Contact	315
	E.4	Computing the Time Derivative of the Tangent Plane	317
	E.5	Computing the Time Derivative of a Rotation Matrix	318
	E.6	Computing the Time Derivative of a Unit Quaternion	320
	E.7	Suggested Readings	323

F Convex Decomposition of 3D Polyhedra — 325

- F.1 Introduction . . . 325
- F.2 Joe's Algorithm . . . 327
  - F.2.1 Determining Candidate Cut Planes . . . 329
  - F.2.2 Computing the Cut Face Associated with a Cut Plane 331
  - F.2.3 Termination Conditions . . . 334
- F.3 Suggested Readings . . . 334

G The Linear-Complementarity Problem — 337

- G.1 Introduction . . . 337
- G.2 Dantzig's Algorithm: The Frictionless Case . . . 339
  - G.2.1 Termination Conditions . . . 347
- G.3 Baraff's Algorithm: Coping with Friction . . . 348
  - G.3.1 Static-Friction Conditions . . . 351
  - G.3.2 Dynamic Friction . . . 361
  - G.3.3 Termination Conditions . . . 362
- G.4 Suggested Readings . . . 363

H Software Implementation — 365

References — 367

Index — 375

1
Computational Dynamics

1.1 Introduction

The quest for realism and precision in computer-graphics simulations of complex systems started decades ago, when engineers realized the importance and cost effectiveness of having reliable computer models for their products. The ability to study the inner workings of a system according to several different scenarios long before the beginning of the manufacture cycle was compelling enough to lead to an impressive amount of work on physically-based simulation and modeling.

The physically based modeling of the interactions between parts in such systems is particularly attractive because they are not limited to a single-domain analysis. On the contrary, the simulations are extremely useful because they can be extended to multiple-domain analysis of the system, such as the combination of thermal and stress-test analysis of the materials used to manufacture the parts, with the forces exerted on them obtained from the mechanical-contact analysis. In this case, the combination can be used to predict the maximum force that can be exerted on each part before it cracks. The set of applications with the potential of benefiting from such work is diverse, ranging from aircraft and automobile design, to structural analysis of buildings, to weather simulations and toxic-plume-spread analyzers, and even to video games.

The challenge in combining multi-domain simulations is that usually each domain of interest being simulated requires the development of specialized mathematical models capable of expressing subtle interactions that match the correct theoretical physical behavior of the system. In many cases, such specialized mathematical models are implemented using different numerical methods that may or may not be compatible with one another. When the numerical methods are compatible, the models can be easily merged and the coupling effects between the different domains can be quickly evaluated. However, there are cases when the numerical methods are incompatible and a direct merge is infeasible. In these cases, the models are usually combined in an interleaved fashion. The interleaved approach consists of solving one method at a time, with their coupling being represented by a set of external forces and constraints that are applied from the system that was just solved to the system that will be solved next. By so doing, each system interacts with the others using its own specialized techniques.

The generation of reliable models for each domain also turns out to be of great interest, since the results that can be drawn from the simulation experiments directly depend on the models used. The models can range from simple first-order approximations for a quick evaluation of the system, to highly complex and accurate models of the theoretical physical behavior, capable of capturing several aspects of the system more realistically. The choice of the model to be used depends on the simulation goals that need to be met, as well as on the computational efficiency required. For example, the dynamic simulation of a wall being blown out by an explosive in a video game does not need to use a highly accurate model of the internal structure of the wall. It suffices to use a simple model that gives a sufficiently accurate feeling of authenticity to the scene. Nevertheless, the same simulation in the context of a miliary operation may need a more accurate model of the wall so that the appropriate weapon can be chosen for the task.

Even though the number of models and specialized mathematical methods that can be used in a physically based simulation is significant, there are two types of multi-domain simulations that represent the most commonly used simulation engines. These simulation engines restrict the general case of physically based modeling to the particular case wherein things interacting are either particles or rigid bodies.

1.2 Particle and Rigid-Body Systems

Arguably, particle and rigid-body systems are the most important and commonly used models in physically based dynamic simulations. They represent a very good trade-off between mathematical complexity and accuracy of the models used to capture the observed real-world behavior of the system.

Particle systems can range from basic implementations of point-mass systems that use discrete particles to represent gaseous or fluid motion[1], to specialized systems that use computational fluid mechanics to simulate turbulent gases and liquids such as swirling steam, gusts of wind, and flooding, to name a few. The former applies standard Newtonian physics to each particle in the system to determine the dynamics of motion, whereas the latter uses sophisticated numerical methods to solve the Navier-Stokes volumetric differential equations of motion.

Rigid-body systems, on the other hand, take into account the shape and mass distribution of the objects being simulated. They are especially suitable to simulate systems where the internal bending, extension or compression of the object can be neglected, meaning, the object does not change its shape during the entire simulation. The rigidity assumption also simplifies the computations, since it makes forces being applied at any point on the object equivalent to a force-torque pair being applied at its center of mass, which can be easily computed.

Rigid bodies can also be combined into systems of articulated rigid bodies, where bodies are attached to each other using joints. There are several types of joints that can be used to connect bodies, and they differ from each other by the degree of freedom of relative motion allowed. An unconstrained rigid body has six degrees of freedom, three degrees to translate and another three degrees to rotate along and about the coordinate axis, respectively.

1.3 Dynamic Simulation Engines

In this section, we shall discuss the general structure of a dynamic-simulation engine for non-penetrating particle and rigid-body systems. Figure 1.1 shows a block diagram representation of the main loop of a dynamic-simulation engine. The engine starts at some time t_0, and advances by some time-step values, executing four high-level steps.

The first step moves all objects from the beginning to the end of the current time step, ignoring any possible collisions that might happen during the movement. This consists of determining the dynamic state of the system at the beginning of the current time step, and using this information to solve the ordinary differential equations (ODEs) of motion for each object. The dynamic state of the system is given by all linear and angular positions, velocities and accelerations, as well as by the net external force-torque pair acting on each object[2] in the system. Since the positions, velocities and

---

[1] In this book, we shall focus the study of particle systems to the case in which particles are approximated by point-mass objects.

[2] In this book, we shall sometimes refer to object as a synonym to particle or rigid body.

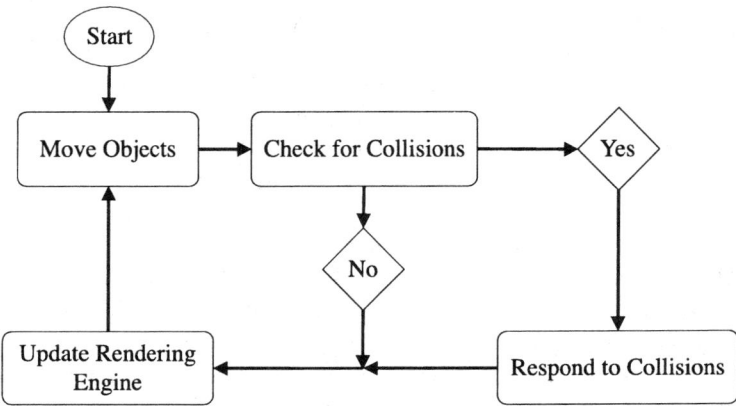

FIGURE 1.1. Block diagram illustrating the general structure of a dynamic-simulation engine.

accelerations at the beginning of the current time step are already known from the previous time step (they are the same as the ones computed at the end of the previous time step, with the exception of the very first time step), the only variables that need to be computed in this task are the net external force-torque pairs. Examples of external forces possibly acting on an object are gravity, contact forces, constraint forces owing to joint attachments if the object is part of an articulated system, and any other external force the environment exerts on the object. The net external force is computed as the combination of each external force acting on the object. If object is a rigid body, each external force is transformed to an external force-torque pair acting on its center of mass, before the combination takes place. Having determined the dynamic state of the system at the beginning of the current time step, we use this information to numerically integrate the differential equations of motion and compute the dynamic state of each object in the system at the end of the current time step.

The second step checks for collisions that may occur between one or more objects during their motion. Collisions are usually detected by checking for geometric intersections between the objects' boundary representations. The collision check can be a very time-consuming task, especially if every object is checked against all others for collisions. In practice, the simulation engine uses auxiliary structures to speed the collision checking.

The first auxiliary structure considered in this book is used to speed the determination of which pairs of objects should be checked for collisions. It consists of some sort of cell decomposition of the simulated world that is guaranteed to contain all objects for the entire simulation. As objects move around the world, the simulation engine keeps track of which of the cells intersect their boundary representations, and checks for collision only between objects that share a common cell.

The second auxiliary structure also covered in this book is used to speed the collision detection between pairs of objects. It consists of decomposing each object in a hierarchical tree of simple structures that can be quickly checked for intersections, such as boxes, spheres or convex polygons. The tree is constructed in a pre-processing step before the simulation engine starts, in such a way that the structure associated with each parent node in the tree contains the structures of all its children. For example, a hierarchical bounding-box representation of an object would consist of one top-level bounding box that contains the entire object, several intermediate levels of possibly overlapping bounding boxes that contain sub-parts of the object, down to leaf bounding boxes that contain one or more faces of the object. The goal is to postpone the more expensive check for collisions between the object's faces as long as possible, substituting them for inexpensive collision checks between their hierarchical representations.

Whenever a pair of objects shares a common cell of the world-cell decomposition, their hierarchical structures are checked for intersections, starting with their top-level structures, and moving down to their intermediate structures until reaching a leaf. Objects that are farther apart, yet in the same cell of the world decomposition, are quickly discarded after checking for intersections between their top-most level representation. Objects that are closer together may require checking for intersections between several intermediate levels of their hierarchical tree representation. Objects that are really close will probably require the more expensive check for collisions between the faces associated with the intersecting leaf nodes. As Figure 1.1 illustrates, if no collisions are detected, then the movement of the objects for the current time step is valid and their new positions and orientations are sent to the rendering engine, which is in charge of updating the display showing the simulation. However, if a collision is detected, the simulation engine moves on to the third step.

The third and most difficult step is to establish the non-penetrating constraints for all collisions and contacts detected in the second step. This consists of responding to collisions that may occur between two or more objects during their motion. Because collisions introduce discontinuities into the accelerations and velocities of the colliding objects, it is necessary to stop the numerical integration of the ODEs of motion just before the collision, resolve the collision to determine the new velocities of the objects just after the collision, and re-start the numerical integration for the remaining time using the updated position and velocity values.

In practice, whenever an intersection is detected, the collision-detection algorithm moves backwards in time to determine an adequate approximation of the exact time just before the collision occurs, that is, just before the colliding objects intersect. Collisions can involve more than two objects, can be coincident in time (i.e., multiple simultaneous collisions), and can be at multiple locations (i.e., several collision points.) The simulation engine keeps track of the most recent simultaneous collisions detected and responds to them as follows.

The collision point, collision normal and relative velocities associated with the most recent collisions detected are computed from the relative geometric displacement of the colliding objects, as well as from their dynamic state just before the collision. This information is passed along to the collision-response algorithm to compute the collision impulses and contact forces, as appropriate. The distinction between contact and collision is usually made by measuring the relative velocity of the colliding objects along the collision normal, just before the collision. If the relative velocity is less than a threshold value, the collision is assumed to be a contact.

Having distinguished between all simultaneous contacts and collisions associated with the most recent collisions detected, the collision-response module computes the collision impulses first, and uses this information to update the relative accelerations and relative velocities of the colliding objects just after the collision. This update may cause some of the contact points to break apart. This happens whenever the relative velocities at contact points after the update become greater than the threshold value. In this case, after the collision impulses are resolved, the relative velocities of the objects at the contact point demonstrate that the objects are now moving away from each other, meaning the contact will break apart and there is no need to take into account the contact force associated with this contact. For the remaining contacts, a contact force is computed to prevent the contacting objects from interpenetrating.

Because resolving collisions requires moving the colliding objects backward in time to place them just before the collision was detected, the simulation engine needs to recompute the dynamic state of the colliding objects for the remaining time until the end of the current time step. This is done by restarting the numerical integration using the new dynamic state of the colliding objects just after the collision. The final position and orientation of the colliding objects is then updated to reflect the changes introduced by the collision. Because these changes affect the path of the objects during time remaining, the simulation engine needs to check again for new collisions. This loop continues until there are no collisions within the current time step.

The forth and final step consists of communicating the final position and orientation of each object in the system to the rendering engine being used. Here, we assume the actual display of the simulation is carried out by the rendering engine, which communicates with the simulation engine through a well defined interface.

1.4 A Computationally Efficient Implementation

Even though the general structure of a dynamic simulation engine is relatively easy to understand, its implementation is usually an involved task. Despite the fact that a naive implementation works, its operation can be frustrating and disappointing, even for a small number of simple rigid bodies and particles being simulated. Naive implementations carry out unnecessary time-consuming computations, since they tend to ignore coherence between time steps, as well as the spatial distribution of the rigid bodies and particles in the scene when computing their interactions. This can easily distort the consumption of important computer resources such as memory and CPU run-time. As it happens, the use of computationally efficient algorithms in dynamic-simulation engines is a necessity than rather a luxury.

In this section, we discuss in detail a possible architectural design of a computationally efficient implementation of a dynamic-simulation engine for non-penetrating particle and rigid-body systems. In doing so, we are mostly interested in studying algorithms that can produce real-time or near real-time performance. The actual description of the algorithms used in each step of the simulation engine proposed in this book, as well as reference software implementations of several modules that make up the entire simulation, are provided in the remaining chapters and appendices. In the particular case of software implementations, we are supporting a Web site with links to the author's own software, as well as other software packages developed by the computer graphics and mechanical engineering industries. The main goal of this section is to justify the book's organization, and describe the high-level steps necessary to implement an efficient dynamic-simulation engine.

1.4.1 Interface with the Rendering Engine

The portability of the simulation engine across a multitude of computer platforms is the first important issue that needs to be addressed. Usually, the results of a dynamic simulation are rendered in a computer display using one of the several sophisticated rendering engines available in the market. Such rendering engines have an internal representation of the scene being rendered, which contains lots of information that is also necessary for the dynamic-simulation engine. At first, the temptation to use the same internal representation of the rendering engine is almost irresistible, since a considerable amount of source-code development can be forgone. However, the fact that there exists a diverse set of internal representations, each tailored to capitalize upon the underlying hardware capabilities, makes it critical to have our own internal representation of objects in the scene that is independent of the specifics of the rendering engine being used. The way this can be done in a dynamic-simulation engine, without sacrificing portability, is to create an interface between the rendering engine and the simulation engine. The number of methods implemented by the interface should be kept small, to avoid unnecessary interactions.

In an efficient implementation, the interface between the rendering and simulation engines should contain at least three basic features. The first feature lets the rendering engine register objects with the simulation engine, and update their status as appropriate. The registration process may consist of passing to the simulation engine some basic information about the object being registered, and receiving back a handle to the internal representation of the object in the simulation engine. This handle (a unique identification number) is used to map the internal representation of objects in the simulation engine to their counterparts in the rendering engine.

The basic information the rendering engine passes to the simulation engine when registering an object should include a list of vertices and faces defining its boundary representation[3]. Upon registering an object, the simulation engine computes its extended representation, which includes the following information in addition to the information already obtained from the rendering engine:

1. Edge list.

2. Face neighbor list.

3. Face normal list.

4. Convex decomposition of the object.

5. Hierarchical representation of the object.

All this information is computed only once, when an object is registered, with respect to the object's local-coordinate system. This information is then used in several modules throughout the simulation engine to speed computations and optimize the overall efficacy, as the object moves within the global-coordinate system.

At the end of each simulation time step, the simulation engine returns to the rendering engine a list of object handles and their updated positions and orientations. The object handles can be used by the rendering engine to quickly get a pointer to the internal representations of the objects, and apply the necessary transformations to position and orient them in the scene. Not all objects registered with the simulation engine are on this list, just those that changed position or orientation since the last time step. As far as efficacy is concerned, it is very important to implement a fast mechanism to retrieve the object structure from its handle (such as a hash table) in both rendering and simulation engines.

---

[3] In this book, we assume the object's geometry is defined by its boundary representation, that is, by its vertices, edges and faces.

The rendering engine should also be capable of updating the status of objects registered with the simulation engine. This consists of carrying out operations ranging from removing the object from the simulation engine, to changing its current position, orientation and velocities, to adjusting its simulation status. Possible values for the simulation status are *inactive*, *static* and *dynamic*. By default, all registered objects can be initially set to the inactive state. In this state, even though the objects exist in the simulation engine, they are ignored during the run-time execution. Static objects, on the other hand, are taken into account during the simulation execution, but are considered fixed objects in the scene. Only dynamic objects have their positions and orientations determined using the physically based computations of the dynamic-simulation engine.

The distinction between static and dynamic objects makes it possible to interface script-based motion with the simulation engine. This can be done by setting the dynamic state of all objects that have a predefined script-based motion to static, and updating their positions between each simulation frame following the script. We just need to be careful when updating the position of each object following the script so as not to overlap any other object. The dynamic-simulation engine enforces the non-penetration constraints on all registered objects for every time step, and if the script-based object is forced to move on top of another object, then the simulation engine will not work properly and will yield an abnormal result for the subsequent simulation time steps. We can overcome this problem by using a kinematically constrained motion based on the scripted position and orientation of the object. By kinematically constrained, we mean that the linear and angular positions, velocities and accelerations of the bodies are obtained from an animation system, possibly by interpolating their values between two consecutive animation frames

The second feature to be implemented in the interface between the rendering and simulation engines lets the rendering engine specify the dimensions of the scene to which the dynamic-simulation engine is constrained. The goal is to impose bounds on the distance an object can move inside the simulation engine. Objects that fall outside these bounds have their status automatically set to inactive, and are left out of the simulation execution for the rest of the simulation.

In order to simplify things, the dimensions of the scene can be given by the coordinates of a bounding box containing the entire simulated world. Once the bounding box is defined, the simulation engine decomposes it into subregions (or cells) that are used to speed the collision-detection phase. The decomposition can be single- or multi-level. In the single-level case, a coarse uniform grid is constructed from the bounding box that contains the entire simulated world. The size of each cell in the grid is determined as a combination of the size of the objects being simulated. In the multi-level case, several uniform grids with cells of different sizes are constructed, forming a coarse-to-fine decomposition of the simulated world. Objects are then assigned to the grids with cells that can completely contain them. The techniques to decompose the simulated world will be explained in more detail in Section 2.3.1.

Last, but not least, the third desired feature of the interface between the rendering and simulation engines should let the rendering engine specify the *sampling time* to be used, that is, the amount of time the dynamic-simulation engine needs to be executed between two consecutive frames. Usually, the sampling time is set to be the inverse of the desired frame rate, so that the dynamic simulation returns the state of the system after each frame, and the rendering engine can update the computer display accordingly. The state of the system can be returned as a list of objects that moved since the last sampling time, with their new positions and orientations given by a translation vector and a rotation matrix, respectively.

It is important to notice that the actual time step used in the simulation engine to numerically integrate the differential equations of motion of each dynamic object may be different from the sampling time. This is most often the case when the numerical method uses adaptive time steps to automatically adjust the current time step used, depending on the value of the estimated integration error.

1.4.2 Moving the Objects

As soon as the simulation engine starts, the first step to be executed consists of moving all dynamic objects from the beginning to the end of the current sampling time, ignoring any collisions that may happen during the movement (see Figure 1.1). For each dynamic object in the system, the simulation engine undertakes the following actions to complete this step.

1. Computes the net force acting on the object. The net force is obtained from the addition of all external forces the environment exerts on the object, such as gravity, contact forces and joint forces, if applicable. If the object is a rigid body or an articulated rigid body, then each external force is transformed to an external force-torque pair acting on its center of mass before they are added to form one net force-torque pair.

2. Numerically integrates the ODEs of motion associated with the object, assuming no collisions occur during the entire movement.

3. Adjusts the cell decomposition of the simulated world to account for the object's movement. This step consists of removing the object from cells it no longer intersects and adding it to the new intersecting cells.

When executing this loop, the simulation engine uses the world-cell decomposition to construct a list of potential collisions. This list contains pairs of objects that are within the same cell or group of cells, and therefore may be colliding with each other during their movements.

1.4.3 Checking for Collisions

The collision check is undertaken only for the pairs of objects that are in the list of potential collisions. Here, the simulation engine builds one global-collision list for the entire world, and one local-collision list for each object involved in a collision. The global-collision list contains information about all collisions, sorted by increasing order of collision time. The local-collision list contains information about the most recent collision associated with its corresponding object. Notice that we should keep a list, as opposed to a single reference to the most recent collision, in order to account for any multiple simultaneous collisions the object might undergo. As new collisions are detected, both the global list and the local references of the objects involved in the collisions are dynamically updated to reflect the changes in the collision times.

For each pair of objects in the list of potential collisions, the simulation engine goes through the following steps before removing the pair from the list.

4. Checks whether the objects' hierarchical representations intersect. The pair of objects is discarded if no intersections are found.

5. If an intersection is detected, the approximate collision time is determined by regressing in time the colliding objects' motion to just before the collision. If either of the colliding objects is already involved in another collision, then the simulation engine needs to check whether the new collision time is smaller than the one just detected, and if so needs to update both global and local lists of collisions to reflect the changes. For example, assume the current pair of objects obtained from the list of potential collisions is (B_1, B_2), and that a new collision C_i was detected between them at time t_i. The simulation engine will then update the local lists of collisions for each object, as follows.

    - Checks whether B_1 is already involved in another collision C_j that occurs *before* C_i. In this case, collision C_i should be discarded, since it may no longer exist after collision C_j is processed. Otherwise, proceed to the next step.
    - Updates B_1's local list of collisions to account for the new collision C_i. There are three possible situations to be considered. If this is the first time object B_1 is involved in a collision, then the simulation engine just adds C_i to B_1's local list of collisions. However, if B_1 was already involved in a collision that happens to have occured simultaneously with collision C_i, then B_1 has multiple simultaneous collisions at time t_i, and collision C_i is also added to B_1's local list of collisions. Finally, if B_1 was already involved in a collision occurring after collision C_i, then the

simulation engine deletes B_1's local list of collisions, and resets it to include C_i only (i.e., resets it to the most recent collision involving B_1.)

The steps above are repeated to B_2. At the end, the list of potential collisions is empty, and the global list of collisions may or may not have entries. If no collisions were detected, then the dynamic state of the system computed in Section 1.4.2 is correct, and the simulation engine returns to the rendering engine the new positions and orientations of the objects that were moved during the current sampling time. Otherwise, the simulation engine has a global list of all collisions detected, ordered by their increasing collision times. Each entry in this list is a list itself reflecting the cases when we have simultaneous collisions between two or more objects. Each individual collision has references to both colliding objects. Also, each object has a reference back to the entry in the global list of collisions that contains the most recent collision it underwent. This flexibility is necessary to properly respond to collisions, since the simulation engine needs to quickly move back and forth between collision structures and the objects involved with them.

1.4.4 Responding to Collisions

The collision response processes only the single or multiple collisions associated with the most recent collision time in the global list of collisions. Here, the simulation engine creates yet another auxiliary list that contains the collisions that are in fact contacts. This distinction is necessary because collisions should be resolved before contacts are resolved. In other words, only after the dynamic state of the system is updated with the collision impulses can we compute the contact forces that will enforce the non-penetration constraints between objects.

The simulation engine executes the following steps for all collisions associated with the most recent collision time in the global list of collisions.

6. Moves all colliding objects backward in time to the moment just before their collisions.

7. Checks whether the collisions are in fact contacts. This check is carried out for each collision and consists of testing whether the module of the relative velocity of the colliding objects at the collision point, along the collision normal, is less than a threshold value. If this is the case, then the collision is said to be a contact and is moved from the global list of collisions to the global list of contacts.

8. Computes the collision impulses associated with each collision. As we shall see later in the chapters to come, this step consists of solving a sparse linear system.

9. Updates the dynamic state of the system with the collision impulses computed in the previous step. This will update the linear and angular velocities of all colliding objects.

10. Checks whether the contacts, if any, are still valid. This check is carried out for each contact associated with an object that was also involved in a collision. This consists of testing one more time whether the module of the relative velocity along the contact normal at the contact point is less than the threshold value. Since the dynamic state of the colliding objects has changed, the contacts that involve any of them will have their relative velocity at the contact point changed as well. In some cases, the contact may beak apart just after the collision impulses are applied, making it necessary to remove them from the global list of contacts.

11. Computes the contact forces at each contact point that prevent the objects from interpenetrating. The contact-force computation involves solving a Linear Complementarity Problem (LCP) that is obtained from the current contact configuration of the system. The general LCP problem formulation, as well as efficient techniques to solve such systems, are addressed in details in Appendix G.

12. Adds the contact forces (and associated torques) to the net external force-torque pair acting on the object. This will enforce the non penetration constraints for the remaining time necessary to reach the end of this sampling time.

After resolving all collisions and contacts, the simulation engine computes the remaining time to reach the end of the current sampling time, and numerically integrates the ODEs of motion *only* for the objects involved in a collision or contact. Again, the simulation engine will move these objects ignoring any collisions that may occur during the movement. Objects that remain in contact for the rest of the current sampling time will not interpenetrate owing to the updated net force and torque computed in step 12.

The final position and orientation associated with each colliding object is updated to reflect the effects of the collision or contact, and a new list of potential collisions is created after the cell subdivision is updated. The simulation engine then loops back to step 4 and continues checking and responding for collisions until it reaches the end of the current sampling time.

At the end of the current sampling time, the simulation engine sends to the rendering engine the list of objects that had their positions or orientations changed since the last update. It then continues moving the objects, checking and responding to collisions, and updating the display until it receives a stop command from the rendering engine.

14   1. Computational Dynamics

1.5   Guide to Readers

The book's organization is justified according to the structure of the efficient dynamic-simulation engine presented in Section 1.4, and the techniques needed to implement it. The book contains five chapters and eight appendices. Each chapter presents many algorithms and covers them in considerable depth, yet makes their design and analysis accessible to all levels of readers, keeping the explanations as elementary as possible without sacrificing depth of coverage or mathematical rigor. The more complex mathematical algorithms and associated implementations are described in details in the appendices. The goal of doing so is to focus the reader's attention to the details of the topic being covered, and not distract him with mathematical issues that can be viewed as "black box" modules that have specific functionality, such as a numerical integrator or a rigid-body mass-properties-computation module.

Chapter 1 introduces the computational-dynamics topic to readers, describing the general structure of a dynamic-simulation engine for non penetrating particle and rigid-body systems. It sets the stage for the remaining chapters of the book by explaining what it takes to design and implement a computationally efficient simulation engine. The following chapters and appendices address the specialized tools and techniques mentioned in this chapter.

Chapter 2 focuses on the problem of computing a hierarchical representation of the geometric description of each simulated object, as well as the simulated world. This representation is used to speed collision-detection checks by taking advantage of the geometric arrangement of the objects in the simulated world, such that collision tests are only carried out on objects that are "close enough" to collide. The hierarchical decomposition of the colliding objects is used to prune unnecessary intersection tests and quickly specify the collision points, or discard the collision if no intersections are found.

Chapter 3 covers the design and implementation of particle systems as a collection of point mass objects that can collide with each other and other rigid-body objects in the simulation. Even though this is one of the simplest models of particle systems that can be used, the computational efficiency and degree of realism that can be attained with these systems is highly attractive. This chapter also discusses in details the use of spring-mass systems to model cloth, and how to implement such systems using point-mass particles connected by a combination of strategically placed springs.

Chapter 4 presents the theoretical and practical aspects of designing and implementing dynamic-simulation engines for rigid-body systems. In this chapter, special attention is given to one of the most difficult and least understood topics in physically based modeling, namely, the computational techniques needed for determining all impulsive and contact forces between bodies with multiple simultaneous collisions and contacts.

Chapter 5 extends the techniques for rigid bodies to include articulated rigid bodies. Here, we shall focus on linking rigid bodies with spherical joints. The goal is to demonstrate and implement techniques that can be used to dynamically simulate articulated rigid bodies. These techniques can be easily applied to include other types of joints suitable to the reader's interests.

The remaining part of the book is devoted to a set of appendices describing the mathematical algorithms used throughout the entire book, each of these standing alone as a broad and complex topic in itself. The appendices focus on the tools being used in the simulation engine. Nonetheless, they provide pointers to the literature, wherein interested readers can get more information about the topic.

Appendix A briefly covers some of the geometric constructions used as building blocks to implement the several intersection tests that are part of the particle-particle, particle-rigid body and rigid body-rigid body collision-detection algorithms. It also discusses how the tangent plane of a collision or contact can be determined given the collision or contact point and normal vector.

Appendix B discusses some of the most common methods used to integrate the differential equations of motion in dynamic simulations. These methods range from simple explicit-Euler, to more sophisticated Runge-Kutta methods, with adaptive time step sizing.

Appendix C presents an alternate representation of rotation matrices using quaternions. This representation is extremely useful in reducing rounding-error problems found when combining rotation matrices. Also, the interpolation between two quaternions representing the orientation of an object is easier than using rotation matrices. This is especially useful when tracing back in time the object's motion to determine the instant just before a collision.

Appendix D shows an efficient algorithm to compute the mass properties of 3D polyhedra. The mass properties include the total volume, total mass, center of mass, and inertia tensor. These quantities are used in the physically based modeling of the dynamics and interactions of objects in the simulated world.

Appendix E presents a detailed description of how the time derivatives of a normal vector, a rotation matrix and a quaternion are computed. These time derivatives are extensively used in Chapters 4 and 5 to describe the dynamics a rigid-body system.

Appendix F addresses the technical barriers to using non-convex polyhedra in a dynamic simulation. Most interesting objects to be simulated are usually non-convex. However, most of the algorithms presented in this book are especially tailored for convex objects. Therefore, it is often necessary to preprocess all objects in the simulation with a convex decomposition module that decomposes the objects into a set of non-overlapping convex parts. The algorithms can then be applied to the convex parts of each object.

Appendix G presents the Linear Complementarity Problem (LCP) in the context of impulse and contact-force computation of multiple simultaneous collisions. This appendix also presents an extension of the original algorithm to cope with static and dynamic friction at the collision or contact point.

Finally, Appendix H discusses architecture principles and design goals for developing dynamic-simulation engines. It also includes information about the Web site supporting this book, which contains pointers to several software packages available for commercial and non-commercial use. The Web site's URL is:

http://hometown.aol.com/animationengine

Readers should be able to use the code (or parts thereof) available from the Web site in their own systems, or develop their own software implementation from scratch using the techniques covered in depth in this book.

2
Hierarchical Representation of 3D Polyhedra

2.1 Introduction

Collision detection is undoubtly the most time-consuming step in a dynamic-simulation engine. In theory, as the simulation evolves, every object needs to be checked for collisions against all other objects in the simulation. Whenever a collision is detected, the simulation engine needs to trace back in time to the instant before the collision, and determine the collision point and collision normal from the relative geometric displacement of the colliding objects.

Usually, collisions are checked by looking for geometric intersections between the objects. When objects are given by their boundary representations, this check can be done by looking for geometric intersections between the primitives of each object, that is, between the polygonal faces defining the boundary of each object. Clearly, checking for collisions between objects this way is a laborious task, and the use of intermediate representations to speed collision checking is critical to achieve real-time performance, especially for simulations involving several hundred objects, each described by several thousand primitives.

In this chapter, we study the use of hierarchical representations to speed the collision-detection phase. Our aim is to compute in a preprocessing stage the hierarchical volumetric decomposition of each object with respect to its local-coordinate system. This usually consists of a tree hierarchy of bounding volumes where the top-most bounding volume bounds the entire object, the intermediate nodes of the tree bound sub-parts of the volume bounded by their parent, and the leaf nodes of the tree bound one or more primitives that lie inside the bounding volume of their parent. Collision checks are then carried out using the objects' hierarchical representations to quickly determine that the objects do not intersect (i.e., are not colliding), or to reduce the number of pair-wise primitive intersection tests needed to check for collision. For example, if the top-most bounding volumes of each object do not intersect, then we can safely conclude that the objects are not colliding. However, if the top-most bounding volumes do intersect, then we have to move down one level in the tree hierarchy to check whether their children intersect. If not, then the objects are not colliding. Otherwise, we move down one more level in the tree hierarchy to the children of the intersecting parents. This process continues until we either reach the leaf nodes of the trees, or detect that the objects do not intersect. Should we reach the leaf nodes of the trees, the collision check proceeds by computing the pair-wise primitive intersections of the primitives bounded by each intersecting leaf.

In practice, there are two important points to consider when using hierarchical representations in a simulation engine. The first is that the intersection test of the bounding volumes must be considerably faster than the intersection test of the primitives. Otherwise, the collision check will take longer using the hierarchical representation than using the original objects' boundary representations. Therefore, our choice of bounding volumes is restricted to simple geometric shapes such as boxes and spheres, which can be quickly tested for intersection against each other. The primitives can also be restricted to convex polygons, or even triangles, to further speed the primitive-primitive intersection tests.

The second point addresses how the hierarchical representation is updated as the object translates and rotates with respect to the world-coordinate system. As mentioned before, the hierarchical decomposition is computed with respect to the object's local-coordinate system. However, all intersection tests should be carried out with respect to the world-coordinate system, thus requiring a coordinate transformation from the object's local-coordinate system to its position and orientation in the world-coordinate system. One solution to this problem would be to transform the entire tree hierarchy of all objects to the world-coordinate system, just before testing for intersection. The drawback of so doing is the substantial waste of time transforming entire tree hierarchies that have only their top-most, or even some of their internal, nodes checked for intersection. All other internal nodes that were transformed but not used in the intersection tests were unnecessary transformed to the world-coordinate system, and the time spent applying the transformation could have been saved. The idea is then to transform only what is absolutely necessary. We can do this as follows.

The simulation engine represents each object in the world-coordinate system by its top-most bounding volume only, and keeps the entire tree hierarchy, as well as the object's boundary representation, in the object's local-coordinate system. At each time step, *only* the top-most bounding volume of each object is moved in the world-coordinate system. The movement consists of updating the position and orientation of the object according to the numerical method being used[1], applying it to the top-most bounding volume of the object.

The collision-detection phase then checks for geometric intersections between the top-most bounding volumes that can potentially collide. Potentially colliding objects are determined from the world-cell decomposition structure, as explained in Section 2.3. Whenever the top-most bounding volumes intersect, the simulation engine transforms only their next-level children from their local-coordinate system to the world-coordinate system. If their next-level children do not intersect, the objects are not colliding and no further transformations are required. Otherwise, the simulation engine keeps transforming only the next-level children of the intersecting bounding volumes until it concludes that the objects are not intersecting, or have intersecting leaf nodes. In that case, each primitive associated with the intersecting leaf node is then transformed from its local-coordinate system to the world-coordinate system before the more expensive primitive-primitive test is carried out. Using this scheme, the simulation engine is guaranteed to transform only the parts of the objects that are absolutely necessary for the collision check, thereby saving substantially on execution time.

Another interesting observation about this scheme is that, because the simulation engine presented in this book is decoupled from the rendering engine, it does not need to position and orient the objects' primitives in the world-coordinate system throughout the simulation. After each time step, the simulation engine just needs to position and orient the top-most bounding volume of each object. Of course, it also needs to advise the rendering engine of the new positions and orientations of the objects that moved since the last simulation time step, so that the rendering engine can itself place and render the objects' primitives at the correct position and orientation. Therefore, as far as the simulation engine is concerned, the cost of moving an object containing several thousand faces, and that does not intersect any other objects in the scene, is the same as moving the top-most bounding volume associated with the object. This in turn reduces even more the simulation engine's execution time.

2.2 Hierarchical Representation of Objects

The hierarchical representations considered in this book are limited to the case when the bounding volumes are either boxes or spheres. Moreover, the object's primitives are assumed to be triangles. This assumption is

---

[1]This is discussed in detail in Appendix B.

used not only to speed the primitive-primitive intersection tests, as discussed in Section 2.4, but also to simplify building Oriented Bounding Boxes (OBB) trees, as explained in Section 2.2.2. The Axis-Aligned Bounding Boxes (AABB) and the Bounding Spheres (BS) representations are not affected by this assumption. These representations are covered in detail in Sections 2.2.1 and 2.2.3, respectively.

In general, it is not clear which hierarchical representation is best, since collision detection is highly dependent on the relative displacement of the objects being considered. For example, if the objects are close enough to each other, the OBB representation usually function better than the others, in the sense that it considerably reduces the number of primitive-primitive intersection tests owing to its tight fit. On the other hand, if the objects are farther apart, the less expensive bounding volume intersection tests of the AABB and BS representations offer a better choice of hierarchical representation.

Independent of the hierarchical representation used, the tree hierarchy can be constructed in a top-down or bottom-up fashion. In the top-down case, the object's primitives are initially assigned to the top-most bounding volume, which in turn is recursively decomposed into sub-volumes according to some partitioning rule until there is only one primitive or group of primitives assigned to each sub-volume. In the latter case, the sub-division ends with the group of primitives if and only if they can no longer be subdivided according to the partitioning rule. Examples of the top-down approach will be discussed in Sections 2.2.1 and 2.2.2. In the bottom-up case, the primitives are individually assigned to an initial bounding volume. These bounding volumes are then merged according to some merging rule, until there is only one top-most bounding volume in the tree containing all primitives.

There are several techniques to partition (or merge) bounding volumes into (or from) two or more sub-volumes to form a tree hierarchy. Examples of such partitions are the binary tree (parent has two children), the quadtree (parent has four children) and the oct-tree (parent has eight children.) In this book we limit our analysis to the most common case of building binary tree hierarchies using the top-down approach. Section 2.5 has references to the literature wherein the other possible partitions are covered in detail.

2.2.1 Axis-Aligned Bounding Boxes

In the Axis-Aligned Bounding Box (AABB) representation, the tree hierarchy is constructed from boxes bounding the primitives associated with them, such that the boxes' axes are aligned to the axis of the object's local-coordinate system. Figure 2.1 illustrates the top-down construction of a binary AABB tree for a simple 2D object.

2.2 Hierarchical Representation of Objects

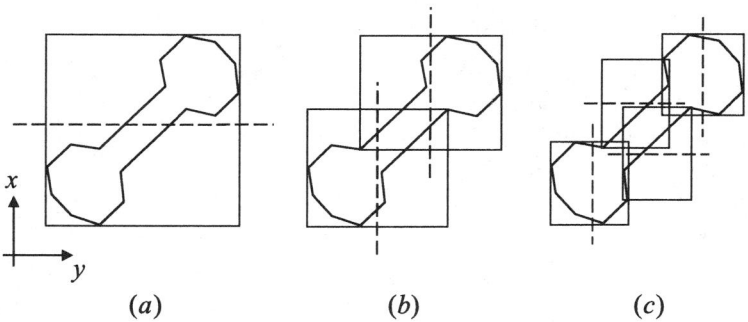

FIGURE 2.1. A 2D example of a binary AABB tree. The boxes at each intermediate level are aligned with the axis of the object's local-coordinate system. The broken lines show the partition plane used at each level.

Initially, the top-most bounding box is constructed by looping through the vertices of all primitives, keeping track of the minimum and maximum values along each axis of the object's local-coordinate system. The minimum and maximum values will define the lower-left and upper-right corners of the of the top-most bounding box, respectively.

A partition plane is then selected such that it splits the top-most bounding box into two regions along its longest axis. The intersection point between the partition plane and the longest axis is chosen such that the two regions will be as balanced as possible, that is, with more or less the same number of primitives assigned to each region of the subdivision. The subdivision rule used here is to pass the partition plane through the mean point of all vertices of all primitives associated with the top-most bounding box. The primitives are then assigned to the region in which their midpoint falls.

At each subsequent level, intermediate bounding boxes are constructed from the primitives associated with them, and new partition planes are created to divide the boxes into two regions. The primitives are then assigned to each region and the process recursively continues until there is only one primitive assigned to each region of the subdivision.

In the event all primitives are assigned to just one region (or the subdivision is unbalanced), another partition plane is chosen such that it divides the second longest axis into two regions, passing through the mean point. If this new partition plane still assigns all primitives to just one region, then a last attempt is made with the partition plane dividing the last axis. In the rare case that the group of primitives is assigned to just one region for all three choices of partition plane, the group is said to be indivisible and the current box containing the group becomes a leaf of the tree. However, in the most common case when the primitives are equally split into both subdivisions, the leaves of the hierarchical tree end up having just one primitive.

2.2.2 Oriented Bounding Boxes

In the Oriented Bounding Box (OBB) representation, the tree hierarchy is constructed from bounding boxes forming a tight fit around the primitives associated with them. In this case, each intermediate bounding box has a different alignment with respect to the object's local-coordinate system, since their orientation depends on the geometric displacement of their primitives. Figure 2.2 illustrates the top-down construction of a binary OBB tree for the same 2D object considered in the AABB case.

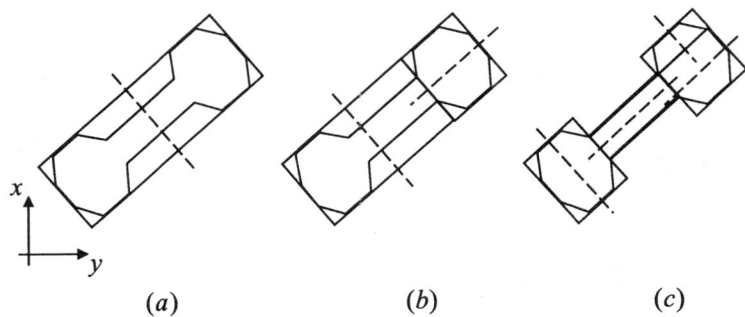

FIGURE 2.2. A 2D example of a binary OBB tree. The boxes at each intermediate level provide a tight fit around their primitives. The broken lines show the partition plane used at each level.

The fact that OBB trees provide a tighter hierarchical representation if compared with AABB trees clearly gives them an advantage over AABB trees when testing for collisions between objects that are close together, since this tightness generally reduces the number of primitive tests to be carried out. However, this comes at the price of having to carry out a more costly overlap test at each intermediate level of the OBB tree, as explained in Section 2.4.

The OBB tree construction is much more complex than the simple AABB tree construction, since the orientation of each intermediate bounding box needs to be computed from the set of primitives associated with it. The OBB tree construction algorithm described in this section assumes that the object's primitives are all triangles, that is, that the object's boundary representation is given by triangular faces. This assumption is especially suited to implementing a simulation engine as described in Appendix H, since by the time an object is registered with the simulation engine its convex decomposition is computed (see Appendix F) and the faces of each convex polyhedron that make up the object are triangulated. The final internal representation of objects in the simulation engine is therefore made up of triangular faces only.

The main difficulty when computing OBB bounding boxes is the determination of the direction of their axes such that the box provides a tight fit around the vertices of the triangle primitives associated with it. This can be done by considering the mean vector and the covariance matrix of the triangle primitives. The mean vector for each triangle primitive T_k is given by

$$\vec{\mu}_k = \frac{1}{3}(\vec{v}_1 + \vec{v}_2 + \vec{v}_3),$$

where $\vec{v}_1$, $\vec{v}_2$ and $\vec{v}_3$ are the vertices defining the triangle. Each vector $\vec{v}_r$ is described by its components $(v_r)_x$, $(v_r)_y$ and $(v_r)_z$. The mean vector of the vertex set is then

$$\vec{\mu} = \frac{1}{n}\sum_{k=1}^{n}\vec{\mu}_k,$$

with n being the total number of triangles being considered when computing the OBB bounding box.

The elements of the 3×3 covariance matrix of each triangle T_k can be computed as

$$C_{ij} = \frac{1}{3}((\bar{p}_1)_i(\bar{p}_1)_j + (\bar{p}_2)_i(\bar{p}_2)_j + (\bar{p}_3)_i(\bar{p}_3)_j),$$

where $i,j \in \{x,y,z\}$ and $\bar{p}_i = (\vec{p}_i - \vec{\mu})$ for $i \in 1,2,3$. The covariance matrix of the vertex set is then

$$C_{ij} = \frac{1}{n}\sum_{k=1}^{n}(C_k)_{ij},$$

with $(C_k)_{ij}$ being the $\{ij\}$ element of the covariance matrix associated with the kth triangle.

Since the covariance matrix is a real symmetric matrix, its eigenvectors are guaranteed to be mutually orthogonal. Moreover, two of its three eigenvectors are the axes corresponding to the maximum and minimum variance of the vertices' coordinates. Therefore, if we use the eigenvectors of the covariance matrix as a base, we can determine a tight-fitting bounding box by transforming all vertices to this base and computing the AABB box of the transformed vertices. In other words, the OBB bounding box has the orientation of the eigenvector base and a size that bounds the maximum and minimum coordinates of the transformed vertices.

It is important to notice that the direction of the eigenvectors of the covariance matrix is influenced, not only by the vertices that define the maximum and minimum coordinates, but by *all* vertices being considered. This may cause problems because interior vertices, which should not affect the bounding-box computation, can influence the direction of the eigenvectors. For example, a large number of interior vertices concentrated in a small area can cause the eigenvectors to align with them, instead of aligning

24   2. Hierarchical Representation of 3D Polyhedra

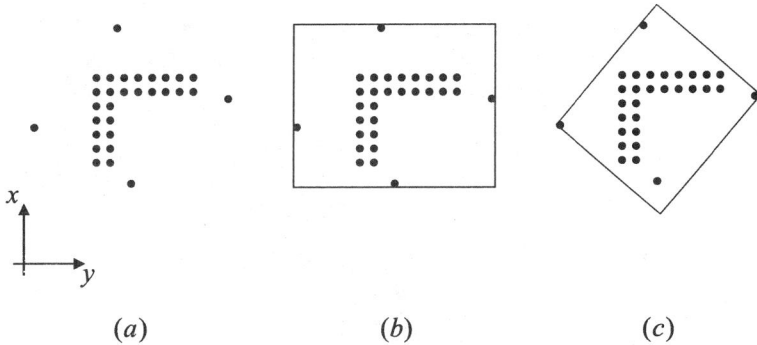

FIGURE 2.3. A 2D example of how interior points can degrade the quality of OBB bounding boxes; (a) The initial set of points; (b) The OBB bounding box created taking all points into account; (c) The OBB bounding box created taking into account only the convex hull points.

with the boundary vertices, thereby creating a low-quality OBB bounding box. This is illustrated in Figure 2.3.

The computation of the covariance matrix should therefore take into account *only* the boundary vertices of the vertex set. It should also be immune to clusters of boundary vertices, since they will tend to influence the direction of the eigenvectors in the same manner clusters of interior vertices do.

Interior vertices can be avoided if we consider only the vertices that are in the convex hull of the vertex set[2]. Clusters of boundary vertices can be ignored if we compute the mean vector and covariance matrix over the surface of the convex hull, as opposed to its vertices. This can be done as follows. The area A_k of each triangular face T_k of the convex hull can be computed directly from its vertices, given by

$$A_k = \frac{1}{2} | (\vec{v}_3 - \vec{v}_1) \times (\vec{v}_3 - \vec{v}_2) | .$$

The total convex hull area A_t is then

$$A_t = \sum_{k=1}^{n_k} A_k ,$$

where n_k is the total number of triangular faces in the convex hull.

The mean vector $\vec{\mu}_t$ associated with the convex hull, weighted by the total convex hull area, is obtained from

$$\vec{\mu}_t = \frac{\sum_{k=1}^{n_k} A_k \vec{\mu}_k}{\sum_{k=1}^{n_k} A_k} = \frac{\sum_{k=1}^{n_k} A_k \vec{\mu}_k}{A_t} .$$

---

[2] The computation of the convex hull of a vertex set is described in Section 2.2.4.

2.2 Hierarchical Representation of Objects

The elements $(C_k)_{ij}$ of the 3×3 covariance matrix of each triangular face T_k, also weighted by the total convex hull area, are given by

$$(C_k)_{ij} = \frac{A_k}{12\,A_t}(9\,(\mu_k)_i\,(\mu_k)_j + (v_1)_i\,(v_1)_j + (v_2)_i\,(v_2)_j + (v_3)_i\,(v_3)_j)\,.$$

Finally, the elements $(C_t)_{ij}$ of the 3×3 covariance matrix associated with the convex hull are computed from the elements $(C_k)_{ij}$ of the covariance matrix of each of its triangular faces as

$$(C_t)_{ij} = \left(\sum_{k=1}^{n_k}(C_k)_{ij}\right) - (\mu_t)_i\,(\mu_t)_j\,.$$

Having determined the covariance matrix, we proceed by computing its associated eigenvectors using one of several methods available for computing eigenvalues and eigenvectors of real symmetric matrices. Section 2.5 has pointers to the literature describing such methods. The OBB axis will be aligned with the direction of the eigenvectors and its dimensions will be given by the extremal vertices along each axis.

2.2.3 Bounding Spheres

In the Bounding Sphere (BS) representation, the tree hierarchy is constructed from minimum-radius bounding spheres encapsulating the primitives associated with them. Figure 2.4 illustrates the top-down construction of a binary BS tree for the same 2D object considered in the AABB and OBB cases.

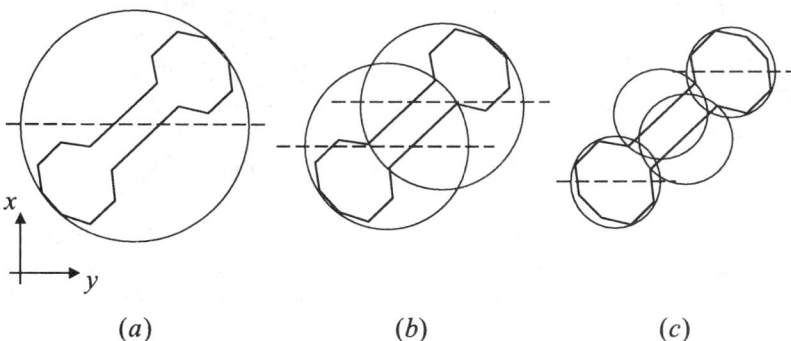

FIGURE 2.4. A 2D example of a binary BS tree. The broken lines represent the partition plane used at each level.

The BS hierarchical tree is usually of poorer quality than its OBB and AABB counterparts, with respect to the tightness of the decomposition. However, its overlap test is undoubtly the easiest and fastest to carry out

26   2. Hierarchical Representation of 3D Polyhedra

(see Section 2.4), thus giving this representation an advantage over the others for quick rejection tests.

In this section we present a method for finding a near-optimal bounding sphere from the set of primitives associated with it. The sphere calculated using this method is usually slightly larger than the minimum-radius sphere, but this inacuracy is offset by the efficiency of the method.

The bounding sphere computation is carried out in two passes through the list of vertices of all primitives associated with it. The first pass is used to estimate the initial center and radius of the sphere. The second pass goes through each vertex in the list and checks whether it is included in the sphere. If it is not included, then the sphere is enlarged to include it. At the end, the center and radius of the near-optimal bounding sphere are determined.

In the first pass, we loop through the list of all vertices to obtain the following six points.

1. The point with maximum x.

2. The point with minimum x.

3. The point with maximum y.

4. The point with minimum y.

5. The point with maximum z.

6. The point with minimum z.

From these six points, we select the two that are farthest apart. These two points will define the first approximation of the diameter of the bounding sphere. The center of the sphere is assumed to be at their midpoint.

In the second pass, we loop again through the list of all vertices, and for each vertex, we compare the square of its distance to the center with the square of the current radius of the bounding sphere. If the distance is smaller than the radius, then the vertex is inside the sphere and we proceed to the next vertex in the list. Otherwise, we adjust the sphere's radius and center as follows.

Let $\vec{v}_i$ be the current vertex being tested against the bounding sphere, and falling outside it. Let $\vec{c}$ be the center of the bounding sphere, r be its radius and $\vec{p}$ be the point in the sphere diametrically opposed to $\vec{v}_i$ (see Figure 2.5(a)).

Let d be the distance between $\vec{v}_i$ and $\vec{c}$, that is

$$d = \sqrt{((v_i)_x - c_x)^2 + ((v_i)_y - c_y)^2 + ((v_i)_z - c_z)^2}.$$

The enlarged sphere is then computed from the current sphere such that $\vec{v}_i$ and $\vec{p}$ become the new diameter, as shown in Figure 2.5(b). The new center $\vec{c}_n$ and radius r_n of the enlarged sphere are given by

2.2 Hierarchical Representation of Objects 27

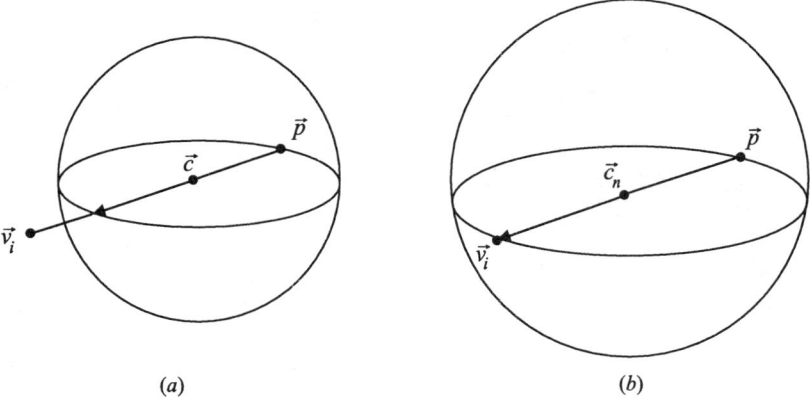

FIGURE 2.5. Efficient, incremental computation of a bounding sphere for a given vertex set; (a) vertex $\vec{v}_i$ falls outside the sphere, and therefore the sphere needs to be enlarged to bound it as well; (b) the sphere is augmented such that $\vec{v}_i$ and $\vec{p}$ define its new diameter.

$$r_n = \frac{r+d}{2}$$
$$\vec{c}_n = \frac{r\vec{c}+(d-r)\vec{v}_i}{d}.$$

This process continues until all vertices are checked for inclusion against the bounding sphere.

Having determined the top-most bounding sphere, a partition plane is chosen such that it passes through the median point of all vertices of all primitives associated with the bounding sphere. The partition plane subdivides the bounding sphere into two regions, and the primitives are assigned to each region following the same rules used on the AABB and OBB cases, namely, the primitive is associated with the region that contains its midpoint. The subdivision continues until there is only one primitive assigned to each bounding sphere, or the primitives cannot be split, in which case the group of primitives is assigned to the bounding sphere.

2.2.4 Convex Hull

The convex hull can be used not only to provide a hierarchical representation of the object's primitives as a tree of convex polyhedra, but also as an intermediate step for computing other types of representations, such as the OBB trees already covered in Section 2.2.2.

The convex hull of a given vertex set S is defined as being the smallest convex set containing S. There are several algorithms and methods that can be used to compute the convex hull in 2D, 3D, or even higher dimensions.

In this section, we focus on the *gift wrapping* method, which is intuitive, easy to visualize in 3D, simple to implement, and applicable to higher-dimensional spaces.

The basic idea of the gift-wrapping method consists of imagining folding a piece of paper around the primitives being considered. We start with a face that is guaranteed to be in the convex hull, and loop through its edges determining its neighbor faces that are also part of the convex hull. The algorithm then proceeds looping through the edges of the new neighbor faces until all faces are discovered and the convex hull is completely determined. All faces will be discovered whenever the list of edges to be checked is empty.

Given a set of vertices $S = \{\vec{v}_1, \ldots, \vec{v}_n\}$, let's assume that the triangular face f_1 defined by vertices $(\vec{v}_1, \vec{v}_2, \vec{v}_3)$ is the starting face guaranteed to be in the convex hull. According to the high-level description of the algorithm presented in the previous paragraph, we need to loop through the edges of face f_1 and determine its associated neighbor faces that are also in the convex hull. A face is said to be in the convex hull if all vertices of S that are not vertices of the face lie on the same side of the plane defined by the face. Since we are using the right-hand coordinate system in our simulation engine, we want all vertices of S to lie on the inside region of the plane. More specifically, we want to construct each face of the convex hull such that its normal is always pointing outwards, as illustrated in Figure 2.6.

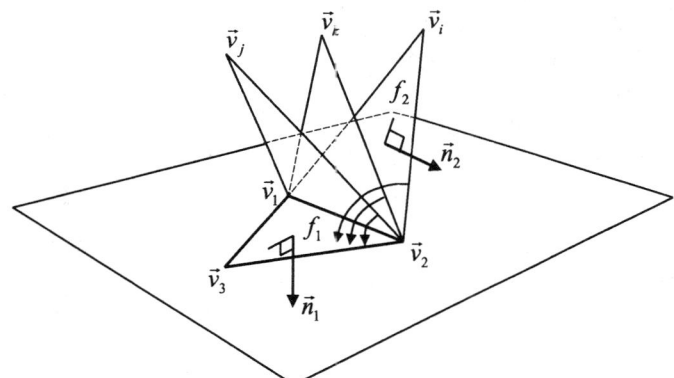

FIGURE 2.6. Determining the neighbor face that shares edge e_1 with f_1. The selected vertex $\vec{v}_i$ will define a face f_2 that forms the largest convex dihedral angle with f_1. The ordering of the vertices defining the new face should be chosen so that the normal vector of the new face always points towards the outside of the object. Because we are using the right-hand coordinate system, the correct order is $(\vec{v}_1, \vec{v}_i, \vec{v}_2)$.

Let's consider, for example, the determination of the neighbor face f_2 that shares edge $e_1 = (\vec{v}_1, \vec{v}_2)$ with f_1. We want to find the vertex $\vec{v}_i \in S$, with $i \neq 1, 2$, such that the triangular face f_2 defined by $(\vec{v}_1, \vec{v}_i, \vec{v}_2)$ forms the largest convex internal dihedral angle at edge e_1. Figure 2.7 shows how the internal dihedral angle can be computed.

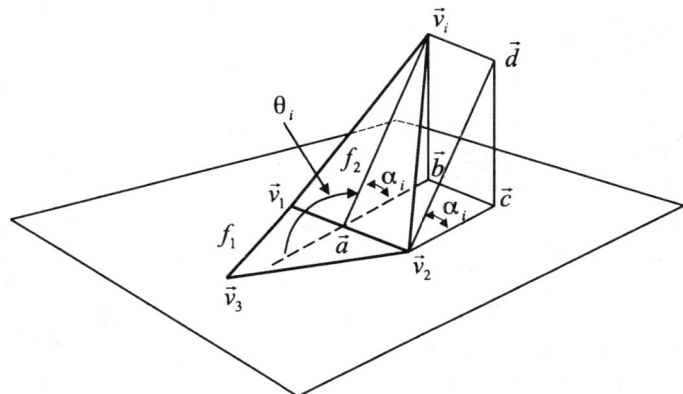

FIGURE 2.7. The internal dihedral angle θ_i associated with vertex $\vec{v}_i$ at edge e_1 defined by vertices $(\vec{v}_1, \vec{v}_2)$, shown as the exterior angle at vertex $\vec{a}$ of triangle $(\vec{a}, \vec{b}, \vec{v}_i)$. Notice that vertex $\vec{b}$ is the projection of $\vec{v}_i$ on the plane of face f_1.

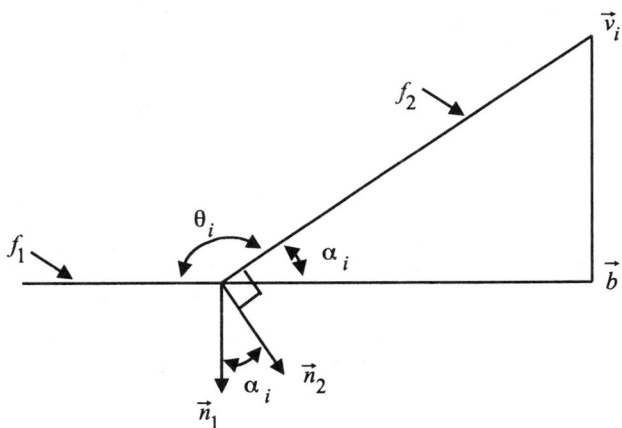

FIGURE 2.8. The interior angle α_1 at vertex $\vec{a}$ can be obtained from the dot product of the face normals $\vec{n}_1$ and $\vec{n}_2$ associated with faces f_1 and f_2, respectively.

Let θ_i be the internal dihedral angle associated with vertex $\vec{v}_i$ at edge e_1. Let $\vec{n}_1$ and $\vec{n}_2$ be the normal vectors of faces f_1 and f_2, respectively. By construction, since the normals at each face are pointing outwards, their dot product gives the cosine of $(\pi - \theta_i)$ (see Figure 2.8). The dihedral angle can then be computed directly from

$$\theta_i = \pi - \arccos(\vec{n}_1 \cdot \vec{n}_2) \ .$$

We select the vertex $\vec{v}_i$ corresponding to the maximum θ_i, and add face f_2 to the list of convex hull faces. The ordering of the vertices defining the new face should be chosen so that the normal vector of the new face always points towards the outside of the object. Because we are using the right-hand coordinate system, the correct order is $(\vec{v}_1, \vec{v}_i, \vec{v}_2)$. The edges of f_2 are then added to the list of edges that need to be checked, so that the algorithm can compute the convex hull faces that share these edges with f_2. It is important to notice that the algorithm assumes each edge is shared by exactly two faces. Therefore, every time a new edge is added to the list of edges that need to be checked, we should first check whether the edge is already in the list. If the edge is already in the list, then one of the faces that contains this edge was already discovered in some previous step, and the other face that contains this edge has just being discovered. In this case, there is no need to check for this edge because both faces that share the edge are already included in the convex hull. Therefore, the edge can be removed from the list. Otherwise, the edge should be added to the list.

Up till now we have assumed the existence of a starting face that is guaranteed to be in the convex hull, and have determined all other faces from it. The only step we still have to describe is how the first face of the convex hull is computed. The computation of the vertices of the first face is incremental, in the sense that we compute one of them at a time. We start with one vertex that is guaranteed to be in the convex hull, then use it to determine the second vertex, thus forming an edge of the starting face. We then use the edge to determine the third vertex that makes up the first face. Having the first face, we proceed as explained before and determine all other convex hull faces.

The first face of the convex hull is computed as follows. Consider the projection of all points on the xy-plane, as shown in Figure 2.9. Let $\vec{a}_1$ be the vertex with the lowest projected y-coordinate value. This vertex is guaranteed to be in the convex hull, since all other points of the vertex set will lie on the same half-space defined by a plane orthogonal to the y-axis (i.e., parallel to the xz-plane), passing through vertex $\vec{v}_1$. Therefore, vertex $\vec{v}_1$ is one of the vertices of the starting face.

The second vertex of the starting face can be found by looping through the projected vertices and selecting a vertex $\vec{a}_2$ such that all other projected vertices lie to the left of the edge $e_p = (\vec{a}_1, \vec{a}_2)$. We can determine whether the projected vertex $\vec{a}_j$ lies to the left or right of edge e_p by considering the sign of the area of the triangle defined by $(\vec{a}_1, \vec{a}_2, \vec{a}_j)$. If the area is positive, then the vertices are in counterclockwise order and the projected vertex $\vec{a}_j$ lies to the left of edge e_1. Otherwise, the projected vertex $\vec{a}_j$ lies to the right of edge e_p.

The area A of the projected triangle $(\vec{a}_1, \vec{a}_2, \vec{a}_j)$ can be quickly computed from the vertices' coordinates

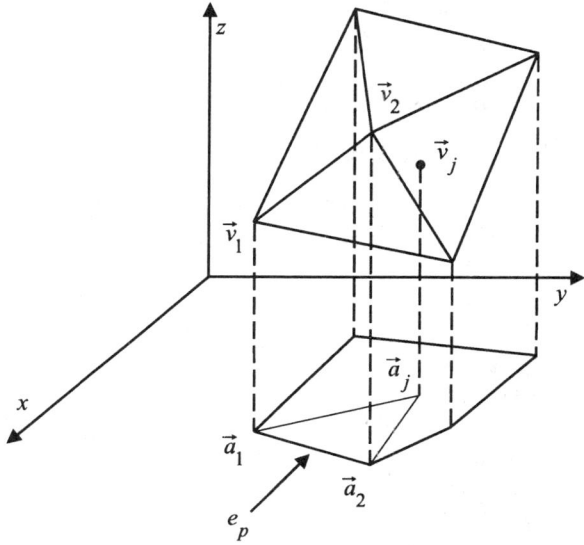

FIGURE 2.9. The first edge of the starting face is computed using the projection of the vertices on the xy-plane. Here the problem is reduced to its 2D counterpart.

$$A = \frac{1}{2} \begin{vmatrix} (\vec{a}_1)_x & (\vec{a}_2)_x & (\vec{a}_j)_x \\ (\vec{a}_1)_y & (\vec{a}_2)_y & (\vec{a}_j)_y \\ 1 & 1 & 1 \end{vmatrix}.$$

Finally, the third vertex of the starting face can be obtained by considering the triangular faces f_j defined by vertices $(\vec{v}_1, \vec{v}_j, \vec{v}_2)$ in 3D space. The order of the vertices defining the triangular face f_j is such that the normal points to the outside of the convex hull[3]. Here, the third vertex $\vec{v}_3$ is selected such that all other vertices lie in the negative half-space defined by the plane that contains the triangular face $(\vec{v}_1, \vec{v}_3, \vec{v}_2)$ (see Figure 2.10).

Let $\vec{n}_j$ be the normal of the plane defined by the triangular face $(\vec{v}_1, \vec{v}_j, \vec{v}_2)$, and let d_j be the plane constant computed as

$$d_j = \vec{n}_j \cdot \vec{v}_1 .$$

Vertex $\vec{v}_p$ will lie on the negative half-space of the plane provided

$$(\vec{n}_j \cdot \vec{v}_p) < d_j .$$

Having the first face, we proceed as explained and compute all other convex hull faces of the polyhedron.

---

[3] Recall that we are using the right-hand coordinate system.

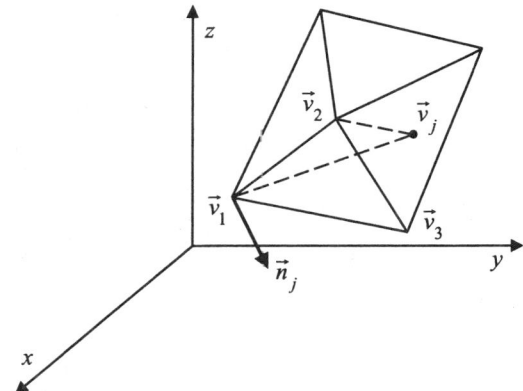

FIGURE 2.10. The starting face is obtained by connecting a third vertex to the starting edge, so that all other vertices lie on the negative half-space defined by the plane that contains the face.

2.3 Hierarchical Representation of the Simulated World

Even though the use of hierarchical representations does speed the collision-detection phase, they themselves do not provide mechanisms to take advantage of the time coherence between consecutive frames in a simulation. For instance, the fact that two or more objects are farther apart, such that their top-most bounding volumes do not intersect, should be exploited in the following simulation time steps to avoid unnecessary collision checks between these objects. The hierarchical representation does minimize the time spent on such unnecessary collision checks, since they are usually dismissed after the top-most bounding volumes are checked against each other. However, the time expended on these unnecessary checks can be significant, especially when the simulation contains several hundred objects.

The idea is then to create a partition of the simulated world into cells, and assign objects to the cells in which their top-most bounding volume intersects. Objects that are assigned to the same cell can be potentially colliding and therefore should be checked for geometric intersections between their hierarchical representations. On the other hand, objects that have no cells in common are clearly distant from each other and should not be checked for collisions at all.

The simulated world considered in this book is assumed to be bounded by a box that defines the maximum and minimum spans along each coordinate axis. The cell decomposition is then a partition of the box into sub-volumes that may or may not contain objects during the simulation.

There are two important issues that should be taken into account when decomposing the simulated world into cells. First, the cell decomposition should be simple, that is, should have simple geometry such that the cost of updating the cells that intersect each moved object is negligible compared with the cost of checking for collisions between their hierarchical representations. This issue is addressed in Section 2.3.1, where the simulated world is subdivided into boxes of uniform size. Second, the size of each cell directly affects the efficacy of the decomposition. For instance, a too-small size will assign several cells to each object, making it more expensive to update the list of occupied cells after each simulation time step. On the other hand, a too-large size will assign several objects to the same cell and a large number of unnecessary collision checks between their hierarchical representations may be carried out. This issue is addressed in Section 2.3.2, where the uniform-grid approach presented in Section 2.3.1 is extended to a multi-level grid that better fits the different sizes of objects being simulated.

2.3.1 Uniform Grid

Uniform-grid decomposition, as its name implies, subdivides the bounding box of the simulated world into cubic cells of same size along each axis of the world-coordinate system. A simple uniform-grid decomposition of an hypothetical world is shown in Figure 2.11.

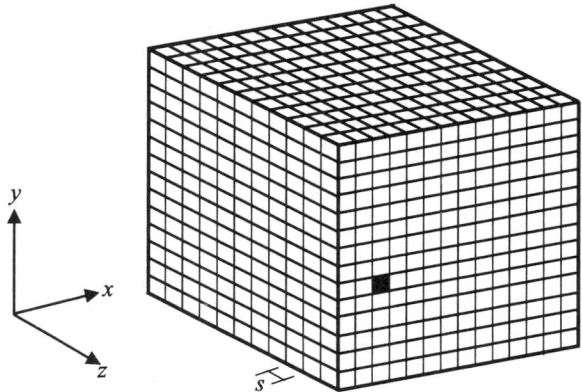

FIGURE 2.11. A simple uniform-grid decomposition of an hypothetical simulated world with a resolution of fourteen boxes along each dimension. Each cell is identified by the index ot its lower-left corner vertex. For example, the first cell is indexed as cell $(0, 0, 0)$. The shaded cell is indexed as cell $(2, 5, 13)$.

In the uniform-grid decomposition, the dimension of the cubic cells plays an important role in minimizing the number of unnecessary objects to cell assignments, and in maximizing the overall efficacy of the simulation. Intuitively, the size of each cell should be:

- Large enough to allow objects to rotate and translate for a while without leaving the cell, thus minimizing the number of dynamic updates of objects assigned to each cell;

- Small enough to have as many objects as possible assigned to different cells, thus minimizing the pair-wise collision checks between the object's hierarchical representations.

Let d_i be the maximum diameter of the top-most bounding volume of object i. For example, if the bounding volume is a box, then the maximum diameter is the distance between the two diagonally opposing vertices defining box. If the bounding volume is a sphere, then the maximum diameter is equal to the diameter of the sphere. The average maximum diameter of the objects being simulated is then

$$\bar{d} = \frac{1}{n} \sum_{i=0}^{n} d_i \; ,$$

where n is the total number of objects[4]. The size b of each cell in the uniform-grid decomposition is also given by the cell's maximum diameter, and can be related to the average maximum diameter of the objects being simulated as

$$\frac{\bar{d}}{b} = k \; , \tag{2.1}$$

with $k \geq 1$. The variable k is used to adjust the size of the cell with respect to the average maximum diameter of the objects being simulated. As a rule, we suggest using $k = 2$, that is, the size of each cell in the uniform grid is twice the average maximum diameter. The rationale behind this choice is as follows. If all objects had the average size, then we could have up to eight objects in each cell (two objects touching each other along each dimension), giving some room to an object to move within the same cell. Also, objects that are farther apart by more than twice their average size are guaranteed not to be in the same cell. On average, this choice gives us a reasonable trade-off between the number of objects assigned to each cell and the number of pair-wise collision checks carried out at each time step if the objects' sizes are close to the average. However, if the objects' sizes vary by orders of magnitude, then a more sophisticated approach, such as the one presented in Section 2.3.2, should be used.

Having selected the size of the boxes in the decomposition, the next step is to provide an efficient mechanism to keep track only of cells that have at

---

[4]The size of particles in a particle system is not taken into account during this computation because particles are usually considered as point mass, as explained in detail in Chapter 3.

2.3 Hierarchical Representation of the Simulated World   35

least one object assigned to them, as opposed to allocating memory to all cells. Clearly, the latter approach is not advisable for cases when the size of each cell is orders of magnitude smaller than the size of the simulated world, since the number of cells along each axis would be huge, and the memory needed for a subdivision containing n cells is n^3. The idea here is to use a hash table to keep track of the occupied cells. There are many ways this hash table of cells can be constructed, and some may be more effective than others, depending on the specifics of the simulation being considered. However, as a rule, we suggest using a hash table of size n, where the key is the sum of the indexes of the cell along each axis. This will have the effect of assigning one slice (plane) of the grid decomposition to each hash-table entry.

The update mechanism using the hash table is used to efficiently detect pairs of potentially colliding objects. Initially, the top-most bounding volume of each object is checked against the cell decomposition. Each cell that intersects the object's bounding volume is added to the hash table of cells. If the cell was already added to the hash table, then there are at least two objects assigned to this cell, and a pointer to this cell is added to a list of cells that need to be checked for collisions. At the end, after all objects are checked against the cell decomposition, another list of potentially colliding objects is created from the list of cells that needs to be checked for collisions. The former list contains pairs of objects that occupy the same cell. For each of these pairs, the more expensive collision check using their hierarchical representation is carried out.

The cells of the uniform-grid decomposition that intersect an object's bounding volume can be efficiently determined if we consider an AABB bounding box of the object's bounding volume in the world-coordinate system. Notice that this selection is independent of the hierarchical tree representation of the objects. The AABB bounding box is aligned with the world coordinate system, as are all cubic cells in the uniform-grid decomposition. The box-box intersection test between boxes that have their axes aligned is extremely fast and can be used to determine the actual cubic cells in the decomposition that need to be checked for intersection with the object's bounding volume. Figure 2.12 illustrates this situation for an object using the bounding-sphere representation. The AABB bounding box of the bounding sphere is used to efficiently locate the cells in the decomposition that need to be checked for intersection with the object's bounding sphere, as opposed to checking the intersection of every cell with the object's bounding sphere.

As objects translate and rotate during the simulation, their top-most bounding volume will move. This movement may cause the bounding volume to no longer intersect some of the cells the object is assigned to, and may also intersect new cells that did not have the object on their list. Therefore, the list of objects assigned to each cell needs to be updated after each simulation time step to reflect these changes. This update can be

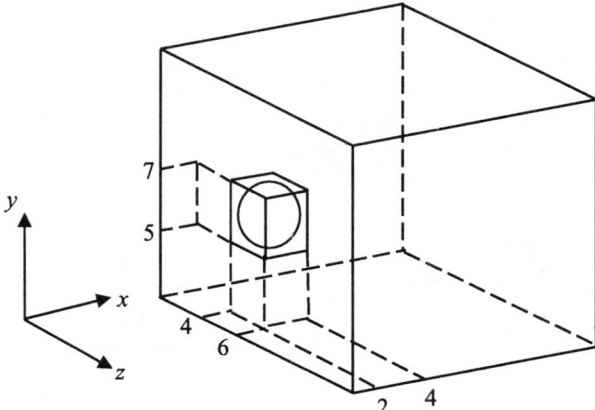

FIGURE 2.12. The AABB bounding box of the object's bounding sphere is used to quickly determine which cells of the decomposition need to be checked for intersection with the bounding sphere. In this case, the bounding sphere will be checked further for intersections with cells (2,5,4), (2,6,4), (3,5,4), (3,6,4), (2,5,5), (2,6,5), (3,5,5) and (3,6,5).

efficiently implemented using coherence between simulation time steps, as follows.

Throughout the simulation, each object keeps track of the indexes of the cells that intersect its bounding volume. At each time step, a new list of indexes of cells that intersect the object's bounding volume is generated. This new list is then compared with the old list. If the new list is the same as the old list, then the object's cell assignment remains the same as in the previous time step, and nothing else needs to be done. If there are cells on the new list that are not on the old list, then we search for these cells in the hash table of cells. If we find the cell in the hash table, then we add a reference to this object and raise the cell's internal counter. Otherwise, we create an entry for the cell in the hash table and set its internal counter to one. Finally, if there are boxes on the old list that are no longer on the new list, then the reference to this object should be removed from the cell's entry in the hash table, and its internal counter is subtracted by one.

The cell's internal counter is used to keep track of the number of objects currently intersecting the cell. The first time the counter is set to two, a reference to this cell is added to the list of cells that contain potentially colliding objects. The counter can be set to values greater than two, but as long as it has at least two, the reference to this cell will be kept in the list. The reference to this cell is removed from the list when the counter first is reduced from two to one. The cell is removed from the hash table of cells when the counter is set to zero.

2.3.2 Multi-Level Grid

If the size of the objects being simulated differs by orders of magnitude, the efficiency of the uniform-grid approach can be improved by extending it to a multi-level grid. The idea is to group at the same level, objects with sizes of the same order of magnitude such that each level can be treated as an uniform grid in itself. The advantage is that each level attempts to maximize the efficiency of its uniform grid, since it is guaranteed to have objects of similar size. There are three important issues that need to be addressed when using this method.

- How many levels should be chosen for a given set of objects?
- What should be the size of the cells at each level?
- How the levels are related so that collisions between objects assigned to different levels can be efficiently detected?

In the uniform-grid case, since we have a single level, the size of the cell is determined from equation (2.1) as a multiple of the average size of the maximum diameter of the objects in the simulation. In the multiple-level case, object i is assigned to level j if

$$k_{min} \leq \frac{d_i}{L_j} \leq k_{max}, \qquad (2.2)$$

where d_i is the maximum diameter of the top-most bounding volume of object i, k_{min} and k_{max} are user-definable constants, and L_j is the size of the cells at level j, such that

$$0 < L_1 < L_2 < \ldots < L_j < \ldots < L_m$$

for $1 \leq j \leq m$. The idea is to assign object i to the largest level j such that equation (2.2) is satisfied. In other words, the largest objects are assigned to the largest boxes (largest levels), so that objects at level $(j+1)$ have diameters greater than objects at level j. The constants k_{min} and k_{max} are used to relate the size of the cells at different levels. They must satisfy

$$\begin{aligned} 0 < \; & k_{min} \; < 1 \\ & k_{max} \; \geq 1 \, . \end{aligned}$$

Let d_{min} and d_{max} be the minimum and maximum diameters of all objects in the simulation. Clearly, the objects associated with d_{min} (the smallest objects) should be assigned to level 1 (the lowest level). This is done by substituting d_{min} and L_1 into equation (2.2), that is

$$k_{min} \leq \frac{d_{min}}{L_1} \leq k_{max} \, .$$

If we make

$$k_{min} = \frac{d_{min}}{L_1},$$

then we have that the size of the cells at the lowest level is given by

$$L_1 = \frac{d_{min}}{k_{min}}. \tag{2.3}$$

The objects associated with d_{max} should be assigned to the largest level m. This is done by substituting d_{max} and L_m into equation (2.2), and making

$$\frac{d_{max}}{L_m} = k_{max},$$

that is

$$L_m = \frac{d_{max}}{k_{max}}. \tag{2.4}$$

Because we want the level assignment to be continuous, we need to make sure the maximum value at level j is equal to the minimum value at level $(j+1)$, that is

$$k_{max} L_j = k_{min} L_{j+1}. \tag{2.5}$$

Equation (2.5) relates the size of the cells at two consecutive levels. We can use this equation to recursively compute the size of the cells at level j as a function of the size of the cells at level 1, as follows:

$$L_2 = \frac{k_{max}}{k_{min}} L_1$$

$$L_3 = \frac{k_{max}}{k_{min}} L_2 = \left(\frac{k_{max}}{k_{min}}\right)^2 L_1$$

$$\ldots$$

$$L_j = \left(\frac{k_{max}}{k_{min}}\right)^{j-1} L_1. \tag{2.6}$$

Because we have the size of the boxes at the first level L_1 and the largest level L_m given by equations (2.4) and (2.5), respectively, we can substitute these equations into (2.6) and compute the number of levels m needed as a function of k_{min}, k_{max}, d_{min} and d_{max}. We have

$$\frac{d_{max}}{k_{max}} = \left(\frac{k_{max}}{k_{min}}\right)^{m-1} \frac{d_{min}}{k_{min}},$$

which gives

2.3 Hierarchical Representation of the Simulated World

$$m = \left\lceil \log_{\left(\frac{k_{max}}{k_{min}}\right)} \left(\frac{d_{max}}{d_{min}}\right) \right\rceil. \tag{2.7}$$

Figure 2.13 shows an example of a multi-level grid assignment. At each level j, the simulated world is subdivided in a uniform grid with boxes of size L_j.

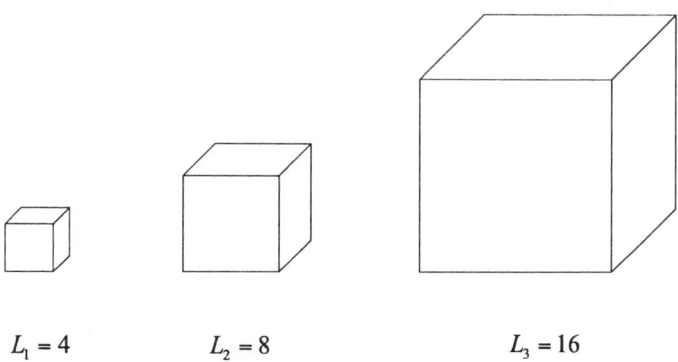

$L_1 = 4$ $L_2 = 8$ $L_3 = 16$

FIGURE 2.13. An example of a multi-level grid assignment for $k_{max} = 1$, $k_{min} = 0.5$, $d_{max} = 16$ and $d_{min} = 2$. The maximum number of levels to be used and their sizes can be directly computed from equations 2.7 and 2.6, respectively. In this case, $m = 3$, $L_1 = 4$, $L_2 = 8$ and $L_3 = 16$.

Therefore, the multi-level grid has one hash table of cells for each level, and the sizes of their cells are given by equation (2.6). The update mechanism for each hash table is the same as that used in the uniform grid, since we do have a uniform grid at each level. Consider, for example, the bounding sphere of an object with center at (5,5,5) and maximum diameter $d = 9.7$ (see Figure 2.14). For the multi-level grid of the simulated world shown in Figure 2.13, according to equation (2.2), the object should be assigned to level 2. Within level 2, the cells that intersect the object's bounding volume are computed the same way as shown in Figure 2.12 for the uniform-grid case.

The only remaining issue is how this scheme can be used to efficiently detect potential collisions between objects assigned to different levels of the grid decomposition. We address this issue by adding a reference to the object, not only to the cells that intersect the object at its level, but also to all other cells that intersect the object at levels greater than its level. For example, an object assigned to level j will have its reference added to all cells that intersect its bounding volume at levels $j, (j+1), \ldots, m$. In the case shown in Figure 2.14, the cells at levels 2 and 3 that intersect the object's bounding volume will keep a reference to this object (see Figures 2.15 and 2.16). Using this scheme, two objects b_1 and b_2 assigned to levels L_{b_1} and L_{b_2} can potentially collide if and only if there exist at least one cell at

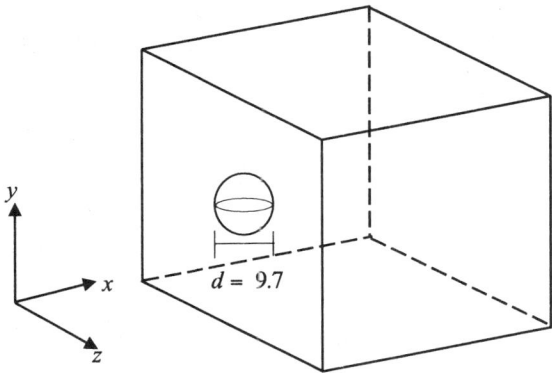

FIGURE 2.14. An example of the multi-level grid assignment of an object to level 2, in the simulated world of Figure 2.13, containing 3 levels.

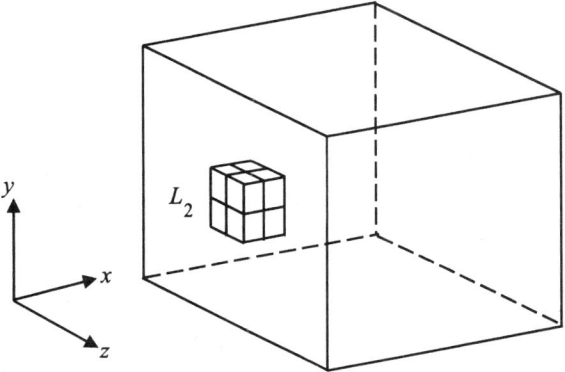

FIGURE 2.15. The cells at level 2 that intersect the object's bounding sphere.

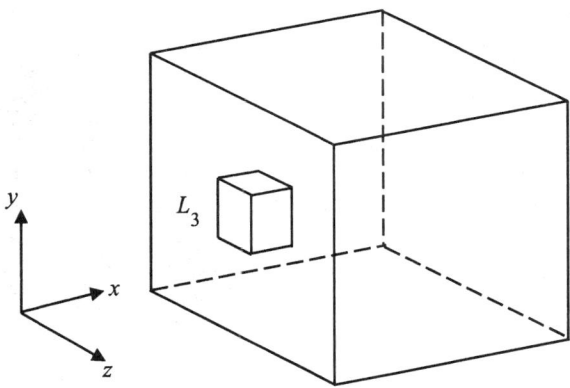

FIGURE 2.16. The multi-level grid assignment makes it necessary to determine the cells at level 3 that intersect the object. This is in order to detect potential collisions between this object and other objects assigned only to level 3.

level max (L_{b_1}, L_{b_2}) that has a reference to both of them. This situation is illustrated in the one-dimensional case shown in Figure 2.17.

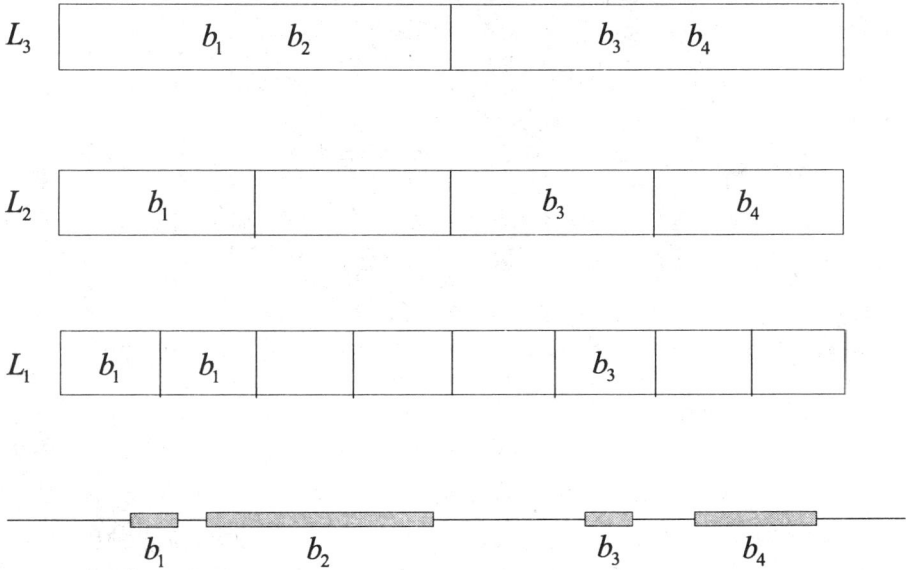

FIGURE 2.17. A one-dimensonal example of how potential collisions are detected between objects assigned to different levels.

In this example, objects b_1 and b_2 are assigned to levels L_1 and L_3, respectively. Since there is a box at level $m = \max(L_1, L_3) = L_3$ that contains a reference to both of them, the objects are added to the list of potentially colliding objects. On the other hand, objects b_3 and b_4 are assigned to levels L_1 and L_2, respectively. Since there are no boxes at level $m = \max(L_1, L_2) = L_2$ that contain a reference to both of them, they are not considered to be on the list of potentially colliding objects, even though there is a cell at level L_3 referring to both of them.

As a rule, we suggest choosing k_{min} and k_{max} such that

$$\frac{k_{max}}{k_{min}} = 2 .$$

This choice means the size of the cells will be a power of two times the minimum diameter d_{min}. With this choice, given an object assigned to level j, it is straightforward to ascertain which cells intersect the object at levels $(j+1), \ldots, m$.

2.4 Collision Detection Between Different Hierarchical Representations

Up till now, we have described several types of hierarchical representations that can be used to speed collision detection between objects in a simulation engine, as well as how we can build them using simple primitives. In this section we shall present efficient algorithms to quickly determine whether two primitives are intersecting.

The primitives of the representations covered in this book are boxes and spheres for the tree hierarchies, and triangles for the faces of the objects. Therefore, we need algorithms for checking the intersection of each possible pair-wise combination of such primitives. Moreover, for the triangle-triangle intersection test, we need to go one step further and save pointers to the intersecting triangles. This information will be used by the collision-response module to estimate the collision (or contact) point between the objects, and determine the collision impulses (or contact forces) needed to prevent their interpenetration.

2.4.1 Computing Box-Box Intersections

The intersection test between two boxes is based on the separating-axis theorem. This theorem states that two boxes A and B are disjoint if and only if there exists a separating plane such that the boxes are located on different sides of the plane.

Let $\vec{n}$ be the normal of a plane P, and let d be its non-negative distance to the origin. The plane P is a separating plane of boxes A and B if

$$\vec{n} \cdot \vec{a} + d \leq 0, \forall \vec{a} \in A \tag{2.8}$$

and

$$\vec{n} \cdot \vec{b} + d > 0, \forall \vec{b} \in B , \tag{2.9}$$

that is, if the projections of A and B along the normal fall on opposite sides of the plane. Equations (2.8) and (2.9) can then be combined into the single equation

$$\vec{n} \cdot \vec{a} < \vec{n} \cdot \vec{b}, \forall \vec{a} \in A, \forall \vec{b} \in B . \tag{2.10}$$

Equation (2.10) states that, if P is a separating plane of boxes A and B, then their images are disjoint under axial projection along an axis parallel to the plane normal $\vec{n}$. In other words, $\vec{n}$ is a separating axis of A and B. It can be shown that the separating-axis candidates are the normals to the faces of A and B, and the normals to the planes defined by one edge of A and one edge of B. This results in 15 potential cases to be tested: 3 different face normals for each box plus 9 pair-wise combinations of edges.

2.4 Collision Detection Between Different Hierarchical Representations

If none of the potential separating axes actually separates the boxes, then the boxes are guaranteed to be overlapping.

Let us first consider the simple intersection case, where the boxes are aligned with each other and are parallel to the world-coordinate system's axis. This case occurs in the hierarchical representation of the simulated world, where all boxes in the uniform or multi-level grids have the same orientation with respect to the world-coordinate system. In such situations, the 15 potential cases are reduced to just 3, since the face normals of each box are the same and the pair-wise combination of their edges always gives another edge. Therefore, the three separating-axis candidates are the axes of the world-coordinate system.

Let each box be represented by its minimum and maximum vertices, as indicated in Figure 2.18.

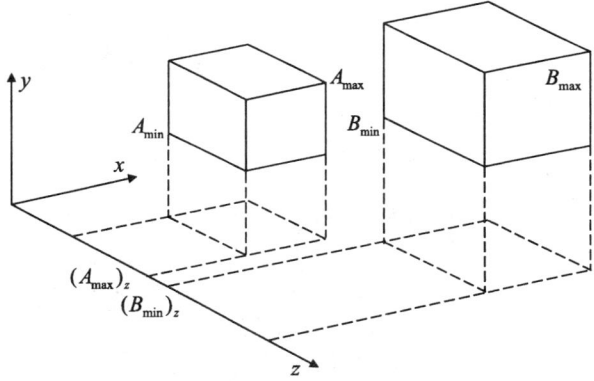

FIGURE 2.18. Axis-aligned box-box intersection test. Each box is defined by its minimum and maximum vertices. The intersection test is then carried out by checking whether their projections along each coordinate axis overlap. In this case, the z-axis is a separating axis and the objects do not overlap.

Let $[(A_{min})_i, (A_{max})_i]$ and $[(B_{min})_i, (B_{max})_i]$ be the projections of boxes A and B along the coordinate axis i, for $i = \{x, y, z\}$. The boxes A and B will *not* overlap if and only if

$$((A_{max})_i < (B_{min})_i) \cup ((B_{max})_i < (A_{min})_i) \qquad (2.11)$$

for at least one projection axis $i \in \{x, y, z\}$. This projection axis is then the separating axis for the boxes. On the other hand, if equation (2.11) is not satisfied for all projection axes, then the boxes are guaranteed to be overlapping.

Having considered the simple axis-aligned case, let's move on to the more complex case wherein the boxes are arbitrarily oriented with respect to each other. This happens when checking for intersections between boxes

44   2. Hierarchical Representation of 3D Polyhedra

in the AABB or OBB hierarchical representations, since they usually have different orientations in the world-coordinate system.

Let $\vec{T}_A$ and $\mathbf{R_A}$ be the translation vector and rotation matrix from A's local-coordinate system to the world-coordinate system. The axis of A in the world-coordinate system will then be given by the columns of $\mathbf{R_A}$, namely $(R_A)_x$, $(R_A)_y$ and $(R_A)_z$. Similarly, let $\vec{T}_B$ and $\mathbf{R_B}$ be the translation vector and rotation matrix from B's local-coordinate system to the world-coordinate system, and the axis of B in the world-coordinate system be $(R_B)_x$, $(R_B)_y$ and $(R_B)_z$. Let $\vec{d}$ be the distance vector between the center of the boxes in the world-coordinate system. The boxes A and B will be disjoint if and only if the sum of the projections of their half-sides along the candidate separating axis $\vec{n}$ is less than the projection of their distance vector $\vec{d}$ along $\vec{n}$, that is

$$(\vec{r_A} \cdot \vec{n} + \vec{r_B} \cdot \vec{n}) < \vec{d} \cdot \vec{n}, \qquad (2.12)$$

where r_A and r_B are the sum of the projections of the half-sides of A and B, respectively. Figure 2.19 illustrates this.

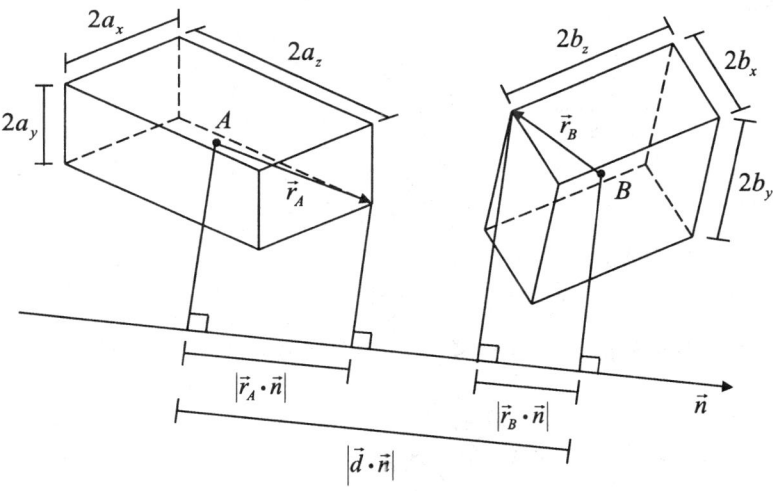

FIGURE 2.19. Arbitrarily oriented box-box test. The boxes will not intersect if the axial projection of the distance between their centers is greater than the sum of the axial projection of their half-sides. There are 15 possible axial directions to be tested.

The distances between boxes A and B to the origin of the world-coordinate system are given by $\vec{T}_A$ and $\vec{T}_B$. Therefore, their distance vector can be directly obtained from

$$\vec{d} = |\vec{T}_A - \vec{T}_B|. \qquad (2.13)$$

2.4 Collision Detection Between Different Hierarchical Representations

The half-sides of each box can be computed from the boxes' minimum and maximum vertices transformed to the world-coordinate system. Let a_x, a_y and a_z be the half-sides of box A along its axes $(R_A)_x$, $(R_A)_y$ and $(R_A)_z$. Similarly, let b_x, b_y and b_z be the half-sides of box B along its axes $(R_B)_x$, $(R_B)_y$ and $(R_B)_z$. The sum of the projections of the half-sides of A and B along $\vec{n}$ are then

$$r_A \cdot \vec{n} = a_x |(R_A)_x \cdot \vec{n}| + a_y |(R_A)_y \cdot \vec{n}| + a_z |(R_A)_z \cdot \vec{n}|,$$
$$r_B \cdot \vec{n} = b_x |(R_B)_x \cdot \vec{n}| + b_y |(R_B)_y \cdot \vec{n}| + b_z |(R_B)_z \cdot \vec{n}|. \quad (2.14)$$

Substituting equations (2.14) and (2.13) into (2.12), we have that $\vec{n}$ is a separating axis if and only if

$$\begin{aligned}(a_x |(R_A)_x \cdot \vec{n}| + a_y |(R_A)_y \cdot \vec{n}| + a_z |(R_A)_z \cdot \vec{n}| \\ + b_x |(R_B)_x \cdot \vec{n}| + b_y |(R_B)_y \cdot \vec{n}| + b_z |(R_B)_z \cdot \vec{n}|) &< \\ |\vec{T}_A - \vec{T}_B| \cdot \vec{n} \end{aligned} \quad (2.15)$$

is satisfied for the 15 possible combinations of $\vec{n}$, namely $\vec{n} = (R_A)_i$, $\vec{n} = (R_B)_i$ or $\vec{n} = (R_A)_i \times (R_B)_j$ for $i, j \in \{x, y, z\}$ and $i \neq j$.

Equation (2.15) can be simplified if we carry out the computations in A's local-coordinate system, as opposed to the world-coordinate system. This can be done by translating all points by $-\vec{T}_A$ and rotating them by $\mathbf{R_A}^{-1} = \mathbf{R_A}^t$. This yields

$$\begin{aligned}\vec{T}_A &= (\vec{T}_A - \vec{T}_A) = (0,0,0) \\ \mathbf{R_A} &= \mathbf{R_A}^t \mathbf{R_A} = \mathbf{I_3} \\ \\ \vec{T}_B &= \mathbf{R_A}^t (\vec{T}_B - \vec{T}_A) \\ \mathbf{R_B} &= \mathbf{R_A}^t \mathbf{R_B}, \end{aligned} \quad (2.16)$$

where $\mathbf{I_3}$ is the 3×3 identity matrix. Substituting equation (2.16) into (2.15), we can explicitly derive the equations for all 15 possible tests for finding a separating axis for boxes A and B with respect to A's local-coordinate system. These results are summarized in Table 2.1, on page 59.

2.4.2 Computing Sphere-Sphere Intersections

The sphere-sphere intersection test is by far the simplest in this chapter. Two spheres are *not* intersecting if and only if the distance between their

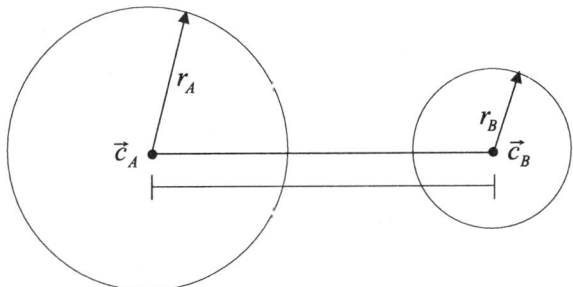

FIGURE 2.20. The sphere-sphere intersection test can be quickly conducted by comparing the distance between the centers of the sphere with the sum of their radii.

centers is greater than the sum of their radii. This is illustrated by Figure 2.20.

Let r_A and $\vec{c}_A$ be the radius and center of sphere A, respectively. Similarly, let r_B and $\vec{c}_B$ be the radius and center of sphere B. The spheres will not overlap if and only if

$$|\vec{c}_A - \vec{c}_B| > (r_A + r_B).$$

2.4.3 Computing Triangle-Triangle Intersections

The triangle-triangle intersection test is considered a primitive-primitive intersection test, since the triangles are in fact faces of the objects in the simulation. Let triangles A and B be defined by vertices $\vec{a}_1$, $\vec{a}_2$, $\vec{a}_3$ and $\vec{b}_1$, $\vec{b}_2$ and $\vec{b}_3$, respectively. The first step of the intersection test is to conduct a quick rejection test. This test consists of determining whether all vertices of one triangle lie on the same side of the plane defined by the other triangle. Let P_a and P_b be the planes defined by triangles A and B, respectively. Let $\vec{n}_a$ and $\vec{n}_b$ be the normal vectors of P_a and P_b. The normals can be directly computed from the vertex list as

$$\begin{aligned} \vec{n}_a &= (\vec{a}_2 - \vec{a}_1) \times (\vec{a}_3 - \vec{a}_1) \\ \vec{n}_b &= (\vec{b}_2 - \vec{b}_1) \times (\vec{b}_3 - \vec{b}_1). \end{aligned}$$

The vertices of triangle B will lie on the same side of P_a if and only if

$$\begin{aligned} &\vec{n}_a \cdot (\vec{b}_1 - \vec{a}_1) \\ &\vec{n}_a \cdot (\vec{b}_2 - \vec{a}_1) \\ &\vec{n}_a \cdot (\vec{b}_3 - \vec{a}_1) \end{aligned} \qquad (2.17)$$

are not zero and have the same sign. If they do not have the same sign, then the following cases can occur.

2.4 Collision Detection Between Different Hierarchical Representations   47

Case 1:
Two of the three equations defined in (2.17) have the same sign and the third evaluates to zero, say that corresponding to $\vec{b}_2$. In this case, the intersection between triangle B and plane P_a is single point, that is, vertex $\vec{b}_2$ (see Figure 2.21). The triangle-triangle intersection test is then reduced to check whether $\vec{b}_2$ lies inside triangle A. This point-in-triangle test can be quickly done by considering the line segment connecting $\vec{b}_2$ to the barycenter of A. If this line segment intersects one of the edges of triangle A, then $\vec{b}_2$ lies outside the triangle. Otherwise, $\vec{b}_2$ lies inside triangle A, and triangle B intersects triangle A. More details on how to implement the point-in-triangle test are given in Section 2.4.4.

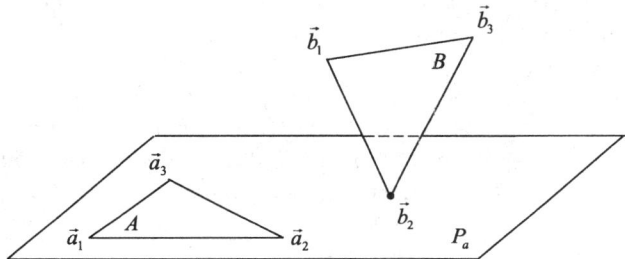

FIGURE 2.21. Case where just one vertex of triangle B touches the plane P_a containing triangle A. As shown, $\vec{b}_2$ lies outside A and the triangles do not intersect.

Case 2:
Two of the three equations defined in (2.17) evaluate to zero, say those corresponding to $\vec{b}_1$ and $\vec{b}_2$. In this case, the intersection between triangle B and plane P_a is the line segment defined by $(\vec{b}_1, \vec{b}_2)$ (see Figure 2.22). We can then apply the point-in-triangle intersection test for each one of the vertices $\vec{b}_1$ and $\vec{b}_2$. If both vertices lie outside triangle A, then triangle B does not intersect triangle A. Otherwise, if one of the vertices lies inside and the other outside, or if both lie inside, then triangle B intersects triangle A.

Case 3:
All three equations defined in (2.17) evaluate to zero. In this case, triangles A and B are coplanar (see Figure 2.23). The intersection test can then be reduced to three point-in-triangle tests by considering the three line segments connecting the vertices of B to the barycenter of A. If at least one of these line segments intersects an edge of A, or the vertex of B associated with this line segment lies on an edge of A, then this vertex is guaranteed to be inside triangle A. In this case, triangle B intersects triangle A. However, if all line segments do not intersect an edge of A,

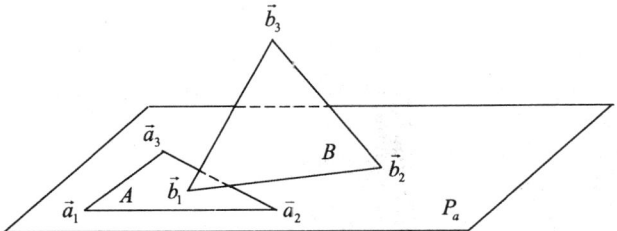

FIGURE 2.22. Case where an edge of triangle B is coplanar with triangle A. As shown, vertices $\vec{b}_1$ and $\vec{b}_2$ lie inside and outside A, respectively, and the triangles intersect.

then triangle B does not intersect the edges of triangle A, but there is still the possibility that A is completely inside B. Therefore, the same test is repeated for A, namely, the vertices of A are connected to the barycenter of B and the line segments are checked for intersection with the edges of B. Again, if at least one of these line segments does not intersect an edge of B, then the vertex of A associated with this line segment lies inside triangle B. In this case, triangle A is inside (i.e., intersects) triangle B.

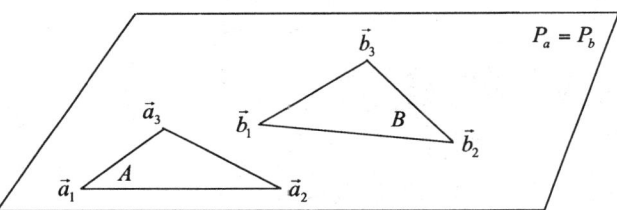

FIGURE 2.23. Case where the triangles A and B are coplanar. As shown, the vertices of B lie outside A, and the triangles do not intersect.

Case 4:
None of the three equations defined in (2.17) evaluates to zero. In this case, we shall have two vertices of B on one side of plane P_a, and the third vertex on the other side of P_a. This is illustrated by Figure 2.24.

Let $\vec{b}_2$ be the vertex that lies on the opposite side of plane P_a. Triangle B will then intersect plane P_a in two points $\vec{p}_1$ and $\vec{p}_2$ defining a line segment on the plane of A. These points can be computed as follows. Consider edge $e_1 = (\vec{b}_1, \vec{b}_2)$ given by its parameterized equation

$$\vec{p} = \vec{b}_1 + t(\vec{b}_2 - \vec{b}_1), \qquad (2.18)$$

where $0 \leq t \leq 1$ and $\vec{p}$ is a point on the edge. The plane equation of P_a is given by

$$\vec{n}_a \cdot \vec{p} = d, \qquad (2.19)$$

2.4 Collision Detection Between Different Hierarchical Representations 49

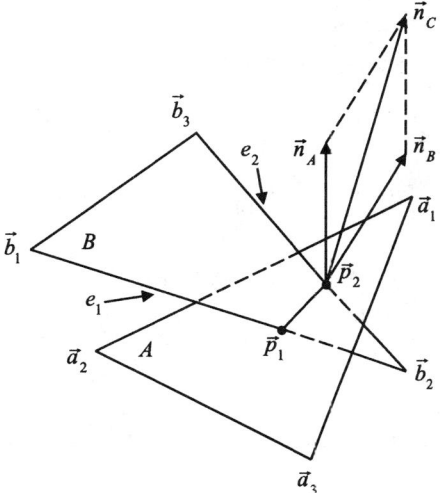

FIGURE 2.24. Triangle-triangle intersection test for the case where no vertices of triangle B are coplanar with triangle A. One vertex of B will lie on the opposite side of the other two vertices, with respect to the plane defined by triangle A.

where $\vec{p}$ is any point on the plane and d is the plane constant given by

$$d = \vec{n}_a \cdot \vec{a}_1 = \vec{n}_a \cdot \vec{a}_2 = \vec{n}_a \cdot \vec{a}_3 \ .$$

Edge e_1 intersects plane P_a at a point $\vec{p}_1$ that satisfies both equations (2.18) and (2.19). Substituting equation (2.18) into (2.19), we can compute the value of t corresponding to the intersection point $\vec{p}_1$, that is

$$t_p = \frac{d - \vec{n}_a \cdot \vec{b}_1}{\vec{n}_a \cdot (\vec{b}_2 - \vec{b}_1)} \ . \tag{2.20}$$

Substituting equation (2.20) back into (2.18), we can immediately find the intersection point $\vec{p}_1$. A similar computation can be done to find the intersection point $\vec{p}_2$ between edge $e_2 = (\vec{b}_2, \vec{b}_3)$ and plane P_a.

The line segment $(\vec{p}_1, \vec{p}_2)$ can then be checked for intersection with triangle A. We can apply the point-in-triangle intersection test for both $\vec{p}_1$ and $\vec{p}_2$. If both points lie outside triangle A, then triangle B does not intersect triangle A. Otherwise, if one of the vertices lies inside and the other outside, or if both lie inside, then triangle B intersects triangle A.

2.4.4 Point-in-Triangle Test

All point-in-triangle tests described in each of the above cases use the segment-segment intersection test, where the line segments are coplanar. This test can be efficiently implemented as follows. Let $s_1 = (\vec{p}_1, \vec{p}_2)$ and $s_2 = (\vec{q}_1, \vec{q}_2)$ be the line segments being tested for intersection, and let

$\vec{n}$ be the normal vector of the plane that contains both segments. The parameterized equations of the segments are then

$$\begin{aligned}\vec{p} &= \vec{p}_1 + t(\vec{p}_2 - \vec{p}_1) \\ \vec{q} &= \vec{q}_1 + m(\vec{q}_2 - \vec{q}_1),\end{aligned}$$

with $0 \le t \le 1$ and $0 \le m \le 1$. The first step of the intersection test consists of carrying out a quick rejection test. This test consists of checking whether the line segments are parallel, that is, checking whether

$$(\vec{p}_2 - \vec{p}_1) \times (\vec{q}_2 - \vec{q}_1) = \vec{0}.$$

If the line segments are not parallel, they will intersect if and only if there exist $t = t_p$ and $m = m_q$ such that

$$\vec{p}_1 + t_p(\vec{p}_2 - \vec{p}_1) = \vec{q}_1 + m_q(\vec{q}_2 - \vec{q}_1), \qquad (2.21)$$

with $0 \le t_p \le 1$ and $0 \le m_q \le 1$. Equation (2.21) can be solved for t_p and m_q if we consider two auxiliary vectors $\vec{k}_p$ and $\vec{k}_q$ given by

$$\begin{aligned}\vec{k}_p &= \vec{n} \times (\vec{p}_2 - \vec{p}_1) \\ \vec{k}_q &= \vec{n} \times (\vec{q}_2 - \vec{q}_1),\end{aligned} \qquad (2.22)$$

that is, $\vec{k}_p$ and $\vec{k}_q$ are non-zero vectors perpendicular to $(\vec{p}_2 - \vec{p}_1)$ and $(\vec{q}_2 - \vec{q}_1)$, respectively. If we apply a dot product by $\vec{k}_p$ on both sides of equation (2.21), then the term multiplying t_p evaluates to zero, and we can therefore determine m_q as

$$m_q = \frac{\vec{k}_p \cdot (\vec{p}_1 - \vec{q}_1)}{\vec{k}_p \cdot \vec{q}_2 - \vec{q}_1)}.$$

If $m_q < 0$ or $m_q > 1$, then the intersection point lies outside the line segment s_2, and the segments do not intersect. Otherwise, we apply a dot product by $\vec{k}_q$ on both sides of equation (2.21), and obtain t_p as

$$t_p = \frac{\vec{k}_p \cdot (\vec{q}_1 - \vec{p}_1)}{\vec{k}_q \cdot \vec{p}_2 - \vec{p}_1)}.$$

Again, if $t_p < 0$ or $t_p > 1$, then the intersection lies outside the line segment s_1, and the segments do not intersect. Otherwise, the segments intersect at the intersection point computed by substituting either t_p or m_q into equation (2.21).

2.4.5 Computing Box-Sphere Intersections

The intersection between a box and a sphere is carried out by considering the point in the boundary of the box that is closest to the sphere, and checking whether its distance to the center of the sphere is greater than the sphere's radius. If the distance is less than or equal to the sphere's radius, then the box intersects the sphere.

Let $\vec{p}$ be a point in the box, and let $\vec{c}$ and r be the sphere's center and radius, respectively. Let x_{min}, x_{max}, y_{min}, y_{max}, z_{min} and z_{max} define the minimum and maximum values of the boundary of the box along each of the coordinated axes, as shown in Figure 2.25.

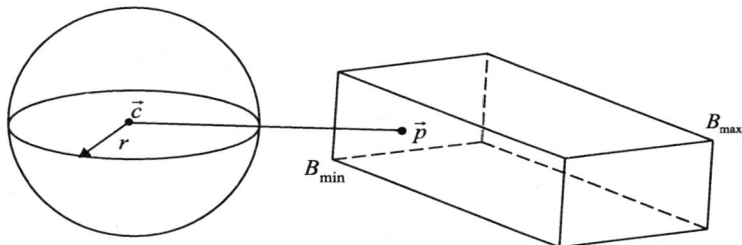

FIGURE 2.25. The closest point to the sphere is on the boundary of the box and minimizes the distance to the center given by equation 2.23.

The square of the distance from $\vec{p}$ to $\vec{c}$ is then given by

$$d^2 = (c_x - p_x)^2 + (c_y - p_y)^2 + (c_z - p_z)^2 . \tag{2.23}$$

The point $\vec{p}$ that is closest to the sphere is that which minimizes equation (2.23), subject to the following constraints:

$$x_{min} \leq p_x \leq x_{max}$$
$$y_{min} \leq p_y \leq y_{max}$$
$$z_{min} \leq p_z \leq z_{max} \quad .$$

Notice that each term of the equation (2.23) is non-negative and can be independently minimized. For example, if $x_{min} \leq c_x \leq x_{max}$, then $p_x = c_x$ minimizes the term $(c_x - p_x)^2$. However, if $c_x < x_{min}$ or $C_x > x_{max}$, then $p_x = x_{min}$ or $p_x = x_{max}$ minimizes the term, respectively. We do a similar analysis for finding the value of p_y and p_z that minimizes their corresponding quadratic terms.

Having determined the coordinates of the closest point to the sphere, we just need to compare its distance to the center of the sphere with the sphere's radius by substituting the coordinates of $\vec{p}$ into equation (2.23), and checking whether

$$d^2 \leq r^2 . \tag{2.24}$$

The box will intersect the sphere if and only if equation (2.24) is satisfied.

2.4.6 Computing Box-Triangle Intersections

The box-triangle intersection test can be quickly carried out in at most three steps. In the first step, we check whether the vertices of the triangle are inside the box. If at least one of the vertices is inside the box, then the triangle intersects the box.

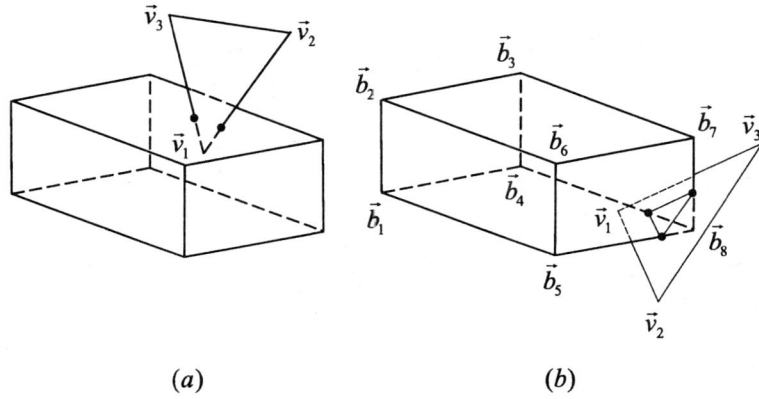

FIGURE 2.26. (a) The triangle intersects the box whenever one of its vertices lies inside the box; (b) The plane containing the triangle intersects the box. Edges of the box that have vertices at opposite sides of the plane, in this case edges $\overline{b_5 b_8}$, $\overline{b_4 b_8}$ and $\overline{b_7 b_8}$, need to be checked for intersection with the triangle.

Let the box be defined by its minimum and maximum vertices, and let $\vec{v}_1$, $\vec{v}_2$ and $\vec{v}_3$ be the vertices of the triangle (see Figure 2.26(a)). Vertex $\vec{v}_i$ is inside the box if and only if

$$\begin{aligned} x_{min} &\leq (v_i)_x \leq x_{max} \\ y_{min} &\leq (v_i)_y \leq y_{max} \\ z_{min} &\leq (v_i)_z \leq z_{max} \end{aligned} \tag{2.25}$$

If at least one of the vertices $\vec{v}_1$, $\vec{v}_2$ or $\vec{v}_3$ satisfies equation (2.25), then the triangle intersects the box. Otherwise, we proceed to the second step.

In the second step, we check whether the plane containing the triangle intersects the box. This can be done by checking whether the eight vertices of the box lie on the same side of the plane (see Figure 2.26(b)). Let $\vec{n}$ be the normal of the triangle and d the plane constant determined from

2.4 Collision Detection Between Different Hierarchical Representations

$$d = \vec{n} \cdot \vec{v}_i, \text{ for } i \in \{1, 2, 3\}.$$

A point $\vec{p}$ is classified with respect to the plane containing the triangle as follows:

$$\begin{aligned} \text{If } \vec{n} \cdot \vec{p} - d > 0 &\Rightarrow \vec{p} \text{ is on positive half-plane} \\ \text{If } \vec{n} \cdot \vec{p} - d = 0 &\Rightarrow \vec{p} \text{ lies on the plane} \\ \text{If } \vec{n} \cdot \vec{p} - d < 0 &\Rightarrow \vec{p} \text{ is on negative half-plane}. \end{aligned} \quad (2.26)$$

Using equation (2.26), we classify each vertex of the box according to its relative position with respect to the plane. If all vertices lie on the same half-space, then we can immediately conclude that the box does not intersect the triangle. Otherwise, we need to consider the edges of the box that intersect the plane, that is, the edges that have vertices at opposite sides of the plane, or one vertex on the plane and another on either side. These edges define line segments and a line segment-triangle intersection test is done for each of them, as explained in detail in Section 2.4.9.

2.4.7 Computing Sphere-Triangle Intersections

The sphere-triangle test is more complex than the box-triangle test, in the sense that it has more steps to be carried out before we can determine whether the sphere is intersecting the triangle.

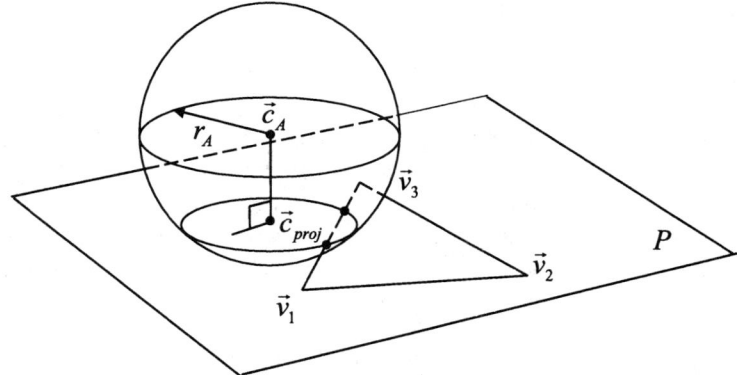

FIGURE 2.27. The triangle intersects the sphere whenever one of its vertices lies inside the sphere, that is, the distance from one of its vertices to the center of the sphere is less than or equal to the sphere's radius.

The first step is to check whether the plane that contains the triangle intersects the sphere. This can be done by comparing the distance of the plane to the center of the sphere with the radius of the sphere. Let r_A and $\vec{c}_A$ be the sphere's radius and center. Let $\vec{v}_1$, $\vec{v}_2$ and $\vec{v}_3$ be the vertices of the triangle defining plane P (see Figure 2.27). Let $\vec{n}$ and d_n be the plane normal and plane constant. The distance between the plane P and the sphere's center is then

$$d_A = |\vec{n} \cdot \vec{c}_A - d_n|.$$

The plane containing the triangle intersects the sphere whenever

$$d_A \leq r_A.$$

If this is the case, then we proceed with the sphere-triangle intersection test by checking whether the vertices of the triangle are inside the sphere. If at least one of the vertices is inside the sphere, then the triangle intersects the sphere. Let d_i be the distance between vertex $\vec{v}_i$ and the sphere's center, that is

$$d_i = |\vec{v}_i - \vec{c}_A|.$$

The sphere will intersect the triangle if

$$d_i \leq r_A \tag{2.27}$$

for at least one vertex $\vec{v}_i$ (see Figure 2.27). If this is not the case, then we proceed to the third step of the sphere-triangle intersection test. In this step, we project the sphere onto the plane containing the triangle, and check whether the projected center lies inside the triangle. The projected center $\vec{c}_{proj}$ is determined from

$$\vec{c}_{proj} = \vec{c}_A - d_n \vec{n}.$$

We can use the point-in-triangle test already explained in Section 2.4.4 to see whether the projected center lies inside the triangle. If the projected center $\vec{c}_{proj}$ lies inside the triangle, then the sphere intersects the triangle (see Figure 2.27). Otherwise, we need to do one more test to check whether the triangle edges intersect the sphere, as explained in the next section.

2.4.8 Computing Line Segment-Sphere Intersections

Let $\vec{p}_1$ and $\vec{p}_2$ define a line segment S, and $\vec{c}_A$ and r_A be the sphere's center and radius, respectively. Consider the line L passing through $\vec{c}_A$ and perpendicular to the line segment S (see Figure 2.28).

Line L will intersect the line segment S at a point $\vec{c}_{proj}$ such that

$$\vec{c}_{proj} = \vec{p}_1 + t(\vec{p}_2 - \vec{p}_1)$$

2.4 Collision Detection Between Different Hierarchical Representations

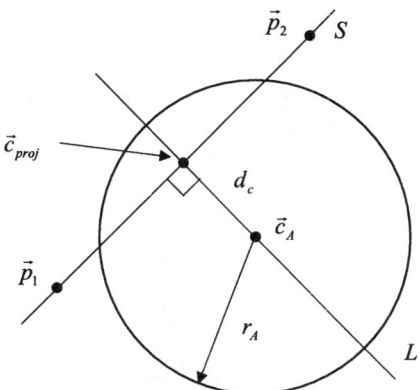

FIGURE 2.28. Intersection test between a sphere and a line segment. We only need consider the intersection of the line segment with the circle resulting from the intersection of the sphere and the plane defined by the center of the sphere, and the end points of the line segment.

for t, given by

$$t = \frac{(\vec{p}_2 - \vec{p}_1) \cdot \vec{c}_A - (\vec{p}_2 - \vec{p}_1) \cdot \vec{p}_1}{(\vec{p}_2 - \vec{p}_1) \cdot (\vec{p}_2 - \vec{p}_1)},$$

that is, $\vec{c}_{proj}$ is the projection of the center of the sphere onto the line segment's supporting line. The distance between the projection point $\vec{c}_{proj}$ and the center of the sphere is directly obtained from

$$d_c^2 = (\vec{c}_{proj} - \vec{c}_A) \cdot (\vec{c}_{proj} - \vec{c}_A).$$

Having determined the projection point $\vec{c}_{proj}$ and its (squared) distance d_c^2 to the center of the sphere, one of the of the following three cases occurs.

1. If $d_c^2 > r_A^2$, then the sphere does not intersect the triangle.

2. If $d_c^2 = r_A^2$, then the supporting line is tangential to the sphere. If the projection point $\vec{c}_{proj}$ is inside the segment, that is, if $0 \leq t \leq 1$, then the line segment intersects the sphere.

3. If $d_c^2 < r_A^2$, then the supporting line intersects the sphere. The segment intersects the sphere if the projection point $\vec{c}_{proj}$ is inside the segment. Otherwise, an intersection occurs if the closest end point to $\vec{c}_{proj}$ is inside the sphere. The closest end point is $\vec{p}_1$ if $t \leq 0$, or $\vec{p}_2$ if $t \geq 1$.

2.4.9 Computing Line Segment-Triangle Intersections

The intersection of a line segment S with a triangle A can be viewed as a subset of the intersection test between two triangles. Let the line segment

56   2. Hierarchical Representation of 3D Polyhedra

be defined by vertices $\vec{s}_1$ and $\vec{s}_2$, and the triangle be defined by vertices $\vec{a}_1$, $\vec{a}_2$ and $\vec{a}_3$.

First, we check whether the vertices defining the line segment lie on the same side of the plane containing the triangle (see Figure 2.29). If this is so, then we can quickly conclude that the segment does not intersect the triangle. Otherwise, we can have one of the following three cases.

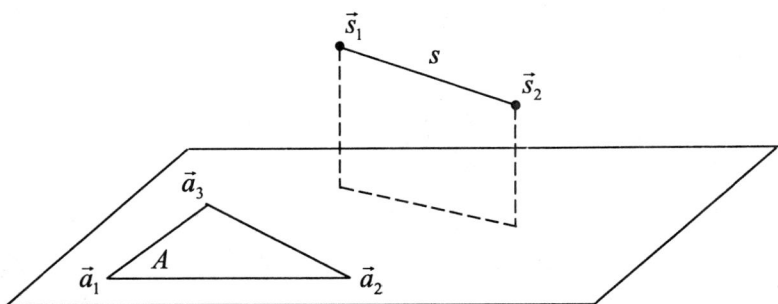

FIGURE 2.29. Case where the vertices defining the line segment lie on the same side of the plane containing triangle A. As shown, the line segment does not intersect the triangle.

Case 1:
One vertex of the line segment, say vertex $\vec{s}_2$, lies on the plane that contains the triangle, and the other lies on either side (see Figure 2.30). In this case, we use the point-in-triangle test for checking whether $\vec{s}_2$ lies inside A. The line segment intersects the triangle only if $\vec{s}_2$ lies inside A.

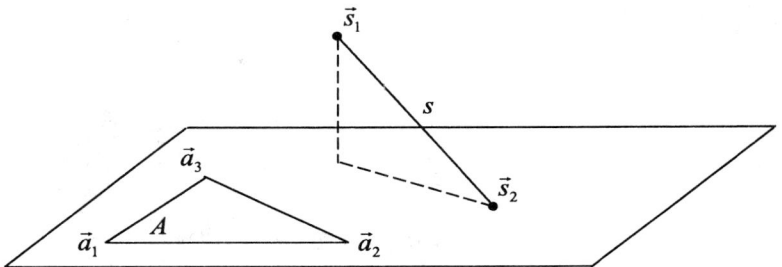

FIGURE 2.30. Case where one vertex of the line segment is coplanar with triangle A. As shown, vertex $\vec{s}_2$ lies outside A, and the line segment does not intersect the triangle.

Case 2:
Both vertices of the line segment lie on the plane containing A (see Figure 2.31). Again, we use the point-in-triangle test for checking the relative

location of the vertices with respect to the triangle. The line segment intersects the triangle only if one of its vertices lies inside the triangle and the other outside, or if both vertices lie inside the triangle.

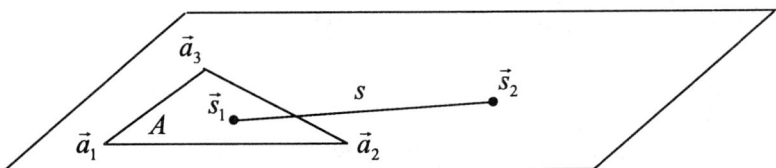

FIGURE 2.31. Case where the line segment is coplanar with triangle A. In the situation shown, vertices $\vec{s}_1$ and $\vec{s}_2$ lie inside and outside A, respectively, and the line segment does intersect the triangle.

Case 3:
The vertices of the line segment lie on opposite sides of the plane containing the triangle (see Figure 2.32). Let $\vec{p}_1$ be the intersection between the line segment and the plane containing the triangle. Using the point-in-triangle test, we can check whether $\vec{p}_1$ lies inside the triangle.

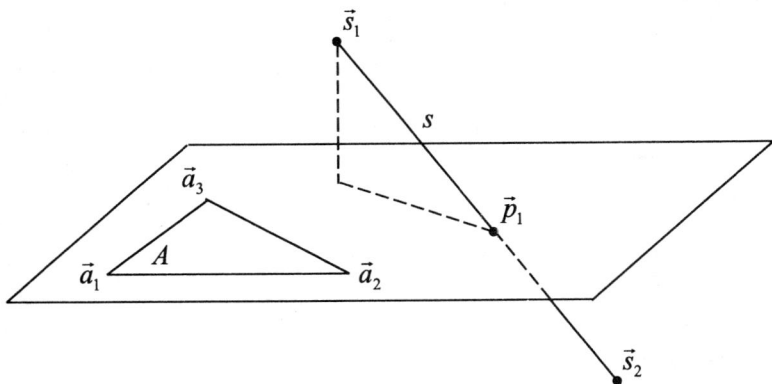

FIGURE 2.32. Case where the vertices defining the line segment lie on opposite sides of the plane containing triangle A. As shown, the intersection point $\vec{p}_1$ between the line segment and the plane lies outside A, and the line segment does not intersect the triangle.

2.5 Notes and Comments

The literature on hierarchical decompositions is extensive, with related publications on several research areas such as computational geometry, computer graphics, robotics and molecular simulations. There are several other representations and variants of the techniques presented in this chapter, specially with respect to implementation.

The OBB tree representation became an option since Gottschalk *et al.* [GLM96, Got96] introduced the separating-axis theorem for carrying out fast interference detection between arbitrarily oriented boxes. Bergen [vdB97] presented a modified interference-detection test using AABB tree representations in which the search for a separating axis considered only the normals of the faces of the boxes. The pair-wise edge-direction tests were ignored, thus reducing the complexity of the test, but consequently being about 6% less accurate.

The use of bounding spheres, instead of bounding boxes, is also popular, owing to the simplicity of its implementation. The efficiency of bounding-sphere representations can be further improved if we use quad-trees or oct-trees instead of binary trees. Samet [Sam89] gives a good introduction to both quad-tree and oct-tree representations. The difficulty in using bounding-sphere representations is to come up with a partition that best approximates the original polyhedra. Hubbard [Hub96] developed a collision-detection algorithm that approximates 3D polyhedra with an oct-tree representation of bounding spheres, using a sophisticated technique based on 3D Voronoi diagrams to construct the spheres at each intermediate level of the decomposition.

The 3D convex-hull computation can be found in Preparata *et al* [PS85], Edelsbrunner [Ede87], and in several other books on computational geometry. In the OBB case, there is also the need to determine the eigenvectors of the covariance matrix of the vertices of the convex hull. Eigenvectors and their associated eigenvalues are covered in detail in Strang [Str91], Golub [GL96] and Horn [HJ91].

The multi-level grid-structure analysis presented in Section 2.3.2 was derived from Mirtich [Mir96b]. Some primitive-primitive tests presented in Section 2.4 were obtained from Gottschalk [Got96] (box-box), Arvo [Arv90] (box-sphere), Ritter [Rit90] (sphere-sphere) and Karabassi *et al.* [KPTB99] (sphere-triangle). The triangle-triangle intersection test presented in Section 2.4.3 is a combination of the three different intersection tests presented in Held [Hel97], Möller [Möl97] and Glaeser [Gla94]. Other interesting primitive-primitive tests involving cones and cylinders can be found in Held [Hel97].

Separating Axis $\vec{n}$	Simplified Overlap Test
$(R_A)_x$	$\|(T_B)_x\| > (a_x + b_x\|(R_B)_{xx}\| + b_y\|(R_B)_{xy}\| + b_z\|(R_B)_{xz}\|)$
$(R_A)_y$	$\|(T_B)_y\| > (a_y + b_x\|(R_B)_{yx}\| + b_y\|(R_B)_{yy}\| + b_z\|(R_B)_{yz}\|)$
$(R_A)_z$	$\|(T_B)_z\| > (a_z + b_x\|(R_B)_{zx}\| + b_y\|(R_B)_{zy}\| + b_z\|(R_B)_{zz}\|)$
$(R_B)_x$	$\|(T_B)_x (R_B)_{xx} + (T_B)_y (R_B)_{yx} + (T_B)_z (R_B)_{zx}\| > (b_x + a_x\|(R_B)_{xx}\| + a_y\|(R_B)_{yx}\| + a_z\|(R_B)_{zx}\|)$
$(R_B)_y$	$\|(T_B)_x (R_B)_{xy} + (T_B)_y (R_B)_{yy} + (T_B)_z (R_B)_{zy}\| > (b_y + a_x\|(R_B)_{xy}\| + a_y\|(R_B)_{yy}\| + a_z\|(R_B)_{zy}\|)$
$(R_B)_z$	$\|(T_B)_x (R_B)_{xz} + (T_B)_y (R_B)_{yz} + (T_B)_z (R_B)_{zz}\| > (b_z + a_x\|(R_B)_{xz}\| + a_y\|(R_B)_{yz}\| + a_z\|(R_B)_{zz}\|)$
$(R_A)_x \times (R_B)_x$	$\|(T_B)_z (R_B)_{yx} - (T_B)_y (R_B)_{zx}\| > (a_y\|(R_B)_{zx}\| + a_z\|(R_B)_{yx}\| + b_y\|(R_B)_{xz}\| + b_z\|(R_B)_{xy}\|)$
$(R_A)_x \times (R_B)_y$	$\|(T_B)_z (R_B)_{yy} - (T_B)_y (R_B)_{zy}\| > (a_y\|(R_B)_{zy}\| + a_z\|(R_B)_{yy}\| + b_x\|(R_B)_{xz}\| + b_z\|(R_B)_{xx}\|)$
$(R_A)_x \times (R_B)_z$	$\|(T_B)_z (R_B)_{yz} - (T_B)_y (R_B)_{zz}\| > (a_y\|(R_B)_{zz}\| + a_z\|(R_B)_{yz}\| + b_x\|(R_B)_{xy}\| + b_y\|(R_B)_{xx}\|)$
$(R_A)_y \times (R_B)_x$	$\|(T_B)_x (R_B)_{zx} - (T_B)_z (R_B)_{xx}\| > (a_x\|(R_B)_{zx}\| + a_z\|(R_B)_{xx}\| + b_y\|(R_B)_{yz}\| + b_z\|(R_B)_{yy}\|)$
$(R_A)_y \times (R_B)_y$	$\|(T_B)_x (R_B)_{zy} - (T_B)_z (R_B)_{xy}\| > (a_x\|(R_B)_{zy}\| + a_z\|(R_B)_{xy}\| + b_x\|(R_B)_{yz}\| + b_z\|(R_B)_{yx}\|)$
$(R_A)_y \times (R_B)_z$	$\|(T_B)_x (R_B)_{zz} - (T_B)_z (R_B)_{xz}\| > (a_x\|(R_B)_{zz}\| + a_z\|(R_B)_{xz}\| + b_x\|(R_B)_{yy}\| + b_y\|(R_B)_{yx}\|)$
$(R_A)_z \times (R_B)_x$	$\|(T_B)_y (R_B)_{xx} - (T_B)_x (R_B)_{yx}\| > (a_x\|(R_B)_{yx}\| + a_y\|(R_B)_{xx}\| + b_y\|(R_B)_{zz}\| + b_z\|(R_B)_{zy}\|)$
$(R_A)_z \times (R_B)_y$	$\|(T_B)_y (R_B)_{xy} - (T_B)_x (R_B)_{yy}\| > (a_x\|(R_B)_{yy}\| + a_y\|(R_B)_{xy}\| + b_x\|(R_B)_{zz}\| + b_z\|(R_B)_{zx}\|)$
$(R_A)_z \times (R_B)_z$	$\|(T_B)_y (R_B)_{xz} - (T_B)_x (R_B)_{yz}\| > (a_x\|(R_B)_{yz}\| + a_y\|(R_B)_{xz}\| + b_x\|(R_B)_{zy}\| + b_y\|(R_B)_{zx}\|)$

TABLE 2.1. The 15 candidate separating axes and their associated tests with respect to A's local-coordinate system. The boxes are overlapping if and only if all tests fail.

3
Particle Systems

3.1 Introduction

Particles are among the simplest and most versatile objects used in dynamic simulations. The fact that their mass is concentrated on a point (i.e., center of mass) considerably simplifies the dynamic equations governing their motion. All interaction forces among themselves and with other objects in the simulation are applied to the points representing each particle, and the rotational motion of a point is undefined, and therefore ignored. The reduced complexity in the dynamic equations allows for an increased number of particles being simulated without significantly imparing operation of the simulation engine. These simplifications make particle systems an extremely attractive option to simulate systems requiring a large number of objects that can be approximated as a collection of point-mass objects. Examples of such systems range from molecules, to smoke, fire, clouds, liquids, and even cloth.

The main difference between the diverse set of particle systems in use nowadays resides in the types of interaction forces considered and the numerical-integration methods used to solve their equations of motion. Since particle systems usually need a large number of particles to achieve their desired effects, the complexity of the computation of the interaction forces between particles, other objects and the simulated environment plays a key role in overall simulation efficiency. Expensive interaction forces such as those that are spatially dependent, that is, their intensity varies with respect to the distance between the particles, can severely impair the efficiency, specially in naive implementations. For example, consider a molecular-dynamics simulation where the spatially dependent Lennard-Jones potential force acts between pairs of particles representing atoms. The computational cost of determining the potential forces between all pairs of particles is therefore $\mathcal{O}(n^2)$ for a particle system containing n particles. Clearly, a naive implementation of the potential-force computation becomes prohibitively expensive, even for a moderate number of particles such as $n = 1000$.

The numerical-integration methods also play an important role in the overall simulation accuracy and robustness. Recall that the actual motion of the particles is determined from the numerical results obtained from the numerical-integration module. A fast but inaccurate method can produce unsatisfactory results that do not reflect the desired behavior of the particles in the system. On the other hand, an accurate numerical method can generate an unsatisfactory performance hit that deems useless any attempts to deliver interactive simulation speeds. In most cases, there is a trade-off between the computational accuracy of the models used to numerically integrate the evolution of particle systems over time, and the simulation's efficiency. Attaining interactive speeds using precise models of the particles' motion often requires computational power that is available only in high-performance computers, such as supercomputers, or a network of parallel computers.

In this book, we focus primarily on particle systems that can be analyzed as classical multibody systems, meaning, the motion of the center of mass of each particle follows the laws of classical mechanics. Such particle systems provide a reasonably good approximation of a wide variety of point-mass systems, such as dust, snow and rain. However, there are yet other types of point-mass systems that are widely used in dynamic simulations and animations that require specialized forms of equations of motion and force interactions to capture the precise physical behavior of the system. Examples of such particle systems include the simulation of turbulent gases (requires solving the volumetric differential Navier-Stokes equations), the simulation of light atoms or molecules (may need to take the quantum effects into account if the translational, vibrational and rotational motions are considered), and the detonation of explosives (requires using the Chapman-Jouget theory, possibly with the equations of the state for the detonated particles derived from laboratory experiments). The detailed explanation of the theoretical framework needed to capture the precise behavior of such specialized systems is beyond the scope of this book. The interested reader is referred to Section 3.10 for pointers to the literature wherein in-depth explanations of such techniques can be found.

Our approach to such particle systems as these, is the same as that commonly used in most animation environments. These software packages usually implement specialized particle systems having a set of user-adjustable parameters used to capture phenomena not otherwise considered in the standard particle-system model derived from classical mechanics. The idea is then to mimic the behavior of the system without having to solve the sophisticated (and computationally demanding) equations of motion associated with it. Section 3.9 presents some specialized particle systems commonly found in animation packages. It is important to notice that this is the *only* section in this book that *does not* consider the mathematically accurate modeling of the dynamics. It was included here to demonstrate the versatility and modeling power that particle systems can have on dynamic simulations and animations.

From the users' perspective, a particle system is defined by both a particle emitter and the particles themselves. The particle emitter, as its name implies, is the source from which particles are created and released in the simulation environment. It can be either attached to other objects in the simulation, or be seem as an object in itself, in which case, the emitter is displayed with its default cubic shape (see Section 3.8). The particle emitter settings are used to control the dynamic behavior of the particles being emitted. These settings define the particle's size, mass, initial velocity and direction of movement, and many other user-adjustable parameters as explained in Section 3.9.1. As soon as the particles are released, their motion is governed by the dynamic equations derived from classical mechanical theory. The particle's parameters give an extra degree of flexibility in governing their motion, providing the added functionality needed to implement the specialized systems covered in this book. Examples of such parameters include the particle's split age, lifespan, color attributes and collision-detection options.

3.2 Particle Dynamics

The dynamic equations that govern the motion of a particle in our standard implementation of a particle system are the same as those governing the motion of a point-mass object in classical mechanics. Let the point-mass be represented by its mass m located at position $\vec{p}(t)$, which varies as a function of time. The velocity of the point is obtained by computing the derivative of its position with respect to time, namely

$$\vec{v}(t) = \frac{d\vec{p}(t)}{dt} \ .$$

Its acceleration is then given by

$$\vec{a}(t) = \frac{d\vec{v}(t)}{dt} \ . \tag{3.1}$$

Let $\vec{F}(t)$ be the net external force acting on the particle at the time instant t. Using Newton's law, we have

$$\vec{F}(t) = \frac{d\vec{L}(t)}{dt}, \qquad (3.2)$$

where $\vec{L}(t)$ is the linear momentum of the particle computed as

$$\vec{L}(t) = m\vec{v}(t). \qquad (3.3)$$

Substituting equation (3.3) into (3.2) and using equation (3.1) we obtain

$$\vec{F}(t) = \frac{d(m\vec{v}(t))}{dt} = m\frac{d\vec{v}(t)}{dt} = m\vec{a}(t). \qquad (3.4)$$

Let $\vec{y}(t)$ denote the dynamic state of the particle at time t, that is, the vector comprising all variables necessary to define the dynamics of the particle at any instant during the simulation. Here, we shall pick the position and linear momentum of the particle to define its dynamic state, namely

$$\vec{y}(t) = \begin{pmatrix} \vec{p}(t) \\ \vec{L}(t) \end{pmatrix}.$$

The dynamic state of the particle at time $t = t_0$ is defined by the particle's position $\vec{p}(t_0)$ and its linear momentum $\vec{L}(t_0)$ computed as $m\vec{v}(t_0)$.

The time derivative of the dynamic state defines how the dynamic state of the particle changes over time, and is given by

$$\frac{d\vec{y}(t)}{dt} = \begin{pmatrix} d\vec{p}(t)/dt \\ d\vec{L}(t)/dt \end{pmatrix} = \begin{pmatrix} \vec{v}(t) \\ \vec{F}(t) \end{pmatrix}.$$

So, the time derivative of the dynamic state at time $t = t_0$ is defined by the particle's velocity $\vec{v}(t_0)$ computed as $(\vec{L}(t_0)/m)$ and the net force $\vec{F}(t_0)$ acting on it.

For a system with N particles, we can combine their individual dynamic states into a single system-wide dynamic-state vector

$$\vec{Y}(t) = \begin{pmatrix} \vec{p}_1(t) \\ \vec{L}_1(t) \\ \ldots \\ \vec{p}_N(t) \\ \vec{L}_N(t) \end{pmatrix}$$

with its corresponding time derivative

$$\frac{d\vec{Y}(t)}{dt} = \begin{pmatrix} \vec{v}_1(t) \\ \vec{F}_1(t) \\ \ldots \\ \vec{v}_N(t) \\ \vec{F}_N(t) \end{pmatrix}. \qquad (3.5)$$

The dynamic simulation of a particle system works as follows. At the beginning of the simulation, we have the dynamic state of every particle, namely their positions and linear momenta defined with respect to the world reference frame. Each simulation time step will then consist of numerically integrating equation (3.5), using the dynamic state of the particles at the beginning of the time step as the initial condition for the numerical integration. There are several numerical methods that can be used to integrate equation (3.5). For example, the Euler method computes a quick (and less accurate) approximation of the time derivative by using only the information about the dynamic state at the beginning of the time step to predict the dynamic state of the system at the end of that time step. Others, such as the several variations of the Runge-Kutta method, use a more sophisticated approach in which the dynamic state at the end of the time step is computed as a weighted sum of the dynamic state of the system at several intermediate positions within it. These and other popular methods are discussed in Appendix B.

The computation of the net external force acting on each particle at each intermediate step of the numerical integrator is determined by summing all external forces acting on the particles. The types of external forces considered in this book range from simple global forces (such as gravity), to point-to-point forces (such as springs), to more computationally demanding spatially dependent forces (such as windy regions). The detailed discussion on how to determine the contribution of each of these forces to the net external force of each particle is presented in Section 3.3.

Initially, the determination of the dynamic state of each particle at the end of the current time step is done without taking into account any possible collisions between particles and other objects in the simulation environment. The information about the initial and final dynamic state of each particle is then used to check for collisions between the particles themselves and with other rigid bodies in the simulation (see Section 3.4). Whenever a collision is detected, the colliding particles have their trajectories traced back in time to the moment just before the collision. The collision point and collision normal are then computed from the relative displacement of the colliding particles. Only then the collision-response module is activated to compute the appropriate impulsive or contact forces that will be applied to change the direction of motion of the colliding particles. This is done slightly differently if we have a particle-particle collision or a particle-rigid body collision (see Sections 3.6 and 3.7 for more details).

The dynamic equations of all particles involved in a collision are then numerically integrated for the remaining period of time, that is, from the collision time to the end of the current time step. This new numerical integration will update the current particles' trajectories to account for all collision forces. Notice that this also requires the numerical integration of the dynamic state of all other particles connected to one or more particles involved in a collision, since the connection usually implies the existence of a force component between the particles. For example, consider a simple particle system consisting of four particles O_1, O_2, O_3 and O_4, and suppose particles O_1 and O_2 are connected by a spring.

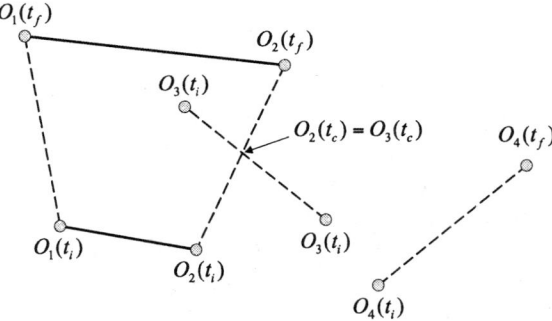

FIGURE 3.1. A simple particle system containing four particles. The dynamic state of the system is numerically integrated from t_i to t_f. A collision between particles O_2 and O_3 is detected at time t_c.

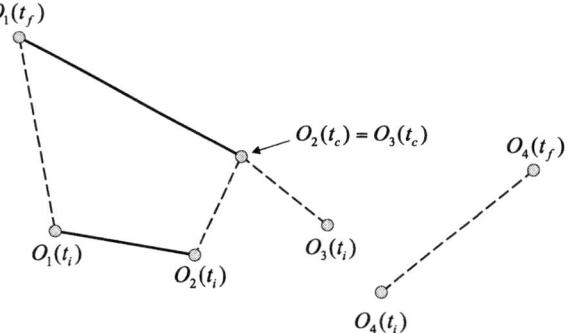

FIGURE 3.2. Particles O_2 and O_3 are traced back in time to the moment just before their collision.

Initially, the dynamic state of the system is numerically integrated from t_i (the beginning of the current time step) to t_f (the end of the current time step). Now, assume the collision-detection module detected a collision between particles O_2 and O_3 at time t_c such that $t_i < t_c < t_f$ (see Figure 3.1). The colliding particles are then traced back in time to the moment before their collision (i.e., traced back to t_c) and the collision impulses are computed so as to prevent their interpenetration. Having applied the collision impulses to both particles, their trajectories are numerically integrated for the remaining period of time, that is, from t_c to t_f. Notice that, if we just trace back in time the trajectories of O_2 and O_3, the spring-force computation between particles O_1 and O_2 will be incorrect in the numerical integration for the remaining period of time. The problem is that, since

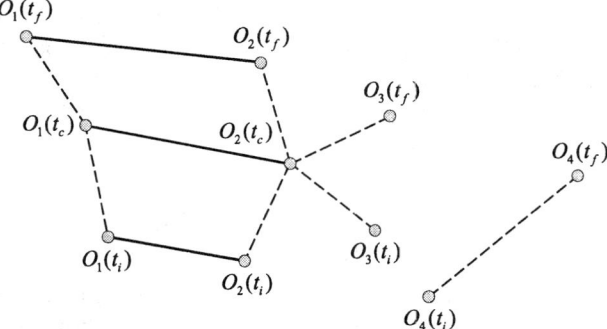

FIGURE 3.3. Particle O_1 is also traced back in time to t_c before the numerical integrator is used to recompute the trajectories of particles O_1, O_2 and O_3. Notice that particle O_4 was not affected by the collision, and therefore remained unchanged throughout the collision-detection and response phases.

particle O_1 was not involved in any collision, its dynamic state corresponds to time t_f, whereas the dynamic state of particle O_2 corresponds to time t_c. So, if we do not trace back in time O_1 too, the spring-force computation will use O_1's position at time t_f when it should use it at the same simulation time of O_2, namely time t_c (see Figure 3.2). In other words, the numerical integration of the dynamic state of all interconnected particles should be synchronized to provide the correct system behavior (see Figure 3.3). On the other hand, particles that are not connected can be asynchronously moved within the same simulation time step. This is the case of particle O_4 as shown in Figure (3.3), since the numerical integration of O_2 and O_3 for the remaining period of time does not affect its dynamic state, already computed.

As far as implementation is concerned, this approach requires some bookkeeping mechanism to efficiently determine which particles are connected to other particles and rigid bodies. The payoff is the significant efficiency gain over the alternative approach of tracing back in time all particles, even those not involved in any collision, to the moment before the most recent collision.

The information about the initial and final dynamic state of each particle along the updated part of its trajectory is used to check again for collisions between all other particles and rigid bodies in the simulation. This process repeats until all particle-particle and particle-rigid body collisions detected within the current time interval have been resolved.

Clearly, the collision check is an intense process that can consume much computational time, especially in a naive implementation. We suggest using the cell decomposition of the simulated world already discussed in Section 2.3 to speed the collision-detection checks. By so doing, as the system evolves, the position of each particle relative to the cell decomposition of the simulated world is tracked by assigning the particle to the cell in which it is included. This dynamic assignment is efficiently implemented by observing that the cell decomposition defines a uniform subdivision of the simulated world into cubic cells (see Section 2.3 for more details). A particle (i.e., point-mass) with coordinates $P = (p_x, p_y, p_z)$ will be included in the cell $C = (c_x, c_y, c_z)$ with

$$c_i = \left\lfloor \frac{p_i}{s_c} \right\rfloor \text{ for } i \in \{x, y, z\},$$

where s_c is the dimension of the cells in the decomposition. Notice that this cell assignment can also be used to speed the computation of spatially dependent forces as explained in Section 3.3.4.

3.3 Interaction Forces

The interaction forces used in most particle-system simulations can be categorized into three different types of forces. The first type considers global interaction forces, that is, forces that are independently applied to all particles in the system. Examples of such forces include gravity and viscous drag (used to simulate air resistance). They are the least expensive interaction forces available in the simulation environment, since their required computational cost is negligible compared with the other types of interaction forces presented in this book.

The second type considers interaction forces between a specific number of particles. Damped springs are a good example of such interaction forces between two given particles. Notice, however, that particles can be attached to more than one particle, and each attachment may use a different interaction force. For example, multiple attachments of springs to pairs of particles can be used to create a mesh of particles (i.e., a spring-mass system) that models cloth, as discussed in Section 3.9.2.

Interactive-user manipulation is also modeled as a point-to-point force between the current mouse position and the selected particle. The goal of using a fictitious interaction force between the mouse and the selected particle is to prevent the introduction of unstable configurations resulting from abrupt mouse movements, as explained in Section 3.3.5.

Finally, the third type of interaction forces considered in this book for particle-system simulations is the spatially dependent force. This comprises forces that depend on the position of the particles, either relative to each other or to the simulated environment. For instance, a gravitational force field depends on the relative positioning of the particles, in the sense that it will have a stronger influence on the motion of nearby particles than on the motion of those farther away. Another example is the interaction force created by defining a windy region on the simulated environment. Particles located on the windy region or passing through it will have to take the wind force into account when computing their net external force, whereas particles that are not in the windy region can ignore that force.

The spatially dependent forces are the most expensive interaction forces considered in particle-system simulations. Approximation methods are generally used to truncate the influence of the force field on particles that are more than one threshold value distant. This truncation technique is discussed in detail in Section 3.3.4.

3.3.1 Gravity

The force contribution of the gravitational force acting on each particle owing to its attraction to the ground (i.e., Earth) is directly obtained as

$$\vec{F} = m\vec{g},$$

where $\vec{g}$ is the gravity acceleration and m is the mass of the particle (see Figure 3.4). The gravity acceleration is in most cases assumed to have constant magnitude and direction pointing downwards (i.e., towards the ground).

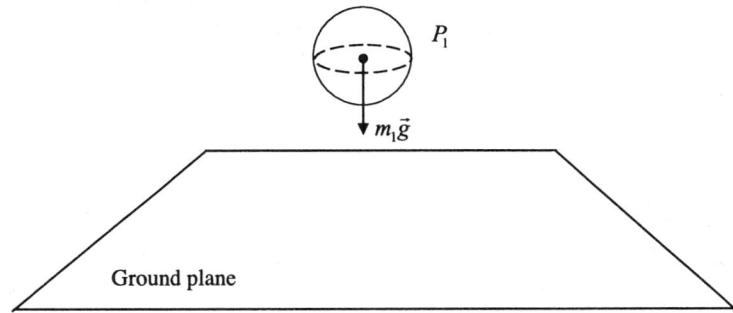

FIGURE 3.4. Gravity pulling particle P_1 with mass m_1 towards the ground plane.

3.3.2 Viscous Drag

The most common use of viscous drag in dynamic simulations of particle systems is to model the air resistance to the particle's movement. The goal is to ensure that particles will eventually come to a rest if there are no other external forces acting on them. Figure 3.5 illustrates this. The force component of the viscous drag is computed as

$$\vec{F} = -k_d \vec{v},$$

where $\vec{v}$ is the velocity vector of the particle and k_d is the coefficient of drag.

70   3. Particle Systems

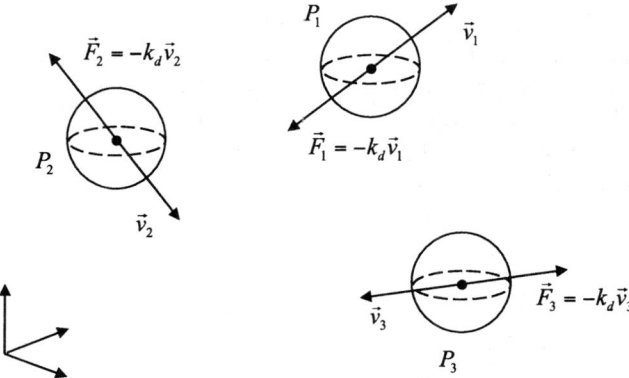

FIGURE 3.5. A set of particles P_i moving in random directions and experiencing air resistance modeled as viscous drag $\vec{F}_i$.

Besides preventing particles from gaining excessive speeds that may introduce instabilities into the numerical-integration method being used, viscous drag can also be used to control the rate at which particles accelerate. For example, a particle system simulating smoke may use a coefficient of drag much greater than that used in a particle system simulating rain. This in turn has the effect of making the smoke particles slowly rise and spread over nearby regions, whereas the rain drops will be allowed to fall at a reasonable speed.

3.3.3 Damped Springs

Springs are mostly used to keep the distance between pairs of particles at a known value. Whenever the particles are pushed apart or pulled together, a spring force is applied to both particles with the same magnitude but opposing direction.

Let P_1 and P_2 be two particles connected by a spring of resting length r_0. Let $\vec{r}_1$, $\vec{v}_1$, $\vec{r}_2$ and $\vec{v}_2$ be the linear position and velocity of particles P_1 and P_2, respectively. The spring-force component acting on both particles is then obtained from

$$\vec{F}_2 = -\left[k_s \left(|\vec{r}_2 - \vec{r}_1| - r_0\right) - k_d \left(\vec{v}_2 - \vec{v}_1\right) \frac{(\vec{r}_2 - \vec{r}_1)}{|\vec{r}_2 - \vec{r}_1|}\right] \frac{(\vec{r}_2 - \vec{r}_1)}{|\vec{r}_2 - \vec{r}_1|}$$

(3.6)

$$\vec{F}_1 = -\vec{F}_2,$$

with $\vec{F}_i$ being the spring force acting on particle P_i for $i \in \{1, 2\}$, k_s being the spring constant and k_d being the damping constant (see Figure 3.6). The damping term of equation (3.6) is used to prevent oscillation, and does not affect the motion of the center of mass of the connected particles.

3.3 Interaction Forces 71

The spring system can be under, over or critically damped, depending on the value of k_d being used. Oscillations occur only when the system is under damped. The interested reader is referred to Section 3.10 for pointers to the literature wherein techniques to compute the value of k_d for under, over and critically damped spring systems can be found.

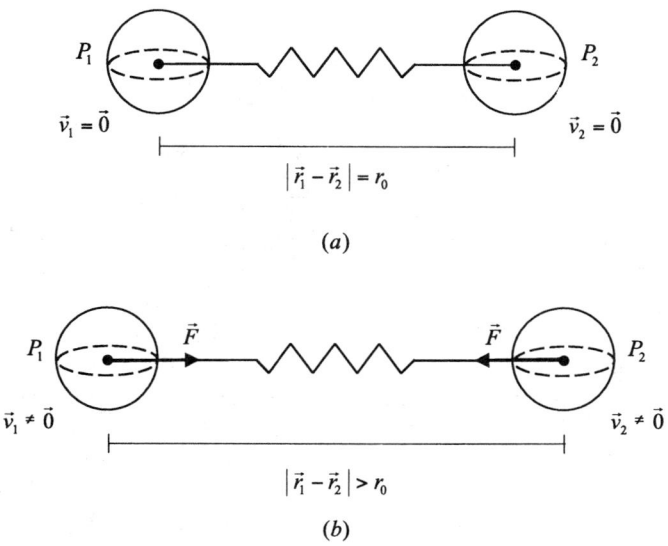

FIGURE 3.6. Particles P_1 and P_2 are connected by a damped spring. (a) Particles at resting position; (b) Spring forces exerted on the particles when they are pulled away from each other.

3.3.4 Spatially Dependent Forces

There are two types of spatially dependent forces considered in this book. The first type, referred to as *constrained force field*, deals with force fields defined over a region of the simulated environment. Such forces interact only with particles located within their region of influence. The second type, referred to as *unconstrained force field*, deals with interaction forces between all particles, which depend on the particles' relative positioning.

Constrained Force Fields
Constrained force fields are defined by their region of influence, force field strength and the drop rate. The region of influence, as its name implies, defines a region of the simulated world to which the force field is constrained. The boundary of the region of influence can be described by a polyhedron in the simulated environment (see Figure 3.7). Particles that are located inside the polyhedron have their trajectories affected by the

72   3. Particle Systems

force-field strength. For efficiency reasons, the region of influence should be represented as a simple polyhedron such as a box or sphere, so that particle-inclusion tests can be efficiently implemented using the algorithms presented in Section 2.4.

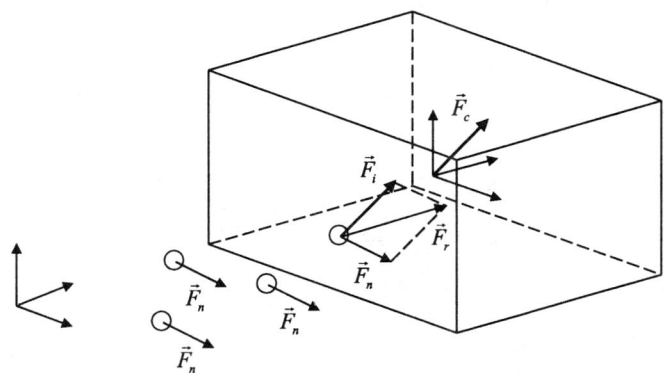

FIGURE 3.7. A windy region defined as a box in the simulated environment. Particles that lie outside the box are not affected by the wind, whereas those inside the box have their net-force vector $\vec{F}_n$ adjusted to $\vec{F}_r$, as the result of taking into account the wind force $\vec{F}_i$ given by equation (3.7).

The force-field strength is defined at the center of the region of influence. As we move away from the center, the strength of the force field is reduced according to its distance to the center. Therefore, particles closer to the center of the region of influence are more affected by the force field than those near the boundary of the region of influence. This is used to provide a smooth transition (i.e., avoid discontinuities) on the dynamics for particles entering and leaving the region of influence.

The drop rate is computed as follows. If the region of influence is a sphere, then the force-field strength at a point $\vec{p}_i$ inside the sphere is

$$\vec{F}_i = \left(1 - \frac{|\vec{r}_i|}{R}\right) \vec{F}_c ,$$

where R is the radius of the sphere, $\vec{r}_i$ is the distance vector from point $\vec{p}_i$ to the center of the sphere $\vec{c}$, and $\vec{F}_c$ is the force-field strength at $\vec{c}$ (see Figure 3.8(a)). When the region of influence is a box centered at $\vec{c}$ with dimensions $B = (b_x, b_y, b_z)$, the field strength at a point $\vec{p}_i$ inside the box is given by

$$\begin{aligned}(\vec{F}_i)_x &= \left(1 - \frac{|(r_i)_x|}{b_x}\right) (\vec{F}_c)_x \\ (\vec{F}_i)_y &= \left(1 - \frac{|(r_i)_y|}{b_x}\right) (\vec{F}_c)_y\end{aligned} \qquad (3.7)$$

$$(\vec{F}_i)_z = \left(1 - \frac{|(r_i)_z|}{b_x}\right)(\vec{F}_c)_z,$$

with $\vec{r}_i$ being the distance vector from point $\vec{p}_i$ to the center of the box. This situation is illustrated in Figure 3.8(b).

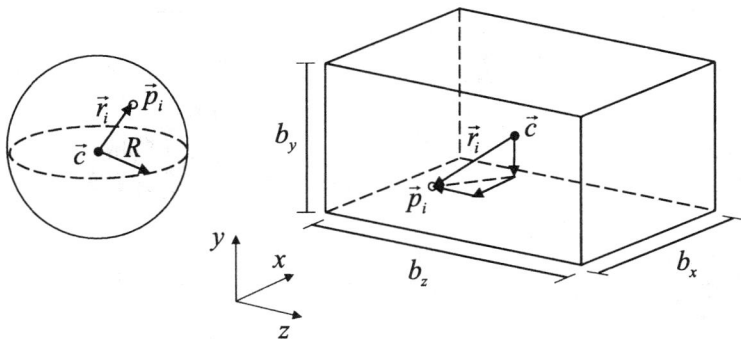

FIGURE 3.8. Taking the drop rate into account when computing the force-field strength at a point $\vec{p}_i$ inside the region of influence defined by: (a) a sphere; (b) a box.

At a preprocessing stage, the region of influence of each constrained force field is intersected with the cell decomposition of the simulated world. By so doing, the simulation engine will know which cells are completely inside, partially covered, or outside the region of influence. This greatly speeds run-time particle-inclusion tests, since such tests are only necessary for particles associated with cells that are partially covered by the region of influence.

Unconstrained Force Fields
Unconstrained force fields are used to specify long-range force interactions between all particles in the system. Each particle influences all others in the system, which in most cases depend on their relative displacement. The farther the particles are apart, the weaker the force interaction affecting their motion. Examples of such force fields include:

- The gravitational potential between two particles P_1 and P_2, computed as

$$\vec{F}_1 = G\frac{m_1 m_2}{|\vec{r}_1 - \vec{r}_2|^2}\frac{(\vec{r}_1 - \vec{r}_2)}{|\vec{r}_1 - \vec{r}_2|}$$
$$\vec{F}_2 = -\vec{F}_1,$$

where m_1, $\vec{r}_1$, m_2 and $\vec{r}_2$ are the mass and position of particles P_1 and P_2, respectively, and $G = 6.672 \times 10^{-11}\ Nm^2 kg^{-2}$ is the universal gravitational constant.

- The Lennard-Jones potential, commonly used for computing non-bounding potentials in molecular-dynamics simulators, and given by

$$\vec{F}_1 = \frac{48}{|\vec{r}_1 - \vec{r}_2|^2} \left(\frac{1}{|\vec{r}_1 - \vec{r}_2|^{12}} - \frac{0.5}{|\vec{r}_1 - \vec{r}_2|^6} \right)$$
$$\vec{F}_2 = -\vec{F}_1 \, .$$

- The Coulomb potential, which is the equivalent of the gravitational potential when the particles have electrical charges

$$\vec{F}_1 = K \frac{q_1 \, q_2}{|\vec{r}_1 - \vec{r}_2|^2} \frac{(\vec{r}_1 - \vec{r}_2)}{|\vec{r}_1 - \vec{r}_2|}$$
$$\vec{F}_2 = -\vec{F}_1 \, ,$$

with q_1 and q_2 being the electrical charges of the particles, and K the Coulomb constant equal to $8.9875 \times 10^9 \; Nm^2 C^{-2}$. In this case, the particles can either repel or attract each other, depending on whether their electrical charges are of the same or opposite signs.

Clearly, the computation of unconstrained forces for a particle system containing n particles has a $\mathcal{O}(n^2)$ computation time complexity, which makes it impractical for interactive simulations, even for a moderate number of particles (i.e., $n \geq 1000$.) Fortunately, there is a workaround to this limitation that reduces the computational complexity to $\mathcal{O}(n)$. It is based on truncating the computation of the interaction forces to particles that are within a cut-off distance from the particle being considered. This can be efficiently implemented if we use the underlying cell subdivision of the simulated world.

As explained in Section 3.2, the simulation engine dynamically assigns particles to the cells in which they are included. Each particle will then interact with other particles in that same cell, and those in neighboring cells that are within a cut-off distance from the cell that contains the particle. Figure 3.9 shows how the neighboring cells are determined for a 2D cell decomposition.

The actual number of neighboring cells to be used depends on the size of each cell and the cut-off distance. Let s be the dimension of each cell along each axis of the world-coordinate frame[1], and let s_c be the desired cut-off distance. Assume we are computing the force interactions for particles contained in cell $C = (c_x, c_y, c_z)$. Now, imagine a cut-off box with its center coincident with the center of cell C, and with sides of length equal to the cut-off distance. This situation is illustrated in Figure 3.10.

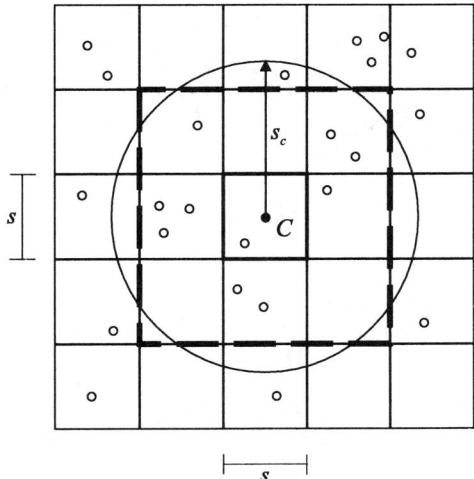

FIGURE 3.9. Particles in cell C will only interact with other particles in C and in neighboring cells that are within the cut-off distance s_c from C.

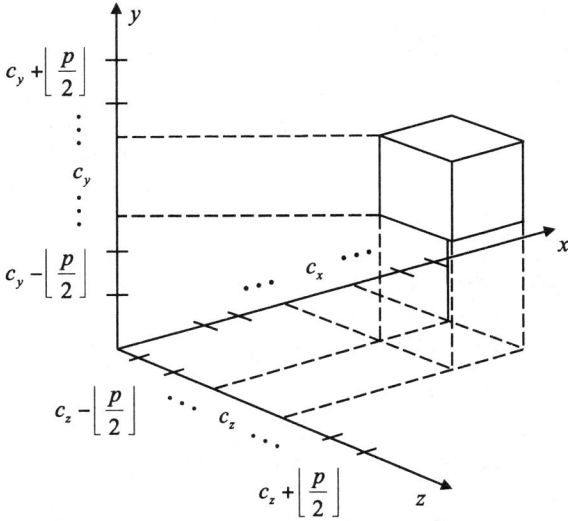

FIGURE 3.10. The computation of the force interactions for particles in cell $C = (c_x, c_y, c_z)$ is limited to the cells with indexes within $\lfloor \frac{p}{2} \rfloor$ from it.

The cut-off box will therefore intersect

$$p = \left\lceil \frac{s_c}{s} \right\rceil \text{ with } p \in \mathbb{N}$$

cells of the subdivision. The particles in $C = (c_x, c_y, c_z)$ will then interact with the particles assigned to cells $C = (i, j, k)$ satisfying

$$0 \leq \left(c_x - \left\lfloor \frac{p}{2} \right\rfloor\right) \leq i \leq \left(c_x - \left\lfloor \frac{p}{2} \right\rfloor\right) \leq c_n$$
$$0 \leq \left(c_y - \left\lfloor \frac{p}{2} \right\rfloor\right) \leq j \leq \left(c_y - \left\lfloor \frac{p}{2} \right\rfloor\right) \leq c_n$$
$$0 \leq \left(c_z - \left\lfloor \frac{p}{2} \right\rfloor\right) \leq k \leq \left(c_z - \left\lfloor \frac{p}{2} \right\rfloor\right) \leq c_n,$$

where c_n is the maximum cell-index value, as discussed in Section 2.3.

3.3.5 User Interaction

The user-interaction force is modeled as a damped spring connecting the current mouse position to the position of the particle being dragged. The goal of using this fictitious spring is to avoid the introduction of unrealistically large external forces acting on the selected particle because of abrupt mouse movements. These large external forces can make the dynamic equations describing the motion of the particle stiff. Stiff systems are more sensible to round-off errors and usually require the use of more elaborated and time-consuming numerical-integration methods, such as the implicit Euler method described in Appendix B.

The main difference between the damped spring described in Section 3.3.3 and the fictitious spring used here is that the resting length of the fictitious spring should be zero. A zero resting length means that the selected particle will only stabilize its motion when its position is coincident with the mouse position. Therefore, as the user drags the particle around, the current mouse position is used to update the actual distance between the particle and the mouse. This distance is then used in equation (3.6) to compute the appropriate spring force to be applied.

3.4 Collision Detection: Overview

Even though particles are modeled as point-mass objects, they are usually represented by simple geometric shapes such as cubes or spheres that can

---

[1] Recall from Section 2.3 that the cell decomposition defines a uniform subdivision of the simulated world.

be rendered in large numbers without affecting too much the overall performance of the rendering engine. The particle's shape can then be used to detect collisions between particles and other objects in the simulated environment.

We say the particle system has *internal collisions* whenever the collisions between its particles are taken into account. The particle system will have *external collisions* if the collisions between its particles and other particles defined by other particle systems are also taken into account. Collisions between rigid bodies and particles will be referred to as *complex collisions*.

In this book, we focus on particle systems that can have internal, external and complex collisions. Also, we assume all particles in the system have a *spherical* shape, possibly with different radii. The main reason for this assumption is efficiency. Since particles are point-mass objects, their trajectories between two consecutive time steps define a straight line segment. When their shape is taken into account, their trajectories will span a volume in 3D space. If we limit our system to one having only spherical particles, the volume spanned by the motion of each particle can be described as a cylinder with a spherical cap (see Figure 3.11). In this context, collision detection between particles and other objects in the simulated environment can be determined using quick and efficient intersection tests.

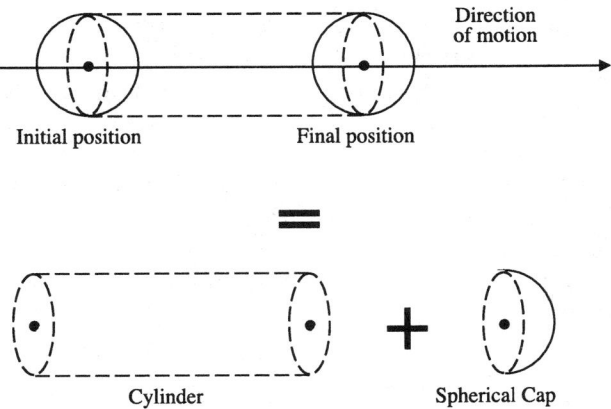

FIGURE 3.11. The trajectory of spherical particles can be described by concatenating simple geometric structures such as a cylinder and a sphere.

The relative displacement of the colliding particles or rigid bodies is used to determine the collision normal and tangent plane at the collision point. The actual computation of the collision normal is slightly different depending on whether we are considering particle-particle or particle-rigid body collisions. This is explained in detail in the following sections. The computation of the tangent plane, on the other hand, is done after the collision normal is determined, and depends strictly on it. Section A.6 of Appendix A presents a thorough derivation of how the tangent plane is obtained given the collision-normal vector $\vec{n}$. Henceforth, we shall take for granted the computation of vectors $\vec{t}$ and $\vec{k}$ that, together with vector $\vec{n}$, form the local frame used to compute the collision and contact forces.

3.4.1 Particle-Particle Collision

The collision detection between particles is usually undertaken by checking whether the particles' trajectories intersect. Because the trajectory of each particle is represented by a cylinder with a spherical cap, the collision-detection check consists of checking for *cylinder-cylinder*, *cylinder-sphere* and *sphere-sphere* intersections. The cylinder-cylinder and cylinder-sphere intersection tests are presented in Sections 3.5.1 and 3.5.3, respectively. The sphere-sphere intersection test was already presented in Chapter 2, in the context of interference detection between hierarchical representations of rigid-body systems.

Unfortunately, transforming the collision-detection problem into an intersection problem does not always give us the correct result. Consider the situation illustrated in Figure 3.12. Clearly, the trajectories of particles P_1 and P_2 do intersect at position I. However, if we take time into account, we quickly conclude that the particles do not collide because the time instant associated with each particle when their trajectories overlap is different. In other words, when particle P_2 is at position I, particle P_1 is still at position A. By the time particle P_1 reaches position I, particle P_2 is already at position B. Therefore, even though their trajectories intersect, they do not really collide.

One way to overcome this problem is to use the geometrical intersection of the trajectories as a reference point to check whether the particles really collide. This is best illustrated using the example shown in Figure 3.12. In this example, we already know that the trajectories of particles P_1 and P_2 intersect at position I. We want to determine the time instant before the intersection happened, that is, just before the particles collide. The idea is then to trace back in time until either both particles are positioned at the moment just before their collision, or their time instants associated with the intersection point I are different and collision is no longer an issue.

The trace back in time can be efficiently implemented using a root-finding algorithm such as the *bisection* method. The bisection method works by moving each particle half way through the current time interval. A sphere-sphere intersection test is then carried out to determine whether the particles intersect at the middle point. If the particles intersect at the middle point, then the current time interval is substituted for its first half (i.e., from the start to middle points) or its second half (i.e., from the middle to end points), depending on the intersection point I being located in the first or second halves, respectively.

If the particles do not intersect at the middle point, then their position at the middle point is compared with the position of the intersection point

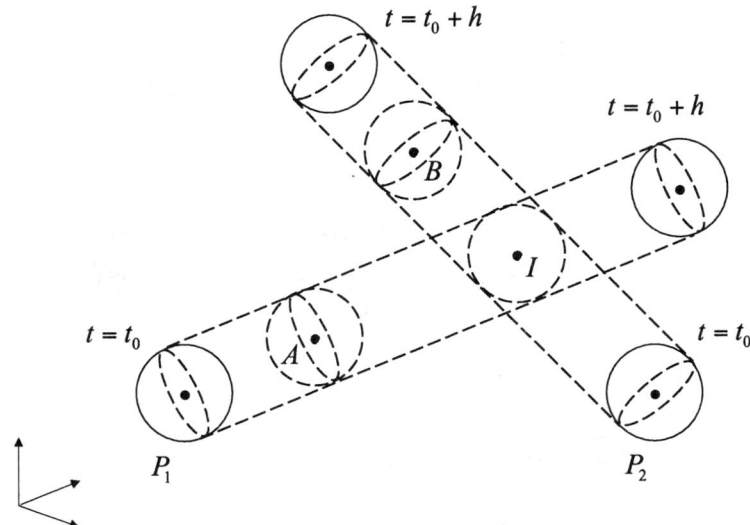

FIGURE 3.12. The geometric intersection of the particle's trajectories may be associated with different simulation times for each particle.

I to help figure out which half should be used to replace the current time interval. Figures 3.13 and 3.14 show the two possible outcomes of this comparison.

In Figure 3.13, the intersection point I is located in the first half of the time interval for particle P_2, but on the second half for particle P_1. In this case, the location of the intersection point I in the trajectory of each particle is associated with different halves of the time interval, that is, the time instant associated with each particle when their trajectories overlap is different. Therefore, they do not collide. In Figure 3.14, the intersection point I is located on the same half of the time interval for both particles. In this case, the current time interval is substituted for the half in which the intersection point I is located.

The current time interval is then recursively subdivided into half parts using the bisection method until either the collision is dismissed as an intersection at different time instants, or the current time interval is shrunk to less than a threshold value. In the latter case, the particles will be intersecting at one end of the time interval, and will not be intersecting at the other end. The collision time (i.e., the instant before the collision) is approximated by the end of the time interval in which the particles are not colliding.

The particles' geometric displacement at the collision time is used to compute the collision normal as the unit vector connecting the center of the spheres representing the particles. The direction of the collision normal is chosen such that the relative velocity of the particles along the collision

80   3. Particle Systems

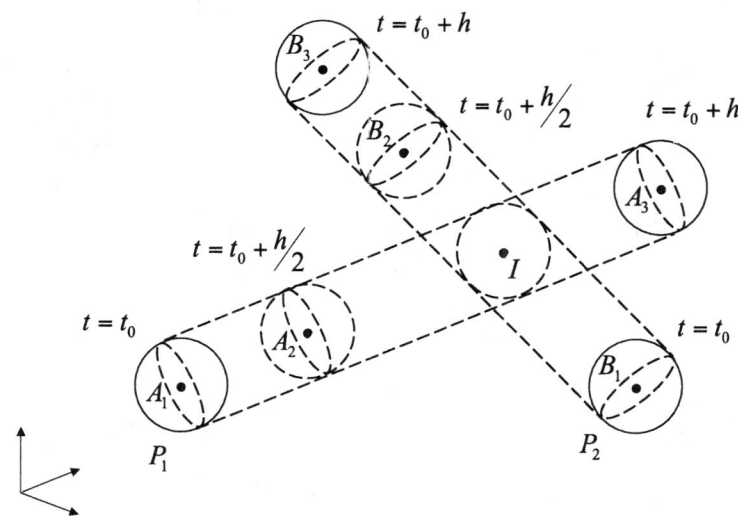

FIGURE 3.13. Applying the bisection method to determine the collision between particles that have their trajectories intersecting. Case in which the intersection point I is assigned to different time intervals of the bisection. The particles do not collide.

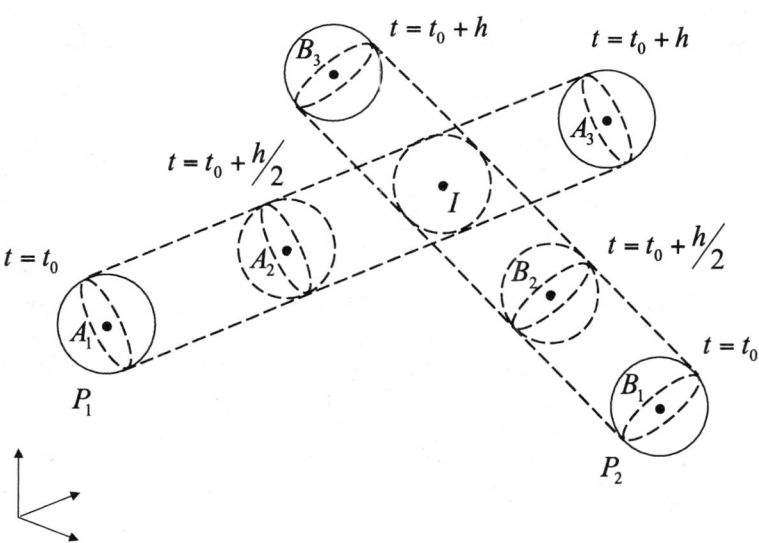

FIGURE 3.14. Case in which the intersection point I is assigned to the same time interval of the bisection. Additional time subdivision is required.

normal is negative, indicating that the particles are moving towards each other.

3.4.2 Particle-Rigid Body Collision

The collision detection between particles and rigid bodies is more complicated than that involving only particles, owing to the use of hierarchical representations for rigid bodies, as discussed in Chapter 2. Here, we have to check first for *particle-hierarchical tree* collisions, and if we end up reaching some leaves of the hierarchical tree during the collision test, we still need to carry out *particle-primitive* collision checks for the primitives associated with each leaf.

Since in this book we are only considering hierarchical representations using either boxes or spheres, the *particle-hierarchical tree* collision checks are transformed into a series of intersection tests between the particle's trajectory and the elements of the hierarchical tree. If the hierarchical tree representation uses boxes (i.e., AABB or OBB tree), then the intersection tests are limited to a sequence of *cylinder-box* and *sphere-box* tests. However, if the hierarchical tree representation uses spheres (i.e., BS tree), then the intersection tests involve *cylinder-sphere* and *sphere-sphere* tests only. Also, since the primitives considered in this book are only triangles, the *particle-primitive* collision checks are transformed to a series of *cylinder-triangle* and *sphere-triangle* intersection tests.

Some of the above intersection tests have already been presented in Chapter 2, when checking for collisions between different hierarchical tree representations of rigid bodies. More specifically, we have covered *sphere-box* in Section 2.4.5, *sphere-sphere* in Section 2.4.2 and *sphere-triangle* in Section 2.4.7. The remaining intersection tests involving cylinders are discussed in Sections 3.5.1 to 3.5.4 of this chapter.

The collision detection between a particle and a rigid body is simplified by ignoring the rigid body's trajectory during the current simulation time step, and carrying out the collision checks only with the rigid body already positioned at the end of the current time step (see Section 4.4 of Chapter 4 for more on rigid body collisions). By so doing, the problem we had when checking for particle-particle collisions with particles that had their trajectories overlapping at different time instants is no longer relevant.

As is the case with most simplifications, there are situations in which the collision detection misses some particle-rigid body collisions using this scheme. However, let's focus first on the cases when the simplification works, and later discuss its drawbacks.

A particle is assumed to collide with a rigid body whenever its trajectory intersects one or more triangle primitives of the rigid body. When an intersection is first detected, the collision-detection module keeps a pointer to the triangle primitive of the rigid body corresponding to the intersection point closest to the particle's initial position (see Figure 3.15).

82   3. Particle Systems

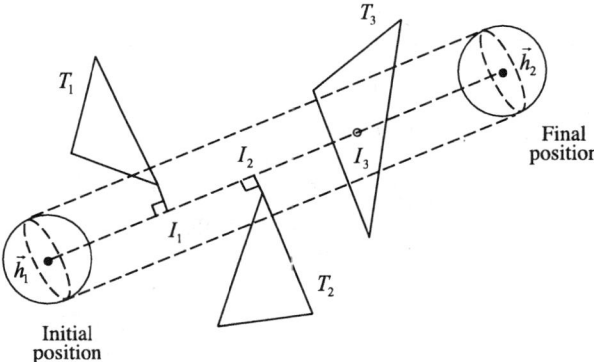

FIGURE 3.15. The particle's trajectory intersects several triangle primitives of a rigid body. The triangle with the closest intersection point to the initial position of the particle is selected as the intersecting triangle. As shown, the intersection point I_1 is closer to H_1 than I_2 and I_3. Therefore, triangle T_1 is selected as the intersecting triangle.

The particle is then traced back in time to the moment just before the collision, using the bisection method to recursively subdivide the time interval into halves. Notice that for the particle-rigid body collision detection there is no need to keep intersecting the particle's trajectory with the rigid body's hierarchical representation at each intermediate step of the bisection method because the rigid body is considered to be "fixed" at the end of the current simulation time step (its trajectory is ignored). Moreover, we already know the triangle primitive that will first come in contact with the particle. Therefore, at each intermediate step of the bisection method, we need only check for intersection between the closest triangle and the particle's trajectory. This process continues until the current time interval is shrunk to less than a threshold value. At this point, the particle will be intersecting the triangle at one end of the time interval, and will not be intersecting the triangle at the other end. The collision time (i.e., time just before the collision) is approximated by the end of the time interval in which the particle does not intersect the triangle. The collision point is assumed to be the center of the sphere representing the particle, and the collision normal can be either a vertex normal, an edge normal or a face normal depending on the particle being closer to a vertex, an edge or a point in the interior of the triangle.

The reaction on the rigid body's motion owing to the impulsive or contact forces of the particles colliding with it will only be applied to the rigid body on the subsequent simulation time step. In other words, even though the rigid body is modeled as a "fixed" object for the current time step, the forces and impulses exerted by particles colliding with it are accumulated as an external force and impulse that will affect the rigid body's motion the next time its dynamic equations are integrated. For instance, consider the situation shown in Figure 3.16, where several particles are moving toward a box at rest.

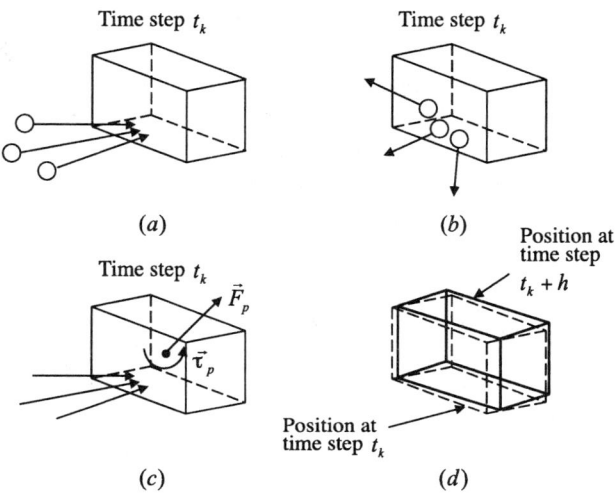

FIGURE 3.16. (a) Several particles hitting a box at rest at time $t = t_k$; (b) The particles' trajectories are updated according to the impulsive forces owing to the collision; (c) The reaction forces owing to the particles' collisions are substituted for a net force-torque pair acting on the box's center of mass; (d) The net force-torque pair is then used in the subsequent simulation time step $t = (t_k + h)$ to update the position and orientation of the box.

The collision-detection module will check for geometric intersections between the particles' trajectories and the box (see Figure 3.16(a)). The collision-response module is then activated to resolve any detected collisions (more details in Section 3.7). The colliding particles have their trajectories updated according to the impulsive forces computed by the collision-response module, so as to prevent interpenetration with the box, as well as other particles during their movement (see Figure 3.16(b)). The reaction impulsive force to the collision of each particle is summed and saved as an external impulsive force-torque pair acting on the box (see Figure 3.16(c)). This external impulsive force-torque pair will then be used to update the box's motion during the next simulation time step (see Figure 3.16(d)).

As mentioned before, there are a couple of drawbacks to ignoring the rigid body's trajectory when checking for particle-rigid body collisions. First of all, collisions that occur at the beginning of the current simulation time step have a higher probability of being missed than collisions that occur near the end of the time step. This owes to the fact that the rigid body is positioned at the end of the current time step before any particle-rigid body collisions are checked. Of course, we could reduce the simulation time step to catch these collision misses, but this would considerably slow the

84   3. Particle Systems

simulation, defeating the purpose of using the simplification in the first place.

Secondly, it is possible to have situations in which, even though the particle's trajectory does not intersect the rigid body's triangular faces, it is completely inside the rigid body (see Figure 3.17). This is a difficult problem to deal with because using standard point-in-polygon tests for checking whether the extreme points of the line defining the particle's trajectory are included in the rigid body's shape can be very time consuming, especially if we don't have a convex decomposition of the rigid body at hand.

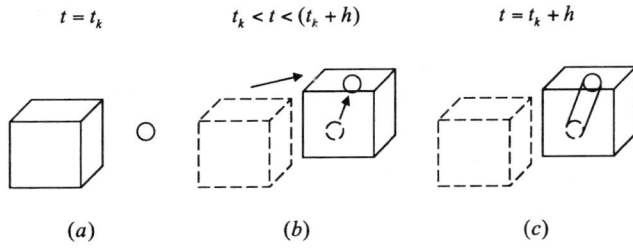

FIGURE 3.17. (a) Rigid body and particle positioned at the beginning of the current simulation time step; (b) Rigid body collides with particle somewhere before the end of the current time step; (c) Rigid body and particle positioned at the end of the current time step. Since the rigid body's trajectory is ignored, no intersections with the rigid body's triangle primitives are detected. The rigid body and particle are assumed to be not colliding, even though the particle's trajectory is completely contained inside the rigid body.

Another way to deal with this problem is to carry out an additional test using the partial intersection information obtained after the rigid body's hierarchical representation is checked for intersection against the particle's trajectory. The fact that the particle's trajectory does not intersect the rigid body means that the intersection between the trajectory and its hierarchical representation is empty at some intermediate level of the hierarchy. This level can be the top-most level containing the parent node, in which case the intersection test is quickly dismissed, or any other intermediate level including the leaf nodes. The extra test then consists of checking for sidedness of the extreme points of the particle's trajectory with respect to the triangle primitives associated with the intermediate node at which the intersection was dismissed. If both extreme points are on the negative side (i.e., inside) of all planes defined by the triangular faces, then the trajectory is completely inside the rigid body. If both extreme points are on the positive side (i.e., outside) of all planes defined by the triangular faces, then the trajectory is completely outside the rigid body. The case in which one point is inside and the other is outside cannot occur at this stage because the trajectory is known not to be intersecting the rigid body.

Even though this extra test would detect the cases in which the particle's trajectory is completely inside the rigid body, it can significantly affect the overall program efficiency. For example, when the intersection is dismissed at the parent node, the extreme points of the trajectory would have to be tested for sidedness against all triangle faces defining the rigid body. Clearly, this is impractical in simulations involving several hundred or more particles.

In this book, we suggest the following approach to efficiently carry out this additional test. Our approach consists of enhancing the AABB and OBB hierarchical representations with some extra information about the relative location of the vertices of their intermediate nodes (i.e., boxes) with respect to the interior and exterior of the rigid body[2]. This is done by testing the relative location of each vertex of the intermediate box B_j with respect to the triangle primitives T_i^j associated with it. Recall that the triangle normals $\vec{n}_{T_i^j}$ are always pointing outwards, that is, towards the exterior of the rigid body. So, vertex $\vec{v} \in B_j$ lies outside the rigid body if and only if

$$\vec{v} \cdot \vec{n}_{T_i^j} > 0, \ \forall T_i^j \text{ associated with } B_j .$$

Conversely, vertex $\vec{v} \in B_j$ lies inside the rigid body if and only if

$$\vec{v} \cdot \vec{n}_{T_i^j} < 0, \ \forall T_i^j \text{ associated with } B_j .$$

We also keep track of the closest triangle face to each vertex of the intermediate boxes. The idea is then to use this inside-outside information of each vertex to classify the six faces of box B_j as pointing to the interior (if all of its vertices are in the interior), exterior (if all of its vertices are in the exterior) or undefined (if some vertices are in the interior and others in the exterior). On a first pass, the extreme points of the particle's trajectory are initially assigned the same label of the faces of the box closest to them. If both extreme points are labeled exterior, then the particle's trajectory lies outside the rigid body and no collisions are reported. However, if one or both are labeled undefined, then we use the closest triangle face to each vertex of the box (adding up to four triangle faces) to improve the sidedness test. If both extreme points are labeled exterior on this second pass, then no collisions are reported. Otherwise, the particle's trajectory is assumed to be inside the rigid body.

Ideally, once the additional test indicates that the particle's trajectory is included in the rigid body, we have to trace back in time both particle and rigid body to the moment before their collision. Unfortunately, this would have the cascade effect of requiring recomputation of all collisions between the rigid body and other particles, since those computed to that

---

[2] This technique does not apply for Bounding Sphere representations.

time considered the rigid body positioned (i.e., "fixed") at the end of the current simulation time step, and not at some intermediate position. In our view, the best way to cope with this tracing problem is to remove the particle from the simulation. This has no side effects with respect to all other collisions already processed, it is easy and fast to implement, it does not affect the efficacy of the simulation, and may even pass unnoticed if there is a large number of particles nearby the one being removed.

Lastly, the third drawback to using the above-mentioned simplification is that all rigid body-rigid body collisions must be processed *before* any particle-particle and particle-rigid body collisions. This is necessary because rigid body-rigid body collisions require that the colliding rigid bodies be traced back in time to the moment just before their collision. Moreover, the impulses and contact forces are immediately applied and their trajectories are updated through the remaining period of time. So, if we compute all particle-particle and particle-rigid body collisions before the rigid body-rigid body collisions are dealt with, there is a chance that some rigid bodies will have their positions at the end of the current time step modified because of some collisions with other rigid bodies. This would require recomputation of all particle-rigid body collisions and particle-particle collisions involving the particles that collided with those rigid bodies that had their trajectories modified by the rigid body-rigid body collisions.

3.5 Collision-Detection Implementation

In this section we shall address all intersection tests necessary to implement the collision detection between particles and rigid-bodies, as explained in Section 3.4. We use the following notation to describe these intersection tests.

The particle's trajectory will be represented by a cylinder H and a sphere S. The sphere S is given by its center $\vec{c}$ and radius r_c, whereas the cylinder H is given by its radius r_h and two points $\vec{h}_1$ and $\vec{h}_2$ defining the direction $\vec{u}_h$ of its axis. The vector $\vec{u}_h$ can be directly obtained from $\vec{h}_1$ and $\vec{h}_2$ as

$$\vec{u}_h = \frac{(\vec{h}_2 - \vec{h}_1)}{|\vec{h}_2 - \vec{h}_1|}.$$

The top and bottom limiting planes of the cylinder are perpendicular to $\vec{u}_h$, passing through points $\vec{h}_2$ and $\vec{h}_1$, respectively. Their equations are

$$\begin{aligned} \vec{p} \cdot \vec{u}_h &= \gamma_2 \text{ (top plane)} \\ \vec{p} \cdot \vec{u}_h &= \gamma_1 \text{ (bottom plane)} \end{aligned} \quad (3.8)$$

for any point $\vec{p}$ in the planes, with

$$\gamma_2 = \vec{h}_2 \cdot \vec{u}_h$$
$$\gamma_1 = \vec{h}_1 \cdot \vec{u}_h$$

being the top and bottom plane constants, respectively.

The rigid-body primitive is represented by a triangle T defined by its vertices $\vec{v}_1$, $\vec{v}_2$ and $\vec{v}_3$. The triangle normal is computed as[3]

$$\vec{n} = (\vec{v}_2 - \vec{v}_1) \times (\vec{v}_3 - \vec{v}_2) .$$

Let P_t be the plane containing the triangle. Any point $\vec{p} \in P_t$ satisfies

$$\vec{p} \cdot \vec{n}_t = \gamma_t ,$$

where $\vec{n}_t$ is the plane normal (i.e., the triangle normal.) and γ_t is the plane constant given by

$$\gamma_t = \vec{v}_1 \cdot \vec{n}_t .$$

A box representing an intermediate node of an AABB or OBB hierarchical tree representation of rigid bodies is defined by its two diagonally opposing extreme points $\vec{b}_{min}$ and $\vec{b}_{max}$.

3.5.1 Computing Cylinder-Cylinder Intersections

The cylinder-cylinder intersection test is primarily used when checking for collisions between the trajectories of two particles. Here, we shall present an intersection test that is custom-tailored for cylinders representing particles' trajectories. Recall that the trajectory is defined by a cylinder with a spherical-cap. Therefore, the most complicated cases that need to be considered in a general cylinder-cylinder intersection test can be avoided. More specifically, the intersection tests between one cylinder and the top or bottom regions of the other cylinder are left unresolved until the cylinder is checked for intersection with the spherical-cap of the other cylinder. The existence of any intersection with the top or bottom regions of the cylinder will necessarily imply the existence of an intersection between their associated spherical-cap and the cylinder (see Figure 3.18), meaning all such intersections will be detected by the cylinder-sphere intersection test discussed in Section 3.5.3. Here, we shall focus on checking for intersections between the cylinders' cap.

---

[3]The simulation engine presented in this book uses the right-hand coordinate system for its internal representation. Therefore, outward normals are computed from the vertex list given in a counterclockwise order.

88   3. Particle Systems

Let H and M represent the two cylinders being checked for intersection. The first step is to carry out some quick rejection tests for M. If the extreme points of cylinder M's axis lie above or below the top or bottom limiting planes of cylinder H, that is, if

$$(\vec{m}_1 \cdot \vec{u}_h - \gamma_2) > 0$$
$$(\vec{m}_2 \cdot \vec{u}_h - \gamma_2) > 0$$

or

$$(\vec{m}_1 \cdot \vec{u}_h - \gamma_1) < 0$$
$$(\vec{m}_2 \cdot \vec{u}_h - \gamma_1) < 0 \quad ,$$

then any cylinder-cylinder intersection will be detected by the cylinder-sphere intersection test, and the cylinders are reported as non-intersecting. Also, if the cylinder M's axis intersects the top limiting plane of cylinder H, that is, if

$$(\vec{m}_1 \cdot \vec{u}_h - \gamma_2) > 0$$
$$(\vec{m}_2 \cdot \vec{u}_h - \gamma_2) < 0$$

or

$$(\vec{m}_1 \cdot \vec{u}_h - \gamma_2) < 0$$
$$(\vec{m}_2 \cdot \vec{u}_h - \gamma_2) > 0 \quad ,$$

then any cylinder-cylinder intersection will contain a region on the top limiting plane, thus it can be detected by the cylinder-sphere intersection test. Again, the cylinder-cylinder intersection test returns false, indicating the cylinders do not intersect. If the above tests fail, then we conduct them again, but for H instead of M. If all rejection tests failed, then the axis of each cylinder is positioned between the limiting planes of the other. The intersection test for this situation can be efficiently carried out as follows.

Consider the line segment defined by cylinder M's axis. Let $\vec{c}_m$ be the closest point of this line segment to the line supporting cylinder H's axis, as shown in Figure 3.19. The determination of the closest point between a line segment and an infinite line is explained in detail in Section A.5. Notice, however, that according to Section A.5 $\vec{c}_m$ can be either a point inside the line segment, or one of its extreme points (i.e., $\vec{m}_1$ or $\vec{m}_2$), depending whether the actual closest point lies inside or outside the line segment. Therefore, $\vec{c}_m$ is guaranteed to lie inside the line segment defining cylinder M's axis.

3.5 Collision-Detection Implementation   89

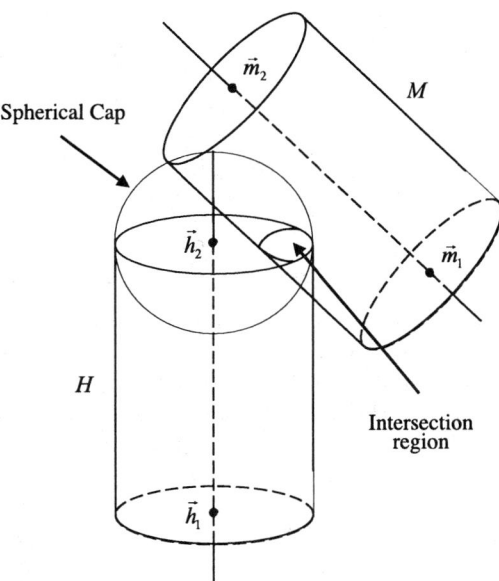

FIGURE 3.18. Cylinder H represents a particle's trajectory from $\vec{h}_1$ to $\vec{h}_2$. Any intersection with the top region of the cylinder (associated with the final trajectory point $\vec{h}_2$) will be detected by the time we check for intersections with the spherical cap.

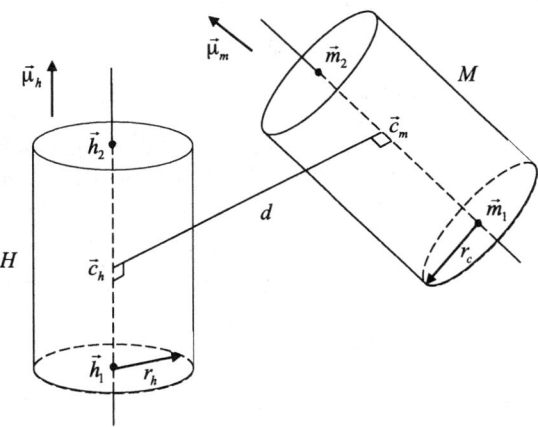

FIGURE 3.19. The cylinder-cylinder intersection test is conducted by comparing the distance d between the closest points $\vec{c}_m$ and $\vec{c}_h$ of their axes with the sum $(r_m + r_h)$ of the their radii.

Let $\vec{c}_h$ be the corresponding point to $\vec{c}_m$, located on the supporting line of cylinder H's axis (see Figure 3.19). The distance d between $\vec{c}_m$ and $\vec{c}_h$ is then

$$d = |\vec{c}_m - \vec{c}_h| . \qquad (3.9)$$

Clearly, the cylinders will not intersect if

$$d > (r_h + r_c) , \qquad (3.10)$$

that is, if the distance between the closest points of the cylinders' axes is greater than the sum of their radii. If equation (3.10) is not satisfied, then the cylinders will intersect only if $\vec{c}_h$ lies within the line segment defined by cylinder H's axis. However, if $\vec{c}_h$ lies outside cylinder H's axis, then we need to test whether the distance between $\vec{c}_m$ and the extreme point of cylinder H's axis that is closest to $\vec{c}_h$ is still less than the sum of the cylinders' radii. In other words, we need to compute

$$d_1 = |\vec{c}_h - \vec{h}_1|$$
$$d_2 = |\vec{c}_h - \vec{h}_2|$$

and select $\vec{c}_h = \vec{h}_1$ if $d_1 < d_2$, or $\vec{c}_h = \vec{h}_2$ otherwise. Then, we need to substitute the new value of $\vec{c}_h$ into equation (3.9), and check whether equation (3.10) is still satisfied. If this is the case, then the cylinders do not intersect. Otherwise, the cylinders do intersect and the intersection point is approximated by their closest points, that is, $\vec{c}_m$ for cylinder M and $\vec{c}_h$ for cylinder H.

3.5.2 Computing Cylinder-Box Intersections

The cylinder-box intersection test is used to detect collisions between particles' trajectories and AABB or OBB hierarchical tree representations. Initially, we conduct a quick rejection test to check whether the box lies completely above or below the top and bottom limiting planes of the cylinder, respectively. The box will lie above the top limiting plane if

$$(\vec{b}_{min} \cdot \vec{u}_h - \gamma_2) > 0 \qquad (3.11)$$

and below the bottom limiting plane if

$$(\vec{b}_{max} \cdot \vec{u}_h - \gamma_1) < 0 . \qquad (3.12)$$

If either equation (3.11) or (3.12) is satisfied, then the box does not intersect the cylinder. Otherwise, we need to conduct the following test for each face of the box.

Let f_i be a face of the box defined by vertices $\vec{v}_1$, $\vec{v}_2$, $\vec{v}_3$ and $\vec{v}_4$. Also, let $\vec{v}_{min}$ and $\vec{v}_{max}$ define the extreme points of face f_i, given by

$$(v_{min})_x = \min((v_1)_x, (v_2)_x, (v_3)_x, (v_4)_x)$$
$$(v_{min})_y = \min((v_1)_y, (v_2)_y, (v_3)_y, (v_4)_y)$$
$$(v_{min})_z = \min((v_1)_z, (v_2)_z, (v_3)_z, (v_4)_z)$$
$$(v_{max})_x = \max((v_1)_x, (v_2)_x, (v_3)_x, (v_4)_x)$$
$$(v_{max})_y = \max((v_1)_y, (v_2)_y, (v_3)_y, (v_4)_y)$$
$$(v_{max})_z = \max((v_1)_z, (v_2)_z, (v_3)_z, (v_4)_z) .$$

We have two possible cases to consider depending on whether the cylinder's axis is perpendicular to the plane containing f_i. Let $\vec{n}_i$ be the outward normal of f_i (i.e., normal pointing from the inside to the outside of the box). The cylinder's axis will be perpendicular to f_i if

$$\vec{u}_h \times \vec{n}_i = \vec{0} \tag{3.13}$$

is satisfied.

Case 1:

If the axis is perpendicular to the plane, we proceed by checking whether the cylinder's axis intersects the plane containing f_i. This line-segment-with-plane intersection test is discussed in Section A.4 of Appendix A.

Let $\vec{g}$ be the intersection point between the infinite line supporting the cylinder's axis and the plane. If $\vec{g}$ lies outside the end points defining the cylinder's axis (see Figure 3.20), then the cylinder does not intersect the box. However, if $\vec{g}$ lies inside the axis, then a point-in-rectangle test is conducted to see whether $\vec{g}$ is actually inside F_i (see Figure 3.21), that is, whether the cylinder and the box do intersect. Otherwise, the cylinder-box intersection test returns false, meaning the cylinder does not intersect the box.

The intersection point $\vec{g}$ will lie inside the rectangular face f_i whenever

$$(v_{min})_x \leq g_x \leq (v_{max})_x$$
$$(v_{min})_y \leq g_y \leq (v_{max})_y$$
$$(v_{min})_z \leq g_z \leq (v_{max})_z .$$

Case 2:

If equation (3.13) is not satisfied, then the cylinder's axis is not perpendicular to the plane containing face f_i. In this case, we need to construct an auxiliary plane P_i to help us determine the closest and farthest points

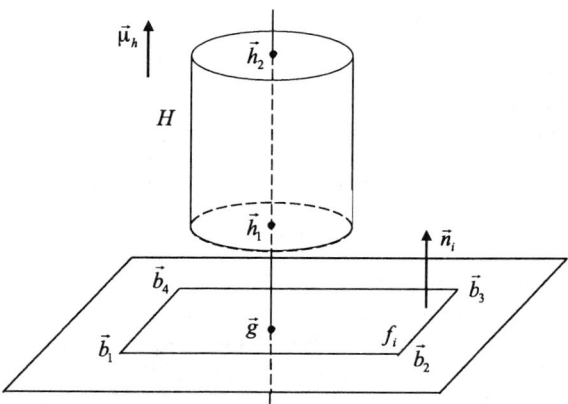

FIGURE 3.20. The intersection point $\vec{g}$ lies outside the cylinder's axis. Therefore, the cylinder does not intersect face f_i.

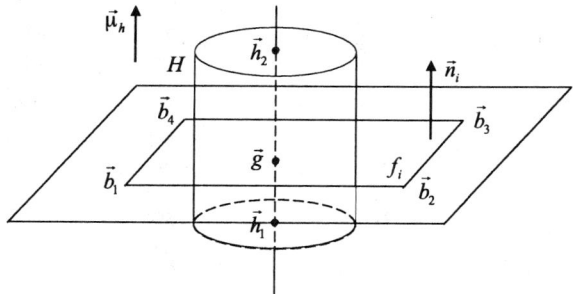

FIGURE 3.21. The intersection point $\vec{g}$ lies inside the cylinder's axis and a point-in-rectangle test is carried out to determine whether $\vec{g}$ lies inside or outside face f_i.

of the cylinder to the plane that contains face f_i. Consider the situation illustrated in Figure 3.22.

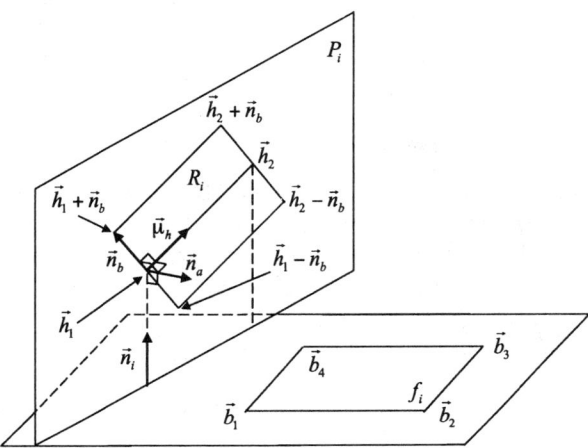

FIGURE 3.22. The auxiliary plane P_i is used to determine the closest and farthest points of the cylinder to the plane containing F_i.

The auxiliary plane P_i is defined as the plane perpendicular to f_i which also contains the cylinder's axis. Its normal vector $\vec{n}_a$ is obtained from

$$\vec{n}_a = \vec{u}_h \times \vec{n}_i .$$

Clearly, since the plane P_i contains the cylinder's axis, its intersection with the cylinder defines a rectangular region R_i. We need to determine the vertices defining R_i. In order to do so, consider the auxiliary vector $\vec{n}_b \in P_i$ given by

$$\vec{n}_b = k_b \frac{(\vec{n}_a \times \vec{u}_h)}{|\vec{n}_a \times \vec{u}_h|} ,$$

where $k_b \in \mathbb{R}$ is a scalar variable that defines the length of vector $\vec{n}_b$. Notice that, by construction, $\vec{n}_b$ is a vector parallel to the top and bottom limiting planes of the cylinder. The vertices of R_i can then be obtained by adding and subtracting $\vec{n}_b$ from the cylinder's axis, with k_b chosen as

$$k_b = r_h \operatorname{sign}(\vec{u}_h \cdot \vec{n}_i) .$$

Thus, the rectangular region R_i is defined by vertices $(\vec{h}_1 - \vec{n}_b)$, $(\vec{h}_2 - \vec{n}_b)$, $(\vec{h}_2 + \vec{n}_b)$ and $(\vec{h}_1 + \vec{n}_b)$ (see Figure 3.22). The cylinder intersects face f_i if and only if f_i intersects R_i. Notice that the closest and farthest points of R_i to the plane containing f_i are $\vec{p}_c = (\vec{h}_1 - \vec{n}_b)$ and $\vec{p}_f = (\vec{h}_2 + \vec{n}_b)$, respectively. So, face f_i intersects R_i if and only if it intersects the line segment defined by $(\vec{p}_c, \vec{p}_f)$. This test is analogous to the intersection test between the cylinder's axis and face F_i when they are perpendicular.

In summary, we decompose the cylinder-box intersection test into six cylinder-rectangle tests. Each of them is implemented as a plane-line segment intersection test, thoroughly explained in Section A.4.

3.5.3 Computing Cylinder-Sphere Intersections

The cylinder-sphere intersection test is used to detect particle-particle collisions, as well as intersections between particles' trajectories and Bounding Sphere hierarchical tree representations. This test can be efficiently implemented if we consider the relative displacement of the center of the sphere with respect to the top and bottom planes limiting the cylinder. One of the following cases occurs:

1. The sphere is completely above the top plane;

2. The sphere intersects the top plane, but its center is above it;

3. The sphere intersects the top plane, but its center is on or below it;

4. The sphere is completely between the top and bottom planes;

5. The sphere intersects the bottom plane, but its center is on or above it;

6. The sphere intersects the bottom plane, but its center is below it;

7. The sphere is completely below the top plane.

Let's start by examining the cases in which the sphere intersects either the top or bottom limiting planes of the cylinder. Figures 3.23 and 3.24 show the situation for cases 2 and 3 of the above list. The analysis of this situation is also applicable to cases 5 and 6.

Let $\vec{c}_1$ and $\vec{c}_2$ be the projection of the center of the sphere on the top plane and on the cylinder axis, respectively[4]. Clearly, the cylinder intersects the sphere if and only if

$$L \leq (r_h + x), \tag{3.14}$$

where x and L are variables to be determined from the relative displacement of the cylinder and the sphere. Consider the triangle defined by vertices $\vec{c}$, $\vec{c}_2$ and $\vec{h}_1$. Applying the Pythagorean theorem, we immediately obtain L as

$$L^2 = |\vec{c} - \vec{h}_1|^2 - |\vec{c}_2 - \vec{h}_1|^2, \tag{3.15}$$

---

[4] See Appendix A for details on how to compute these projections.

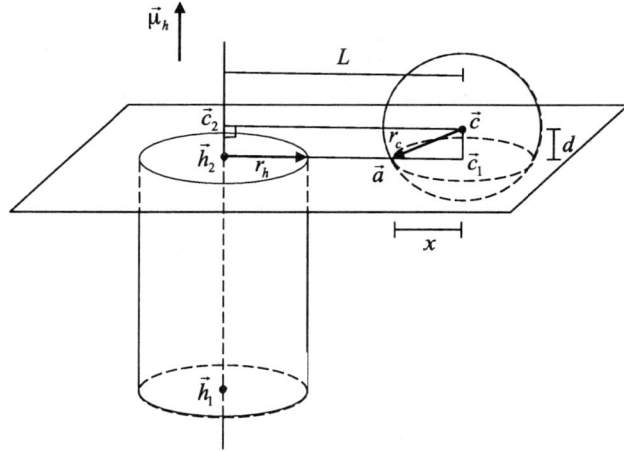

FIGURE 3.23. The sphere intersects the top plane, with the center of the sphere being above the plane. A similar figure can be drawn to depict the sphere's intersecting the bottom plane.

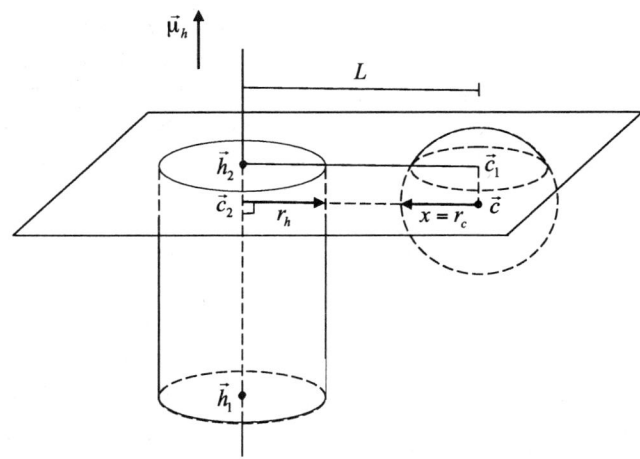

FIGURE 3.24. The sphere intersects the top plane, with the center of the sphere being below the plane. A similar figure can be drawn to depict the sphere's intersecting the bottom plane.

where the vector $(\vec{c}_2 - \vec{h}_1)$ can be efficiently computed by projecting the vector $(\vec{c} - \vec{h}_1)$ onto the cylinder axis, that is

$$(\vec{c}_2 - \vec{h}_1) = (\vec{c} - \vec{h}_1) \cdot \vec{u}_h . \tag{3.16}$$

If the center of the sphere is on or below the top plane, then we immediately have (see Figure 3.24)

$$x = r_c . \tag{3.17}$$

Otherwise, the variable $x = |\vec{a} - \vec{c}_1|$ can be obtained by applying the Pythagorean theorem to triangle $\vec{c}, \vec{c}_1$ and $\vec{a}$, that is

$$x^2 = r_c^2 - d^2 . \tag{3.18}$$

Notice from Figure 3.23 that

$$d = (\vec{c} - \vec{c}_1) = (\vec{c}_2 - \vec{h}_2) = (\vec{c}_2 - \vec{h}_1) - (\vec{h}_2 - \vec{h}_1) .$$

Using equation (3.16), we have

$$d = (\vec{c}_2 - \vec{h}_1) - (\vec{h}_2 - \vec{h}_1) = (\vec{c} - \vec{h}_1) \cdot \vec{u}_h - (\vec{h}_2 - \vec{h}_1) . \tag{3.19}$$

The variable x is then determined by substituting equation (3.19) into (3.18). We can immediately verify whether the cylinder intersects the sphere by plugging x and L into equation (3.14).

Notice that this intersection test is used only for cases 2, 3, 5 and 6. The other three possible cases, namely cases 1, 4 and 7, are addressed in a different way, as shown in Figure 3.25.

In cases 1 and 7, the distances of the center of the sphere to its projection on both the top and bottom planes are greater than the radius of the sphere, that is

$$d > r_c . \tag{3.20}$$

Substituting equation (3.20) into (3.18), we obtain

$$x^2 = r_c^2 - d^2 < 0 ,$$

which has no real solution. Therefore, the cylinder does not intersect the sphere in cases 1 and 7.

Finally, if the relative displacement corresponds to case 4, we have

$$x = r_c$$

and the intersection test to be used is the same as that in Figure 3.24.

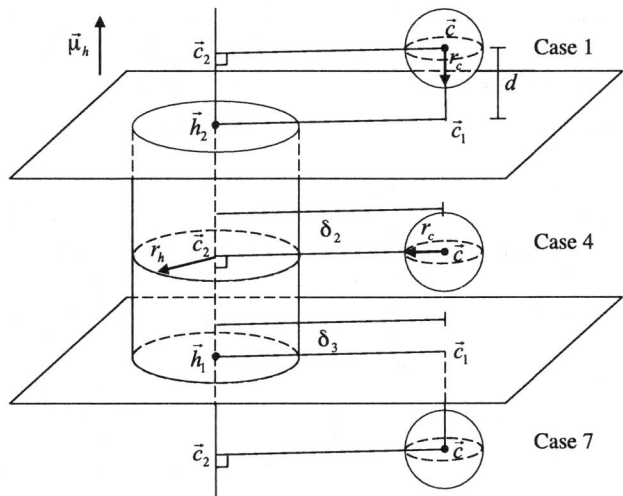

FIGURE 3.25. Relative displacement of the cylinder and the sphere for cases 1, 4 and 7.

At this point, we have covered the intersection tests for all seven possible cases of relative displacement between the cylinder and the sphere. However, we still need to address how to determine in which of the seven cases we are given the initial cylinder and sphere positions.

In order to do so, we start by computing the value of d, using equation (3.19). If $|d| > r_c$, then we are either at case 1 or 7, and the cylinder does not intersect the sphere. Else, if $d > 0$, then we are at case 2 and the variable x is computed using equation (3.18). If our case does not fit in any of the above situations, then we need to check whether

$$(\vec{c} - \vec{h}_1) \cdot \vec{u}_h < 0 . \tag{3.21}$$

If equation (3.21) evaluates to true, then we are at case 6 and the variable x is also computed using equation (3.18). Otherwise, we are at case 3, 4 or 5, and the variable x is obtained from equation (3.17).

3.5.4 Computing Cylinder-Triangle Intersections

The cylinder-triangle intersection test presented in this section is the first of a two-part collision-detection test between particles' trajectories and triangle primitives forming a rigid body. For efficiency reasons, the algorithm described here is not a general cylinder-triangle intersection test because the check for intersections between triangles and the limiting planes defining the cylinder is left partly unresolved until the second part of the collision-detection test is performed. The second part of the detection test, already explained in Section 2.4.7, consists of checking for sphere-triangle intersections, with the sphere representing the final position of the particle. Notice that there is no need to carry out a sphere-triangle intersection test at the initial position of the particle because that was the particle's final position at the previous simulation time step, and therefore has been checked already.

As mentioned in Section 3.4, the cylinder represents the particle's trajectory for the current time step. In this context, the point $\vec{h}_1$ is associated with the initial position of the particle and the point $\vec{h}_2$ to its final position. The cylinder's axis represents the actual trajectory of the point-mass particle. Therefore, collision points between the cylinder and the triangle should be located on the cylinder axis, representing the position of the particle at the moment of collision.

The first step in the cylinder-triangle intersection test consist of carrying out a quick rejection test. This test is used to check whether all vertices of T lie either above the top or below the bottom limiting planes of the cylinder. If so, then we can immediately conclude that the cylinder does not intersect the triangle (see Figures 3.26 and 3.27).

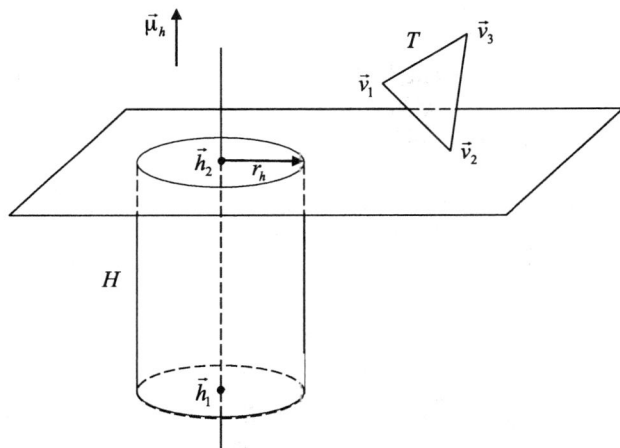

FIGURE 3.26. The triangle does not intersect the cylinder whenever its vertices lie above the top limiting plane of the cylinder.

The vertices of triangle T will lie above the top limiting plane of the cylinder if

$$(\vec{v}_i \cdot \vec{u}_h - \gamma_2) > 0, \ \forall\, i \in \{1, 2, 3\} \tag{3.22}$$

and they will lie below the bottom limiting plane of the cylinder if

$$(\vec{v}_i \cdot \vec{u}_h - \gamma_1) < 0, \ \forall\, i \in \{1, 2, 3\}\,. \tag{3.23}$$

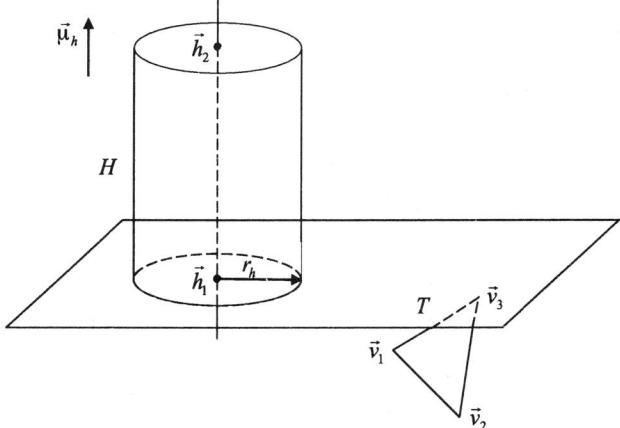

FIGURE 3.27. The triangle does not intersect the cylinder whenever its vertices lie below the bottom limiting plane of the cylinder.

The cylinder does not intersect the triangle if either equations (3.22) or (3.23) is satisfied for all vertices. Otherwise, more tests are needed to determine whether triangle T is in fact intersecting cylinder H.

The next step consists of checking whether the cylinder axis is parallel to P_t, the plane that contains the triangle. That is, we need to check whether

$$\vec{u}_h \cdot \vec{n}_t = 0 , \qquad (3.24)$$

where $\vec{n}_t$ is the plane normal. For the sake of clarity, let's subdivide our analysis at this point into two main cases, namely: the cylinder axis does not intersect P_t (i.e., the axis is parallel to the plane); the cylinder axis does intersect P_t.

Case 1: Axis is parallel to plane.

This is the case in which equation (3.24) is satisfied. Since the axis is parallel to the plane, we need to compute ther distance and compare it with the cylinder's radius. The distance between an axis parallel to a plane and the plane itself is the same as the distance from any point in the axis, say $\vec{h}_1$, to the plane, and is given by

$$d = (\vec{h}_1 \cdot \vec{n}_t) - \gamma_t .$$

Clearly, if

$$|d| > r_c ,$$

then the cylinder does not intersect the triangle (see Figure 3.28).

However, if this is not the case, then the plane P_t is not only parallel to the cylinder's axis, but also intersects the cylinder. This is illustrated in Figure 3.29.

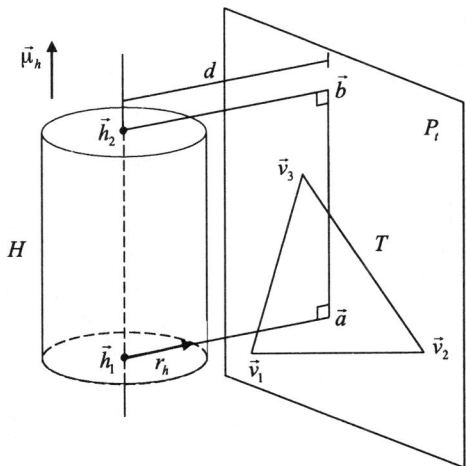

FIGURE 3.28. The cylinder axis is parallel to the plane containing the triangle. In this case, there will be no intersections if the distance d between the cylinder axis $(\vec{h}_1, \vec{h}_2)$ and the plane is greater than the radius r_c of the cylinder.

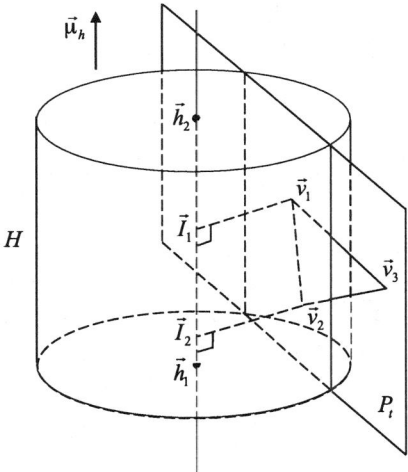

FIGURE 3.29. Plane P_t containing the triangle is parallel to, and intersects, the cylinder. We project onto the cylinder's axis the triangle vertices that are inside the cylinder, and set as intersection point I the projection closer to the initial point $\vec{h}_1$. In the situation shown, I_2 is selected as the intersection point between the particle and the triangle.

At this point, we need to conduct several inclusion tests to see whether the intersection region between P_t and the cylinder contains the triangle. First, we need to check whether one or more vertices of the triangle lie inside the cylinder. Vertex $\vec{v}_i \in T$, for $i \in \{1, 2, 3\}$, lies inside the cylinder if it is positioned between the top and bottom limiting planes and its distance to the cylinder is less than the radius $\vec{r}_c$. Putting into equations, we need to verify that

$$\begin{aligned}(\vec{v}_i \cdot \vec{u}_h - \gamma_2) &< 0 \\ (\vec{v}_i \cdot \vec{u}_h - \gamma_1) &> 0 \\ |d_i| &\leq r_c\end{aligned}$$

is satisfied for at least one vertex of T. If this is so, then the cylinder intersects the triangle (see Figure 3.29). The intersection point is chosen by projecting the triangle vertices inside the cylinder onto its axis, and selecting the projected vertex closest to $\vec{h}_1$, that is, closest to the initial point in the trajectory.

In the event all vertices of T lie outside the cylinder, then we need to check whether the intersection region between the cylinder and the plane containing the triangle is completely or partly inside the triangle. This can be efficiently done if we project the cylinder axis onto the plane that contains the triangle, as shown in Figure 3.30.

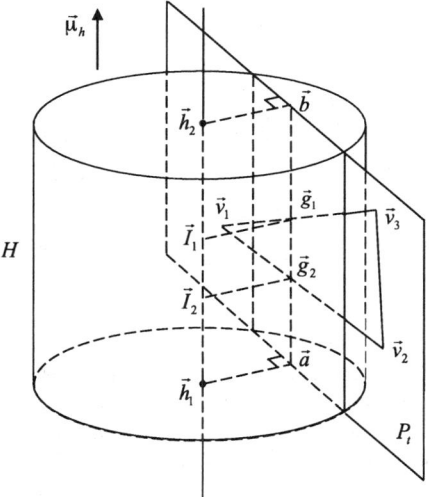

FIGURE 3.30. The projected cylinder axis $(\vec{a}, \vec{b})$ is checked for intersections with the triangle edges. Edges $(\vec{v}_1, \vec{v}_2)$ and $(\vec{v}_3, \vec{v}_1)$ intersect the projected axis at points $\vec{g}_2$ and $\vec{g}_1$, respectively. The point $\vec{g}_2$ is selected as the intersection point because it is closer to the initial position $\vec{h}_1$.

Let $\vec{a}$ and $\vec{b}$ be, respectively, the projection points of $\vec{h}_1$ and $\vec{h}_2$ on the plane P_t. The projected axis is then checked for intersection with the triangle edges. This *segment-triangle* intersection test is explained in detail in Section 2.4.9. If intersections are found, then we choose as the intersection point the intersection closest to $\vec{h}_1$ (see Figure 3.30). If the projected axis does not intersect the edges of the triangle, then the triangle does not intersect the cylinder. However, it can still intersect the top or bottom limiting planes of the cylinder. The intersection with the top plane is left unresolved at this point because it will be carried out later during the *sphere-triangle* intersection test (the top plane is inside the sphere representing the particle at the end point of its trajectory.) On the other hand, the intersection test with the bottom plane (i.e., the plane associated with the starting point of the trajectory) is completely ignored, since it was already carried out during the previous simulation time step (the starting point of the trajectory was the ending point of the previous time step.) Therefore, in both cases, the cylinder-triangle intersection test returns false, meaning the cylinder does not intersect the triangle.

There is, however, a special case to be considered before we compute the intersection between the projected axis $(\vec{a}, \vec{b})$ and the triangle edges. If the particle is in contact with the rigid body, then the intersection point between the particle and the triangular faces representing the object can be its initial point $\vec{h}_1$. This happens whenever the particle is at rest or rolling on top of a triangular face, with an external force (such as gravity) pulling along the triangle normal, as shown in Figure 3.31.

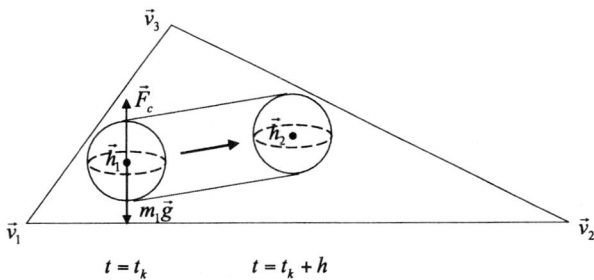

FIGURE 3.31. The particle is rolling on top of a triangular face. At the start of the current time step, gravity pulls the particle down and a collision is detected at the beginning of the movement. The collision-response module then computes the contact force $\vec{F}_c$ that will prevent the particle from moving downwards. The particle's trajectory is then recomputed taking the reaction force into account, and the particle moves parallel to the triangular face, from its initial position $\vec{h}_1$ to its final position $\vec{h}_2$.

In order to deal with this special case, we need first to test whether $\vec{a}$ lies inside the triangle. If so, then the cylinder intersects the triangle, and the intersection point is set to $\vec{h}_1$ (this point-in-triangle test is covered in

Section 2.4.4). Otherwise, the projected axis is checked for intersections with the triangle edges, as explained above.

Case 2: Axis intersects the plane.

Here, equation (3.24) is not satisfied. We first need to compute the intersection point $\vec{g}$ between the plane and the line supporting the axis.

The intersection point $\vec{g}$ will be given by

$$\vec{g} = \vec{h}_1 + k_g (\vec{h}_2 - \vec{h}_1) \qquad (3.25)$$

for some $k_g \in \mathbb{R}$ (see Section A.4 of Appendix A for details on how to determine k_g). If $k_g < 0$ or $k_g > 1$, then the cylinder does not intersect the triangle. However, if $0 \leq k_g \leq 1$, then we need to check whether the intersection point $\vec{g}$ lies inside or outside the triangle (see Figure 3.32.) This test can be efficiently implemented using the point-in-triangle test already discussed in Section 2.4.4. If $\vec{g}$ lies inside the triangle, then the cylinder intersects the triangle and we set $\vec{g}$ as their intersection point.

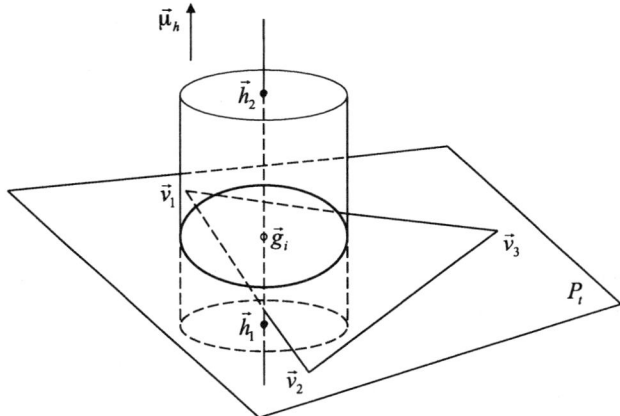

FIGURE 3.32. The intersection point $\vec{g}$ between the cylinder axis and the plane that contains the triangle lies inside or outside the triangle. In the situation shown, point $\vec{g}$ lies inside the triangle.

Otherwise, if $\vec{g}$ lies outside the triangle, then one of the following cases can occur:

1. One or more edges of the triangle intersect the cylinder cap;

2. The triangle is intersecting the top limiting plane of the cylinder;

3. The triangle is intersecting the bottom limiting plane of the cylinder;

4. And, of course, the triangle is not intersecting the cylinder at all.

104   3. Particle Systems

Case 2 is left unresolved for the moment because the top limiting plane is also part of the spherical cap, and therefore its intersection with the triangle is left to the sphere-triangle test. Also, case 3 is ignored because the bottom limiting plane at the current time step was the top limiting plane at the previous time step and was already checked at that time. In both cases, the cylinder-triangle intersection test returns false, meaning the cylinder does not intersect the triangle.

Case 1 requires some extra steps, as illustrated in Figure 3.33. For each edge of the triangle, we determine its closest point and distance to the cylinder's axis. The closest point of an edge to a line is computed as indicated in Section A.5 of Appendix A.

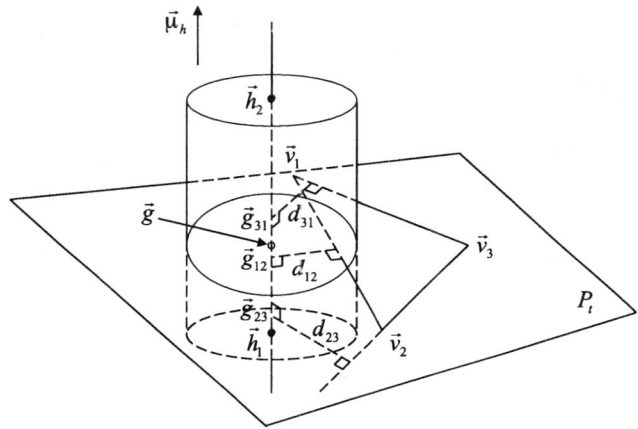

FIGURE 3.33. The distance d_{12} is less than r_c, meaning $\vec{g}_{12}$ is approximated as the intersection point between the cylinder and the triangle.

Let d_{ij} be the distance from edge $(\vec{i}, \vec{j})$ to the cylinder's axis. If the distance from all edges is such that

$$d_{ij} > r_c, \text{ for } i, j \in \{1, 2, 3\}, i \neq j,$$

then, at first, none of the edges intersects the cylinder cap. The triangle may still intersect the top limiting plane, possibly passing through the cylinder cap (see Figure 3.34). However, as mentioned, checking whether the triangle intersects the top limiting plane is left to the sphere-triangle intersection test, and no further tests are required at this stage.

However, if one or more distances d_{ij} are less than or equal to r_c, then we check whether their associated closest point $\vec{g}_{ij}$ is inside the cylinder. If this is the case, then the cylinder intersects the triangle and the intersection point is set to the $\vec{g}_{ij}$ closest to $\vec{h}_1$ (the initial position of the particle). The point-in-cylinder test is covered in Section 3.5.5.

Finally, if all closest points $\vec{g}_{ij}$ are outside the cylinder, then the triangle may still be intersecting the top limiting plane of the cylinder (see Fig-

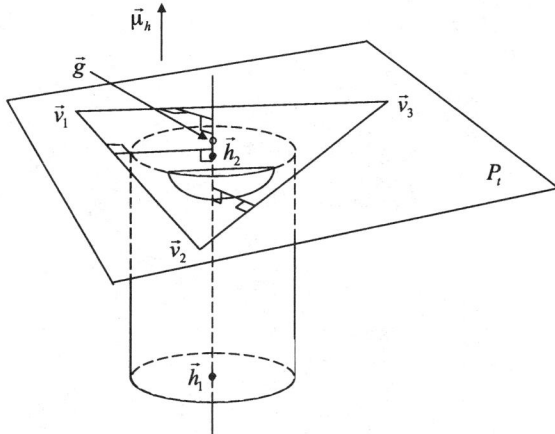

FIGURE 3.34. The distances between the triangle edges and the cylinder axis are greater than r_c, but the triangle still intersects the cylinder's top limiting plane. The point $\vec{g}$ is the intersection between the cylinder's axis and the triangle plane.

ure 3.35). Again, this case is dealt with in the sphere-triangle intersection test.

3.5.5 Point-in-Cylinder Test

The point-in-cylinder test can be carried out in three easy steps. First, we determine the relative position of the point $\vec{p}$ with respect to the top and bottom limiting planes of the cylinder. If $\vec{p}$ lies above or below the top and bottom limiting planes, respectively, then it is not included in the cylinder. Otherwise, if $\vec{p}$ lies on either of the planes, then we need to compute its distance to $\vec{h}_1$ or $\vec{h}_2$ as appropriate, and compare it with the radius of the cylinder. For instance, if $\vec{p}$ lies on the top limiting plane, then we need to check whether

$$|\vec{p} - \vec{h}_2|^2 \leq r_h^2 . \tag{3.26}$$

If equation (3.26) is satisfied, then the $\vec{p}$ lies inside the cylinder.

Finally, if $\vec{p}$ lies between the top and bottom limiting planes of the cylinder, then we need to compute its distance to the cylinder's axis and compare it with the cylinder's radius. Let d be the distance between $\vec{p}$ and the cylinder's axis (see Section A.2 of Appendix A for details on how to compute the distance between a point and a line segment). The point is inside the cylinder if and only if

$$d \leq r_h .$$

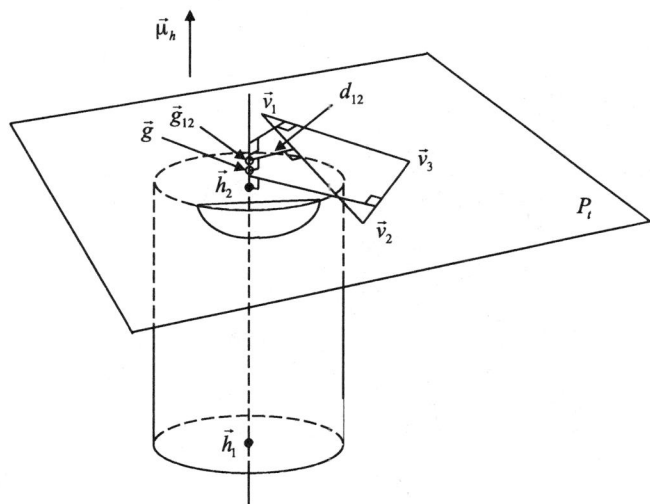

FIGURE 3.35. Distance d_{12} is less than r_c, but the closest point $\vec{g}_{12}$ lies outside the cylinder axis. Nonetheless, the triangle still intersects the cylinder at its top limiting plane.

3.6 Particle-Particle Collision Response

Whenever a particle-particle collision is detected, the collision-response module is invoked to compute the appropriate collision impulses or contact forces that will prevent interpenetration between the colliding particles. As explained in Section 3.4, the trajectories of the colliding particles are traced back in time to the moment before their collision. The collision point and normal are then determined from their geometric displacement.

In the case of particle-particle collisions, the collision point is the actual particles' position, since the particles are treated as point-mass objects and all collision forces are directly applied to the center of the sphere representing each particle. The particles' shape is only used to compute the normal direction of the collision. As indicated in Figure 3.36, the collision normal is determined by connecting the center of the particles' spheres.

The colliding particles are arbitrarily assigned indexes 1 and 2, and the normal direction is selected such that the relative velocity $(\vec{v}_1 - \vec{v}_2)$ of the particles along the normal is negative just before the collision, that is, we choose $\vec{n}$ such that

$$(\vec{v}_1 - \vec{v}_2) \cdot \vec{n} < 0 \qquad (3.27)$$

is satisfied just before the collision. This assignment is critical, since from Newton's principle of action and reaction the collision impulses and contact forces between the particles have the same magnitude, but opposite directions. Following our convention, a positive impulse should be applied to

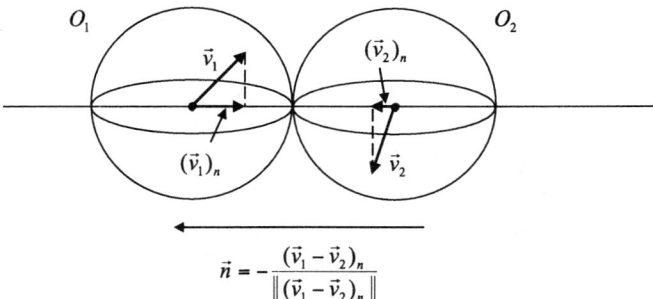

FIGURE 3.36. Particles O_1 and O_2 are traced back in time to the moment before their collision. The collision normal $\vec{n}$ is defined by the line connecting the center of the spheres. Its direction should be the opposite of the relative velocity $(\vec{v}_1 - \vec{v}_2)$ of the particles along the normal.

the particle with index 1, whereas a negative impulse should be applied to the particle of index 2. Therefore, it is very important to keep track of the index assigned to each particle so as to later apply the collision impulses and contact forces on the correct direction (i.e., with the correct sign) to each particle. Also, notice that in the case of multiple particle-particle collisions, a particle might be assigned different indexes for each collision it is involved.

The difference between a collision and a contact is determined from the module of the relative velocity along the collision normal, at the collision point. If the relative velocity of the particles along the normal, at the moment before the collision, is less than a threshold value, then the particles are said to be in contact and a contact force is computed to prevent their interpenetration. Otherwise, the particles are said to be in collision and an impulsive force is applied to instantaneously change the direction of motion of the particles to avoid the imminent interpenetration.

It may happen that several particles are involved in multiple collisions and contacts. If so, the collision-response module should first resolve all collisions by simultaneously computing all impulsive forces. Having determined all impulsive forces, the collision-response module proceeds by applying the impulses to the appropriate particles. By the time the impulses are applied, some of the contacts may break, depending on whether the relative acceleration of the particles at their contact point, along the contact normal, is positive, zero or negative. A contact force is then simultaneously computed for all contacts that have a negative relative acceleration along their contact normal.

3.6.1 Computing Impulsive Forces for a Single Collision

Let's start by examining the case where we have one or more simultaneous collisions, each involving two different particles. Here, each collision can be

108   3. Particle Systems

dealt with independently of the others, since they do not have particles in common.

Let collision C, involving particles O_1 and O_2, be defined by its collision normal $\vec{n}$ and tangent axes $\vec{t}$ and $\vec{k}$, as indicated in Figure 3.37. Let $\vec{v}_1 = ((v_1)_n, (v_1)_t, (v_1)_k)$ and $\vec{v}_2 = ((v_2)_n, (v_2)_t, (v_2)_k)$ be the particles' velocities just before the collision, and $\vec{V}_1 = ((V_1)_n, (V_1)_t, (V_1)_k)$ and $\vec{V}_2 = ((V_2)_n, (V_2)_t, (V_2)_k)$ be their velocities just after the collision. We need to compute their velocities just after the collision, along with the impulsive force $\vec{P} = (P_n, P_t, P_k)$. This yields a total of nine unknowns ($\vec{P}$, $\vec{V}_1$ and $\vec{V}_2$), thus requiring the solution of a system with nine equations.

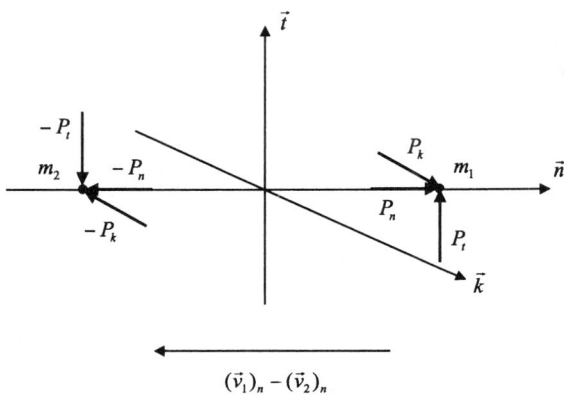

FIGURE 3.37. Particles O_1 and O_2 just before their collision. The impulsive force is applied to each particle, with same magnitude, but opposite direction.

Applying the principle of impulse and momentum for each particle along the three collision axes, we obtain six out of the nine equations needed:

$$m_1 (V_1)_n - m_1 (v_1)_n = P_n \tag{3.28}$$
$$m_1 (V_1)_t - m_1 (v_1)_t = P_t \tag{3.29}$$
$$m_1 (V_1)_k - m_1 (v_1)_k = P_k \tag{3.30}$$
$$m_2 (V_2)_n - m_2 (v_2)_n = -P_n \tag{3.31}$$
$$m_2 (V_2)_t - m_2 (v_2)_t = -P_t \tag{3.32}$$
$$m_2 (V_2)_k - m_2 (v_2)_k = -P_k \ . \tag{3.33}$$

The next equation is obtained by the empirical relation involving the coefficient of restitution and the relative velocity of the particles along the normal direction of collision. Let e denote the coefficient of restitution along the normal direction. We have

$$(V_1)_n - (V_2)_n = -e\left((v_1)_n - (v_2)_n\right) . \tag{3.34}$$

The remaining two equations are obtained from the Coulomb friction relations at the collision point. If the relative motion of the particles along $\vec{t}$ and $\vec{k}$ is zero just before the collision, that is, if

$$(v_1)_t - (v_2)_t = 0$$
$$(v_1)_k - (v_2)_k = 0,$$

then their relative motion will remain zero after the collision. More specifically, we use

$$(V_1)_t = (V_2)_t \quad (3.35)$$
$$(V_1)_k = (V_2)_k \quad (3.36)$$

as the two remaining equations to solve the system. However, if the relative motion is not zero, then the particles are sliding along $\vec{t}$ and $\vec{k}$ at the collision point. The collision impulse will then act on the opposite direction of motion, trying to prevent the sliding. If it succeeds, then equations (3.35) and (3.36) should be used. Otherwise, the particles continue sliding throughout the collision, and we use

$$P_t = (\mu_d)_t P_n \quad (3.37)$$
$$P_k = (\mu_d)_k P_n \quad (3.38)$$

as the two remaining equations to solve the system. Notice that $(\mu_d)_t$ and $(\mu_d)_k$ are the dynamic Coulomb friction coefficients along the $\vec{t}$ and $\vec{k}$ axes, respectively. Since P_t and P_k are always opposing the sliding motion, the coefficients of friction can be either positive or negative to reflect that condition. The actual signs of the coefficients depend on the relative velocity of the particles along their associated axes, just before the collision. The signs are directly computed from

$$\text{sign}((\mu_d)_t) = \frac{(v_2)_t - (v_1)_t}{(v_2)_n - (v_1)_n} \quad (3.39)$$

$$\text{sign}((\mu_d)_k) = \frac{(v_2)_k - (v_1)_k}{(v_2)_n - (v_1)_n}. \quad (3.40)$$

This directional-friction model is a generalization of the widely used model of relating the tangential and normal impulses using just one omnidirectional coefficient of friction μ_d, as in

$$P_{tk} = \mu_d P_n, \quad (3.41)$$

where P_{tk} is the impulse on the tangent plane given by

$$P_{tk} = \sqrt{P_t^2 + P_k^2} \ .$$

For example, if friction is isotropic, that is, independent of direction, then we can write

$$\begin{aligned}(\mu_d)_t &= \mu_d \cos \phi \\ (\mu_d)_k &= \mu_d \sin \phi\end{aligned}$$

for some angle ϕ, so

$$\begin{aligned}P_{tk} &= \sqrt{P_t^2 + P_k^2} \\ &= \sqrt{(\mu_d)^2 \, P_n^2 \, \cos\phi^2 + (\mu_d)^2 \, P_n^2 \, \sin\phi^2} \\ &= \mu_d \, P_n \ ,\end{aligned}$$

which is the same result obtained using the omnidirectional-friction model of equation (3.41). The main advantage of using the directional-friction model is that the non-linear equation

$$|P_{tk}| = \sqrt{P_t^2 + P_k^2} \le \mu_d \, P_n \ ,$$

which needs to be enforced when the particles are not sliding at the collision point, can be substituted for two linear equations

$$\begin{aligned}|P_t| &\le \mu_t \, P_n \\ |P_k| &\le \mu_k \, P_n\end{aligned}$$

which are equivalent to the non-linear equation if friction is isotropic, and, most important, are easier to handle in matrix form, as we shall see shortly.

So, as far as friction is concerned, we have to consider two possible cases. In the first case, we assume the particles continue sliding on the tangent plane after collision, and we use equations (3.28) to (3.34) with equations (3.37) and (3.38) to compute the collision impulse and velocities after the collision. In the second case, the particles are not sliding on the tangent plane after the collision, and we use equations (3.28) to (3.34) with equations (3.35) and (3.36) instead. For now, let's focus on the solution corresponding to the first case. Later, we consider the required modifications needed to address the second case.

Summing equations (3.28) and (3.31), (3.29) and (3.32), (3.30) and (3.33), we get

$$m_1 (V_1)_n + m_2 (V_2)_n = m_1 (v_1)_n + m_2 (v_2)_n \qquad (3.42)$$
$$m_1 (V_1)_t + m_2 (V_2)_t = m_1 (v_1)_t + m_2 (v_2)_t \qquad (3.43)$$
$$m_1 (V_1)_k + m_2 (V_2)_k = m_1 (v_1)_k + m_2 (v_2)_k . \qquad (3.44)$$

Subtracting the same equations pair-wise, we obtain

$$P_n = \frac{(m_1 (V_1)_n - m_2 (V_2)_n) - (m_1 (v_1)_n - m_2 (v_2)_n)}{2} \qquad (3.45)$$
$$P_t = \frac{(m_1 (V_1)_t - m_2 (V_2)_t) - (m_1 (v_1)_t - m_2 (v_2)_t)}{2} \qquad (3.46)$$
$$P_k = \frac{(m_1 (V_1)_k - m_2 (V_2)_k) - (m_1 (v_1)_k - m_2 (v_2)_k)}{2} . \qquad (3.47)$$

Substituting equations (3.45) to (3.47) into (3.37) and (3.38) gives

$$(\mu_d)_t\, m_1 (V_1)_n - m_1 (V_1)_t - (\mu_d)_t\, m_2 (V_2)_n + m_2 (V_2)_t =$$
$$(\mu_d)_t\, m_1 (v_1)_n - m_1 (v_1)_t - (\mu_d)_t\, m_2 (v_2)_n + m_2 (v_2)_t \qquad (3.48)$$
$$(\mu_d)_k\, m_1 (V_1)_n - m_1 (V_1)_k - (\mu_d)_k\, m_2 (V_2)_n + m_2 (V_2)_k =$$
$$(\mu_d)_k\, m_1 (v_1)_n - m_1 (v_1)_k - (\mu_d)_k\, m_2 (v_2)_n + m_2 (v_2)_k . \qquad (3.49)$$

Making a change of variables such that

$$(U_i)_j = m_i ((V_i)_j - (v_i)_j), \text{ for } i = \{1, 2\}, j = \{n, t, k\} ,$$

the system defined by equations (3.42) to (3.44), (3.34), (3.48) and (3.49) can then be written as

$$(U_1)_n + (U_2)_n = 0$$
$$(U_1)_t + (U_2)_t = 0$$
$$(U_1)_k + (U_2)_k = 0$$
$$\frac{(U_1)_n}{m_1} - \frac{(U_2)_n}{m_2} = -(1+e)((v_1)_n - (v_2)_n)$$
$$(\mu_d)_t (U_1)_n - (U_1)_t - (\mu_d)_t (U_2)_n + (U_2)_t = 0$$
$$(\mu_d)_k (U_1)_n - (U_1)_k - (\mu_d)_k (U_2)_n + (U_2)_k = 0 .$$

Solving for the $(U_i)_j$ we obtain

$$(U_1)_n = m_1 ((V_1)_n - (v_1)_n)$$

$$\begin{aligned}
&= m_{12}\,(1+e)\,((v_2)_n - (v_1)_n) & (3.50)\\
(U_1)_t &= m_1\,((V_1)_t - (v_1)_t)\\
&= (\mu_d)_t\,m_{12}\,(1+e)\,((v_2)_n - (v_1)_n) & (3.51)\\
(U_1)_k &= m_1\,((V_1)_k - (v_1)_k)\\
&= (\mu_d)_k\,m_{12}\,(1+e)\,((v_2)_n - (v_1)_n) & (3.52)\\
(U_2)_n &= m_2\,((V_2)_n - (v_2)_n)\\
&= -m_{12}\,(1+e)\,((v_2)_n - (v_1)_n) & (3.53)\\
(U_2)_t &= m_2\,((V_2)_t - (v_2)_t)\\
&= -(\mu_d)_t\,m_{12}\,(1+e)\,((v_2)_n - (v_1)_n) & (3.54)\\
(U_2)_k &= m_2\,((V_2)_k - (v_2)_k)\\
&= -(\mu_d)_k\,m_{12}\,(1+e)\,((v_2)_n - (v_1)_n)\,. & (3.55)
\end{aligned}$$

where

$$m_{12} = \frac{m_1\,m_2}{(m_1 + m_2)}\,.$$

The particle velocities $\vec{V}_1$ and $\vec{V}_2$ just after the collision are directly obtained from equations (3.50) to (3.55). Substituting their values into equations (3.45) to (3.47), we immediately get the impulse $\vec{P}$.

All derivations up till now considered the case wherein the colliding particles continue sliding throughout the collision. If the particles are not sliding after collision, either because they were not sliding before the collision or the sliding motion stopped during the collision, then equations (3.35) and (3.36) should be used instead of equations (3.37) and (3.38), which are repeated here for convenience:

$$\begin{aligned}
(V_1)_t &= (V_2)_t\\
(V_1)_k &= (V_2)_k\,.
\end{aligned}$$

Notice that the sliding motion on the tangent plane is directly affected by the coefficients of restitution and friction, as well as by the relative velocities of the particles just before the collision. Intuitively, for a given coefficient of restitution and relative velocities, the sliding motion will continue if the coefficient of friction is small, or will stop if the coefficient of friction is sufficiently large. Therefore, there exists a critical coefficient of friction value associated with a given coefficient of restitution and relative velocity of the particles. If the actual coefficient of friction is less than the critical coefficient of friction, then slide continues throughout the collision and the system equations associated with the first case should be considered. However, if the actual coefficient of friction is greater than or equal to the critical coefficient of friction, then slide stops somewhere during the collision and the system equations associated with the second case should be considered instead.

Let's derive an expression for computing the critical coefficient of friction. If we substitute back $(V_1)_t$, $(V_1)_k$, $(V_2)_t$ and $(V_2)_k$ obtained from equations (3.51), (3.52), (3.54) and (3.55), respectively, into equations (3.37) and (3.38), we get

$$\frac{(v_1)_t) + (\mu_d)_t\, m_{12}\,(1+e)\,((v_2)_n - (v_1)_n)}{m_1} =$$
$$\frac{(v_2)_t) - (\mu_d)_t\, m_{12}\,(1+e)\,((v_2)_n - (v_1)_n)}{m_2}$$
$$\frac{(v_1)_k) + (\mu_d)_k\, m_{12}\,(1+e)\,((v_2)_n - (v_1)_n)}{m_1} =$$
$$\frac{(v_2)_k) - (\mu_d)_k\, m_{12}\,(1+e)\,((v_2)_n - (v_1)_n)}{m_2}.$$

Solving for $(\mu_d)_t$ and $(\mu_d)_k$, we obtain

$$(\mu_d)_t = (\mu_d)_t^c = \frac{1}{(1+e)} \frac{((v_2)_t - (v_1)_t)}{((v_2)_n - (v_1)_n)} \quad (3.56)$$

$$(\mu_d)_k = (\mu_d)_k^c = \frac{1}{(1+e)} \frac{((v_2)_k - (v_1)_k)}{((v_2)_n - (v_1)_n)}, \quad (3.57)$$

where $(\mu_d)_t^c$ and $(\mu_d)_k^c$ are the critical values of the coefficient of friction such that the sliding motion stops exactly at the end of the collision.

We do the following in practice. First, compute the critical coefficient of friction using equations (3.56) and (3.57). Then, compare the actual coefficient of friction $(\mu_d)_t$ and $(\mu_d)_k$ to their associated critical values. If $(\mu_d)_t < (\mu_d)_t^c$, then sliding continues along $\vec{t}$ and we use equation (3.37). Else, if $(\mu_d)_t \geq (\mu_d)_t^c$, then sliding along $\vec{t}$ stops during the collision and we use equation (3.35). The same analysis is used for comparing $(\mu_d)_k$ with $(\mu_d)_k^c$ and selecting the appropriate system equation.

Also, notice that there is no need to derive a new set of solutions to the system equations for the case in which sliding stops during the collision; we just need to use the critical value of the coefficient of friction instead of the actual friction value in the solution equations already obtained. Recall that, if we set $(\mu_d)_t = (\mu_d)_t^c$, then we immediately obtain the desired condition $(V_1)_t = (V_2)_t$. By analogy, if we set $(\mu_d)_k = (\mu_d)_k^c$, we get $(V_1)_k = (V_2)_k$.

An alternate representation of the system equations for computing the collision impulses commonly found in the literature is the partitioned-matrix representation. Even though this representation is not particularly useful for the singe-collision case because we were able to compute the impulse and final velocities, it proves to be extremely useful when dealing with

114   3. Particle Systems

multiple simultaneous collisions. Here, we shall focus on the partitioned-matrix representation of a single collision. In Section 3.6.2, we extend it to the multiple-collision case.

As previously explained, we use equations (3.28) to (3.34) with equations (3.37) and (3.38) whenever the particles continue sliding on the tangent plane after collision. In this case, the system equations can be put into the following matrix format:

$$\begin{pmatrix} 0 & 0 & 0 & 1 & 0 & 0 & -1 & 0 & 0 \\ -(\mu_d)_t & 1 & 0 & 0 & 0 & 0 & 0 & 0 & 0 \\ -(\mu_d)_k & 1 & 0 & 0 & 0 & 0 & 0 & 0 & 0 \\ -1 & 0 & 0 & m_1 & 0 & 0 & 0 & 0 & 0 \\ 0 & -1 & 0 & 0 & m_1 & 0 & 0 & 0 & 0 \\ 0 & 0 & -1 & 0 & 0 & m_1 & 0 & 0 & 0 \\ 1 & 0 & 0 & m_2 & 0 & 0 & 0 & 0 & 0 \\ 0 & 1 & 0 & 0 & m_2 & 0 & 0 & 0 & 0 \\ 0 & 0 & 1 & 0 & 0 & m_2 & 0 & 0 & 0 \end{pmatrix} \begin{pmatrix} (P_{1,2})_n \\ (P_{1,2})_t \\ (P_{1,2})_k \\ (V_1)_n \\ (V_1)_t \\ (V_1)_k \\ (V_2)_n \\ (V_2)_t \\ (V_2)_k \end{pmatrix} = \begin{pmatrix} -e\left((v_1)_n - (v_2)_n\right) \\ 0 \\ 0 \\ m_1(v_1)_n \\ m_1(v_1)_t \\ m_1(v_1)_k \\ m_2(v_2)_n \\ m_2(v_2)_t \\ m_2(v_2)_k \end{pmatrix} \quad (3.58)$$

where $\vec{P}_{1,2} = ((P_{1,2})_n, (P_{1,2})_t, (P_{1,2})_k)$ is the collision impulse between particles O_1 and O_2.

If sliding along the $\vec{t}$ direction stops by the end of the collision, then we need to use equation (3.35) instead of equation (3.37). This in turn has the effect of substituting the second row of the system matrix for

$$\begin{pmatrix} 0 & 0 & 0 & 0 & 1 & 0 & 0 & -1 & 0 \end{pmatrix}.$$

Conversely, if sliding along the $\vec{k}$ directions stops by the end of the collision, then we need to substitute the third row of the system matrix for

$$\begin{pmatrix} 0 & 0 & 0 & 0 & 0 & 1 & 0 & 0 & -1 \end{pmatrix}.$$

The partitioned-matrix representation of the system can be directly obtained from equation (3.58), and is given by

$$\begin{pmatrix} \mathbf{A}_{1,2} & \mathbf{B}_{1,2} & -\mathbf{B}_{1,2} \\ -\mathbf{I} & m_1 \mathbf{I} & 0 \\ \mathbf{I} & 0 & m_2 \mathbf{I} \end{pmatrix} \begin{pmatrix} \vec{P}_{1,2} \\ \vec{V}_1 \\ \vec{V}_2 \end{pmatrix} = \begin{pmatrix} \vec{d}_{1,2} \\ m_1 \vec{v}_1 \\ m_2 \vec{v}_2 \end{pmatrix}, \quad (3.59)$$

3.6 Particle-Particle Collision Response

where **I** is the 3×3 identity matrix, **0** is the 3×3 zero matrix, and

$$\vec{d}_{1,2} = \begin{pmatrix} -e\left((v_1)_n - (v_2)_n\right) \\ 0 \\ 0 \end{pmatrix}.$$

The matrices $\mathbf{A}_{1,2}$ and $\mathbf{B}_{1,2}$ are chosen depending on whether sliding continues along the tangent plane after the collision ends. We have four possible cases to consider.

1. If $(\mu_d)_t < (\mu_d)_t^c$ and $(\mu_d)_k < (\mu_d)_k^c$, then:

$$\mathbf{A}_{1,2} = \begin{pmatrix} 0 & 0 & 0 \\ -(\mu_d)_t & 1 & 0 \\ -(\mu_d)_k & 1 & 0 \end{pmatrix} \text{ and } \mathbf{B}_{1,2} = \begin{pmatrix} 1 & 0 & 0 \\ 0 & 0 & 0 \\ 0 & 0 & 0 \end{pmatrix}$$

2. If $(\mu_d)_t \geq (\mu_d)_t^c$ and $(\mu_d)_k < (\mu_d)_k^c$, then:

$$\mathbf{A}_{1,2} = \begin{pmatrix} 0 & 0 & 0 \\ 0 & 0 & 0 \\ -(\mu_d)_k & 1 & 0 \end{pmatrix} \text{ and } \mathbf{B}_{1,2} = \begin{pmatrix} 1 & 0 & 0 \\ 0 & 1 & 0 \\ 0 & 0 & 0 \end{pmatrix}$$

3. If $(\mu_d)_t < (\mu_d)_t^c$ and $(\mu_d)_k \geq (\mu_d)_k^c$, then:

$$\mathbf{A}_{1,2} = \begin{pmatrix} 0 & 0 & 0 \\ -(\mu_d)_t & 1 & 0 \\ 0 & 0 & 0 \end{pmatrix} \text{ and } \mathbf{B}_{1,2} = \begin{pmatrix} 1 & 0 & 0 \\ 0 & 0 & 0 \\ 0 & 0 & 1 \end{pmatrix}$$

4. If $(\mu_d)_t \geq (\mu_d)_t^c$ and $(\mu_d)_k \geq (\mu_d)_k^c$ then $\mathbf{A}_{1,2} = \mathbf{0}$ and $\mathbf{B}_{1,2} = \mathbf{I}$.

Notice that the first row of the partitioned matrix shown in (3.59) contains the coefficient of restitution and friction equations, and it is associated with the state variable $\vec{P}_{1,2}$. The second and third rows contain the conservation-of-linear-momentum equations associated with the final velocities $\vec{V}_1$ and $\vec{V}_2$, respectively. This ordering is extremely important because it can significantly simplify the updates required for the multiple-collision case.

3.6.2 Computing Impulsive Forces for Multiple Simultaneous Collisions

During a simulation, there might be situations in which three or more particles are simultaneously colliding with each other. In these cases, instead of resolving one collision at a time not taking into account the presence of the others, the simulation engine needs to group the particles into clusters that

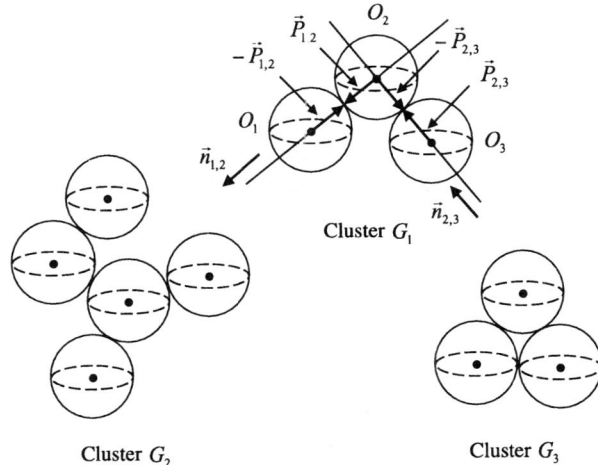

FIGURE 3.38. Multiple particle collisions separated into three clusters. Particle O_i is added to cluster G_j if it is colliding with at least one particle already in G_j. The collision-response module resolves each cluster in parallel, since they have no collisions in common and therefore can be viewed as independent groups of collisions.

share at least one collision. The collisions within each cluster can then be simultaneously resolved independent of all other clusters (see Figure 3.38).

Consider the computation of the collision impulses associated with cluster G_1, as shown in Figure 3.38. Let collisions C_1 and C_2 be the collisions $(O_1 - O_2)$ and $(O_2 - O_3)$, respectively. As far as particle O_2 is concerned, the linear momentum equation owing to both collisions becomes

$$m_2 (\vec{V}_2 - \vec{v}_2) = -\vec{P}_{1,2} + \vec{P}_{2 \mapsto 1, 3 \mapsto 2} , \qquad (3.60)$$

where $P_{2 \mapsto 1, 3 \mapsto 2}$ is the impulse $\vec{P}_{2,3}$ of collision C_2 expressed in the local coordinate frame[5] associated with collision C_1. The minus sign on $\vec{P}_{1,2}$ shows that particle O_2 has index 2 with respect to collision C_1.

Clearly, the impulse owing to collision C_2 will also affect the computation of the impulse owing to collision C_1, and vice-versa. Therefore, the correct way to compute the collision impulses is to take both collisions into account when solving the system of equations. Recall from Section 3.6.1 that we adopted the convention that a positive impulse is applied to the particle with index 1, and a negative impulse is applied to the particle with index 2. The choice of indexes is related to the relative velocities of the particles along the collision normal, such that equation (3.27) is satisfied at the moment just before the collision.

---

[5]The local-coordinate frame is defined by the collision normal and tangent plane.

If a particle is involved in multiple collisions, it is possible to have it assigned to different indexes for each collision. For the particular situation of cluster G_1, particle O_2 has index 2 with respect to its collision with particle O_1, and index 1 with respect to its collision with particle O_3. This in turn affects the choice of sign when combining the multiple-collision impulses in the system equations (i.e., the minus sign of $\vec{P}_{1,2}$ and the plus sign of $P_{2\mapsto 1, 3\mapsto 2}$ in equation (3.60)). Moreover, the collision normal and tangent plane are different for each collision. So, we also need to implement a change of base between the collision impulses before combining them (i.e., the vector $P_{2\mapsto 1, 3\mapsto 2}$ used in equation (3.60)).

The best way to deal with multiple collisions is to represent the system equations associated with each cluster in its partitioned matrix form

$$\mathbf{A}\,\vec{x} = \vec{b},$$

where $\vec{x}$ is the state vector containing the variables that need to be determined. In the case of a single collision, the state vector is defined by the collision impulse and final velocities of the colliding particles. However, when dealing with multiple collisions, the state vector can be viewed as the concatenation of several single-collision state vectors, with the added complexity that no variables should be accounted for more than once. For instance, Figure 3.39(a) shows the result of a naive concatenation of state vectors for the multiple collisions associated with cluster G_1 of Figure 3.38.

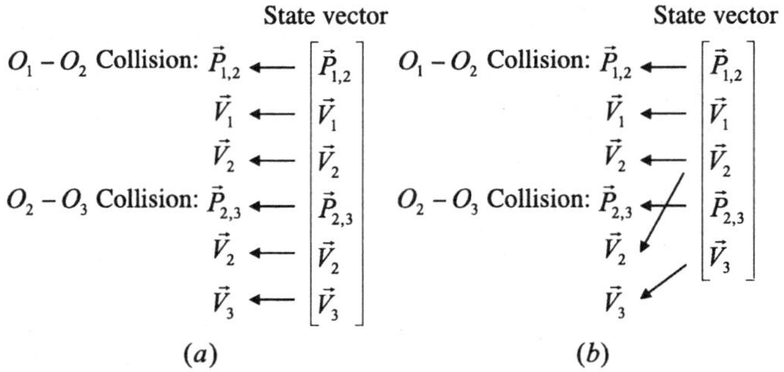

FIGURE 3.39. (a) A naive concatenation creates multiple entries for the final velocities of particles that are involved in more than one collision; (b) The state-vector variables should have a link back to their collisions. More than one link is used for multiple collisions, as in the case of $\vec{V}_2$.

Since particle O_2 is involved in both collisions, its final velocity $\vec{V}_2$ is counted twice. The correct way to create the state vector is then to keep track of the variables already added, and mark as "common" those added more than once. This is illustrated in Figure 3.39(b).

Having determined the state vector associated with a cluster, the next step is to fill in the rows of matrix $\mathbf{A}$ and vector $\vec{b}$. This can be done by considering the equation associated with the first link of each variable in the state vector. For example, for the cluster G_1 of Figure 3.38, the first variable of the state vector is $\vec{P}_{1,2}$. This variable is linked to the $(O_1 - O_2)$ collision. Its associated equations are the coefficient of restitution and friction equations. Therefore, the first row of matrix $\mathbf{A}$ and vector $\vec{b}$ is

$$\begin{pmatrix} \mathbf{A}_{1,2} & \mathbf{B}_{1,2} & -\mathbf{B}_{1,2} & 0 & 0 \\ x & x & x & x & x \\ x & x & x & x & x \\ x & x & x & x & x \\ x & x & x & x & x \end{pmatrix} \begin{pmatrix} \vec{P}_{1,2} \\ \vec{V}_1 \\ \vec{V}_2 \\ \vec{P}_{2,3} \\ \vec{V}_3 \end{pmatrix} = \begin{pmatrix} \vec{d}_{1,2} \\ x \\ x \\ x \\ x \end{pmatrix}.$$

The second variable of the state vector is $\vec{V}_1$. This variable is also linked to the $(O_1 - O_2)$ collision. Its associated equations are the conservation of linear momentum for particle O_1. So, the second row of matrix $\mathbf{A}$ and vector $\vec{b}$ is

$$\begin{pmatrix} \mathbf{A}_{1,2} & \mathbf{B}_{1,2} & -\mathbf{B}_{1,2} & 0 & 0 \\ -\mathbf{I} & m_1 \mathbf{I} & 0 & 0 & 0 \\ x & x & x & x & x \\ x & x & x & x & x \\ x & x & x & x & x \end{pmatrix} \begin{pmatrix} \vec{P}_{1,2} \\ \vec{V}_1 \\ \vec{V}_2 \\ \vec{P}_{2,3} \\ \vec{V}_3 \end{pmatrix} = \begin{pmatrix} \vec{d}_{1,2} \\ m_1 \vec{v}_1 \\ x \\ x \\ x \end{pmatrix}.$$

Doing the same for all other state variables, we obtain

$$\begin{pmatrix} \mathbf{A}_{1,2} & \mathbf{B}_{1,2} & -\mathbf{B}_{1,2} & 0 & 0 \\ -\mathbf{I} & m_1 \mathbf{I} & 0 & 0 & 0 \\ \mathbf{I} & 0 & m_2 \mathbf{I} & 0 & 0 \\ 0 & 0 & \mathbf{B}_{2,3} & \mathbf{A}_{2,3} & -\mathbf{B}_{2,3} \\ 0 & 0 & 0 & \mathbf{I} & m_3 \mathbf{I} \end{pmatrix} \begin{pmatrix} \vec{P}_{1,2} \\ \vec{V}_1 \\ \vec{V}_2 \\ \vec{P}_{2,3} \\ \vec{V}_3 \end{pmatrix} = \begin{pmatrix} \vec{d}_{1,2} \\ m_1 \vec{v}_1 \\ m_2 \vec{v}_2 \\ \vec{d}_{2,3} \\ m_3 \vec{v}_3 \end{pmatrix}.$$

(3.61)

Notice the difference between rows 1 and 4 of the system matrix shown in equation (3.61). Since $\vec{V}_2$ is common to both $(O_1 - O_2)$ and $(O_2 - O_3)$ collisions, the matrices $\mathbf{A}_{2,3}$ and $\mathbf{B}_{2,3}$ were rearranged to correctly multiply their associated state-vector variables. The correct order is $\mathbf{B}_{2,3}$ multiplying the linear velocity of the particle with index 1, $\mathbf{A}_{2,3}$ multiplying the impulse associated with collision $(O_2 - O_3)$ and $(-\mathbf{B}_{2,3})$ multiplying the linear velocity of the particle with index 2. Because for collision $O_2 - O_3$ particle O_2 is associated with index 1 and particle O_3 with index 2, the arrangement of these block matrices in row 4 of equation (3.61) is different from the one obtained for row 1.

Also, notice that equation (3.61) was built following *only* the first link of each state-vector variable. Now, we need to update equation (3.61) with the multiple-collision terms. This can be done by considering the state variables that have more than one associated link. The first link was used to define the row. The following links are used to update some elements of the row with the multiple-collision terms.

In general, if particle O_i is involved in more than one collision, then the row associated with $\vec{V}_i$, that is, the row associated with its final velocity, needs to be updated. Say for example that particle O_i has a second link to particle O_j. Let $\vec{P}_{i,j}$ designate the state-vector variable corresponding to the impulse associated with this collision. So, the index of $\vec{V}_i$ in the state vector defines the row of the system matrix to be updated, and the index of $\vec{P}_{i,j}$ in the state vector defines the column of the system matrix that needs to be updated. Therefore, we need to update the element

$$[\text{index of } \vec{V}_i][\text{index of } \vec{P}_{i,j}]$$

of the system matrix given in equation (3.61).

The actual update consists of accounting for $\vec{P}_{i,j}$ in the linear momentum equations associated with particle O_i. This can be done by expressing $\vec{P}_{i,j}$ with respect to the local-coordinate frame of the collision corresponding to the first link of the state variable $\vec{V}_i$.

Say for example that the first link of the state variable $\vec{V}_i$ is associated with collision C_m involving particles O_i and O_m. Let the local-coordinate frame $\mathcal{F}_{m,i}$ of collision $(O_m - O_i)$ be defined by vectors $\vec{n}_{m,i}, \vec{t}_{m,i}$ and $\vec{k}_{m,i}$.

The second link of the state variable $\vec{V}_i$ is associated with collision C_j involving particles O_i and O_j. Let the local-coordinate frame $\mathcal{F}_{i,j}$ of collision $(O_i - O_j)$ be defined by vectors $\vec{n}_{i,j}, \vec{t}_{i,j}$ and $\vec{k}_{i,j}$. The collision impulse $\vec{P}_{i,j}$ defined in the local frame $\mathcal{F}_{i,j}$ is expressed in the local frame $\mathcal{F}_{m,i}$ as

$$\vec{P}_{i \mapsto m, j \mapsto i} = \mathbf{M}_{i \mapsto m, j \mapsto i} \vec{P}_{i,j} ,$$

with

$$\mathbf{M}_{i \mapsto m, j \mapsto i} = \lambda \begin{pmatrix} \vec{n}_{i,j} \cdot \vec{n}_{m,i} & \vec{n}_{i,j} \cdot \vec{t}_{m,i} & \vec{n}_{i,j} \cdot \vec{k}_{m,i} \\ \vec{t}_{i,j} \cdot \vec{n}_{m,i} & \vec{t}_{i,j} \cdot \vec{t}_{m,i} & \vec{t}_{i,j} \cdot \vec{k}_{m,i} \\ \vec{k}_{i,j} \cdot \vec{n}_{m,i} & \vec{k}_{i,j} \cdot \vec{t}_{m,i} & \vec{k}_{i,j} \cdot \vec{k}_{m,i} \end{pmatrix} .$$

The variable λ can be either 1 or -1, depending whether particle O_i is assigned to index 2 or 1 in collision O_j. The necessary multiple-collision term update is then

$$[\text{index of } \vec{V}_i] \, [\text{index of } \vec{P}_{i,j}] = \vec{P}_{i \mapsto j, i \mapsto m} .$$

As an example, let's apply this multiple-collision term update to the cluster G_1 of Figure 3.39. In this example, the second link of $\vec{V}_2$ points to the collision of particle O_2 with particle O_3. Therefore, we need to update the element at

$$[\text{index of } \vec{V}_2]\,[\text{index of } \vec{P}_{2,3}] = [3,4]$$

in the system matrix of equation (3.61). The actual update will be to substitute the current **0** element at position $[3,4]$ for

$$\mathbf{M}_{2 \mapsto 1, 3 \mapsto 2} = \lambda \begin{pmatrix} \vec{n}_{2,3} \cdot \vec{n}_{1,2} & \vec{n}_{2,3} \cdot \vec{t}_{1,2} & \vec{n}_{2,3} \cdot \vec{k}_{1,2} \\ \vec{t}_{2,3} \cdot \vec{n}_{1,2} & \vec{t}_{2,3} \cdot \vec{t}_{1,2} & \vec{t}_{2,3} \cdot \vec{k}_{1,2} \\ \vec{k}_{2,3} \cdot \vec{n}_{1,2} & \vec{k}_{2,3} \cdot \vec{t}_{1,2} & \vec{k}_{2,3} \cdot \vec{k}_{1,2} \end{pmatrix}, \quad (3.62)$$

where frame $\mathcal{F}_{1,2}$ is defined by vectors $\vec{n}_{1,2}$, $\vec{t}_{1,2}$ and $\vec{k}_{1,2}$, and frame $\mathcal{F}_{2,3}$ is defined by vectors $\vec{n}_{2,3}$, $\vec{t}_{2,3}$ and $\vec{k}_{2,3}$. Also, since particle O_2 is assigned to index 1 in its collision with particle O_3 (see Figure 3.38), we should use $\lambda = +1$ in equation (3.62). That is, we need to set

$$\text{element } [3,4] = \mathbf{M}_{2 \mapsto 1, 3 \mapsto 2}.$$

The final system matrix for this particular example is then:

$$\begin{pmatrix} \mathbf{A}_{1,2} & \mathbf{B}_{1,2} & -\mathbf{B}_{1,2} & 0 & 0 \\ -\mathbf{I} & m_1 \mathbf{I} & 0 & 0 & 0 \\ \mathbf{I} & 0 & m_2 \mathbf{I} & \mathbf{M}_{2 \mapsto 1, 3 \mapsto 2} & 0 \\ 0 & 0 & \mathbf{B}_{2,3} & \mathbf{A}_{2,3} & -\mathbf{B}_{2,3} \\ 0 & 0 & 0 & \mathbf{I} & m_3 \mathbf{I} \end{pmatrix} \begin{pmatrix} \vec{P}_{1,2} \\ \vec{V}_1 \\ \vec{V}_2 \\ \vec{P}_{2,3} \\ \vec{V}_3 \end{pmatrix} = \begin{pmatrix} \vec{d}_{1,2} \\ m_1 \vec{v}_1 \\ m_2 \vec{v}_2 \\ \vec{d}_{2,3} \\ m_3 \vec{v}_3 \end{pmatrix}.$$

In summary, for each state-vector variable with more than one link, we need to update the elements of the system matrix corresponding to each of these collisions. When all elements are updated, we solve the resulting linear system using, for example, the Gaussian elimination method. Another option would be to use specialized methods to solve sparse linear systems, since the system matrix is often sparse. The solution would then give the correct values of the state-vector variables to be used by the collision-response module.

3.6.3 Computing Contact Forces for a Single Contact

Two particles are said to be in contact whenever their relative velocities along the collision normal are either zero, or less than a threshold value. In these cases, a contact force should be applied, instead of the impulsive force described in Section 3.6.1.

The contact-force computation is considerably different from the impulsive-force computation. In the latter case, we have the equations of conservation of linear momentum and the coefficients of friction and restitution defining the system equations. Here, we need to derive other conditions to compute the contact forces, based on the contact geometry[6] and dynamic state of each particle.

The first condition states that the relative acceleration of the particles at the contact point, along the contact normal, should be greater than or equal to zero, assuming that a negative value indicates that the particles are accelerating towards each other. In this case, if the computed contact force is such that the relative acceleration at the contact point along the contact normal is zero, then the particles remain in contact. However, if their relative acceleration is greater than zero, then contact is about to break.

The second condition implies that the contact-force component along the contact normal should be greater than or equal to zero, indicating that the particles are being pushed away from each other. The contact force is not allowed to have a negative value, that is, is not allowed to keep the particles connected to each other, preventing their separation.

The third and final condition states that the contact force should be set to zero if the contact between the particles is about to break. In other words, if the relative acceleration at the contact point, along the contact normal, is greater than zero, then contact is about to break and the contact force should be set to zero.

Let's translate these three conditions into meaningful equations that can be used to compute the contact force. Figure 3.40 illustrates a typical situation in which particles O_1 and O_2 are shown at the moment before contact, in contact, and interpenetrating each other in case the contact force is not applied.

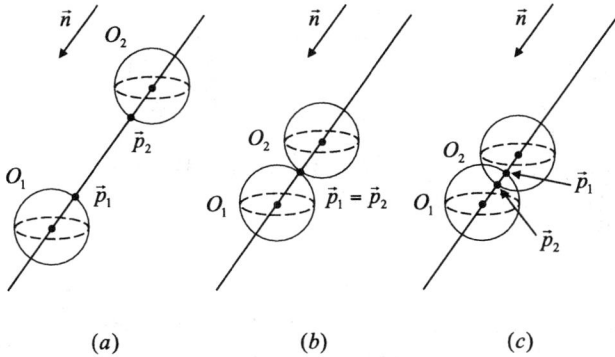

FIGURE 3.40. (a) Particles O_1 and O_2 are about to make contact with other at points $\vec{p}_1$ and $\vec{p}_2$; (b) Contact is established whenever $\vec{p}_1 = \vec{p}_2$; (c) Interpenetration occurs if $(\vec{p}_1 - \vec{p}_2) \cdot \vec{n} < 0$, where $\vec{n}$ is the contact normal.

---

[6]When collision becomes a contact, the collision normal will also be referred to as the contact normal.

Let $\vec{p}_1(t)$ and $\vec{p}_2(t)$ be the position of particles O_1 and O_2 that are about to be in contact. Consider the vector $\vec{q}(t)$ defined as

$$\vec{q}(t) = \begin{pmatrix} q_n(t) \\ q_t(t) \\ q_k(t) \end{pmatrix} = \begin{pmatrix} (\vec{p}_1(t) - \vec{p}_2(t)) \cdot \vec{n}(t) \\ (\vec{p}_1(t) - \vec{p}_2(t)) \cdot \vec{t}(t) \\ (\vec{p}_1(t) - \vec{p}_2(t)) \cdot \vec{k}(t) \end{pmatrix}, \quad (3.63)$$

where $\vec{n}(t)$ is the contact normal, pointing from particle O_2 to particle O_1, and $\vec{t}(t)$ and $\vec{k}(t)$ are vectors defining the tangent plane at the contact. Clearly, $q_n(t)$ defines a distance measure between points $\vec{p}_1(t)$ and $\vec{p}_2(t)$, along the contact normal, as a function of time. We have $q_n(t) > 0$ if the particles are separated, $q_n(t) = 0$ if the particles are in contact, and $q_n(t) < 0$ if the particles are interpenetrating (see Figure 3.40). Let t_c be the instant at which contact is established, that is

$$\vec{q}(t_c) = \vec{0} .$$

The first condition states that the relative acceleration at the contact point, along the contact normal, should be greater than or equal to zero. This is equivalent to enforce that

$$\left. \frac{d^2 q_n(t)}{dt^2} \right|_{t=t_c} \geq 0 . \quad (3.64)$$

If we let $\vec{a}(t) = (a_n(t), a_t(t), a_k(t))$ be the relative acceleration at the contact point, we can rewrite equation (3.64) as

$$a_n(t_c) \geq 0 . \quad (3.65)$$

The components $a_t(t)$ and $a_k(t)$ define the relative acceleration at the contact point on the tangent plane of the contact. They are only used if static or dynamic friction is considered at the contact point, as will be explained later in this section.

The second condition states that the contact-force component along the contact normal should be non-negative, that is

$$F_n \geq 0 , \quad (3.66)$$

where $\vec{F} = (F_n, F_t, F_k)$ is the contact force to be determined. If friction is taken into account, then the tangent components F_t and F_k of the contact force are computed following the Coulomb friction model. To elaborate, if the relative velocity of the points $\vec{p}_1$ and $\vec{p}_2$ along $\vec{t}$ is zero or less than a threshold value, then there is no sliding at the contact point. In this case, the component F_t will assume values in the range of

$$-(\mu_s)_t F_n \leq F_t \leq (\mu_s)_t F_n ,$$

depending on the relative-acceleration component $a_t(t)$ being positive or negative. In other words, F_t will do its best to prevent the particles from sliding at the contact point by always opposing the relative acceleration $a_t(t)$[7]. On the other hand, if the relative velocity along $\vec{t}$ is greater than the threshold value, then the particles are sliding at the contact point and

$$F_t = +(\mu_d)_t F_n \text{ or } F_t = -(\mu_d)_t F_n ,$$

depending on the relative acceleration $a_t(t)$ being negative or positive. Here, $(\mu_d)_t$ is the dynamic coefficient of friction along the $\vec{t}$ direction. A similar analysis holds for $\vec{k}$.

The third and last condition states that the contact force is zero if the contact is breaking away, that is, if the relative acceleration along the contact normal is positive. Equivalently, we have

$$F_n \, a_n(t_c) = 0 , \tag{3.67}$$

meaning that, if F_n is greater than zero, then the particles are in contact and the relative acceleration is zero. Otherwise, if a_n is greater than zero, then the contact is about to break and the contact force should be zero. Putting it all together, we see that the computation of the contact force involves solving the following system of equations:

$$\begin{aligned} a_n(t_c) &\geq 0 \\ F_n &\geq 0 \\ F_n \, a_n(t_c) &= 0 . \end{aligned} \tag{3.68}$$

Here, we adopt the convention that a positive contact force $+\vec{F}$ is applied to particle O_1 (i.e., the particle with index 1) and a negative contact force $-\vec{F}$ is applied to particle O_2 (i.e., the particle with index 2).

According to equation (3.64), the relative acceleration $a_n(t)$, along the contact normal, can be obtained by differentiating equation (3.63) twice with respect to time. By so doing, we see that the first time derivative of equation (3.63) gives

$$\frac{d\,q_n(t)}{dt} = \left(\frac{d\vec{p}_1(t)}{dt} - \frac{d\vec{p}_2(t)}{dt} \right) \cdot \vec{n}(t) + \\ (\vec{p}_1 - \vec{p}_2) \cdot \frac{d\vec{n}(t)}{dt} , \tag{3.69}$$

or equivalently

---

[7] Notice that F_t is zero if $a_t(t)$ is zero.

$$v_n(t) = (\vec{v}_1(t) - \vec{v}_2(t)) \cdot \vec{n}(t) +$$
$$(\vec{p}_1 - \vec{p}_2) \cdot \frac{d\vec{n}(t)}{dt}, \quad (3.70)$$

where $\vec{v}_1(t)$ and $\vec{v}_2(t)$ are the velocity vectors of points $\vec{p}_1(t)$ and $\vec{p}_2(t)$. This gives us an expression for the relative velocity $v_n(t) = dq(t)/dt$ of points $\vec{p}_1(t)$ and $\vec{p}_2(t)$ along the contact normal, as a function of their velocities and collision normal. The time derivative of the collision normal indicates its rate of change in direction as a function of time.

Differentiating equation (3.69) once again with respect to time

$$\frac{d^2 q(t)}{dt^2} = \left(\frac{d^2 \vec{p}_1(t)}{dt^2} - \frac{d^2 \vec{p}_2(t)}{dt^2} \right) \cdot \vec{n}(t) +$$
$$2 \left(\frac{d\vec{p}_1(t)}{dt} - \frac{d\vec{p}_2(t)}{dt} \right) \cdot \frac{d\vec{n}(t)}{dt} +$$
$$(\vec{p}_1 - \vec{p}_2) \cdot \frac{d^2 \vec{n}(t)}{dt^2}, \quad (3.71)$$

or equivalently

$$a_n(t) = (\vec{a}_1(t) - \vec{a}_2(t)) \cdot \vec{n}(t) + 2(\vec{v}_1(t) - \vec{v}_2(t)) \cdot \frac{d\vec{n}(t)}{dt} +$$
$$(\vec{p}_1 - \vec{p}_2) \cdot \frac{d^2 \vec{n}(t)}{dt^2}, \quad (3.72)$$

where $\vec{a}_1(t)$ and $\vec{a}_2(t)$ are the acceleration vectors of points $\vec{p}_1(t)$ and $\vec{p}_2(t)$, respectively. This gives us an expression for the relative acceleration $a_n(t) = d^2 q(t)/dt^2$ of points $\vec{p}_1(t)$ and $\vec{p}_2(t)$ along the contact normal, as a function of their accelerations, velocities, contact normal and rate of change in direction of the contact normal.

At the instant of contact $t = t_c$, points $\vec{p}_1(t)$ and $\vec{p}_2(t)$ are coincident, that is

$$\vec{p}_1(t_c) = \vec{p}_2(t_c). \quad (3.73)$$

Substituting equation (3.73) into (3.72), we obtain an expression of the relative acceleration along the contact normal at the instant of contact:

$$a_n(t_c) = (\vec{a}_1(t_c) - \vec{a}_2(t_c)) \cdot \vec{n}(t_c) + 2(\vec{v}_1(t_c) - \vec{v}_2(t_c)) \cdot \frac{d\vec{n}(t_c)}{dt}. \quad (3.74)$$

According to equation (3.74), the relative acceleration at the instant of contact has two terms. The first term depends on the accelerations of the

3.6 Particle-Particle Collision Response

contact points, which in turn are related to the contact force using Newton's law. The second term depends on the velocities of the contact points and the rate of change in direction of the collision normal.

For now, let's assume the contact to be frictionless[8], that is

$$\vec{F} = F_n \vec{n} \;.$$

If we separate the terms that depend on the contact force from those that do not, we can rewrite equation (3.74) as

$$a_n(t_c) = (a_{11})_n F_n + b_1 \;. \tag{3.75}$$

Substituting equation (3.75) in (3.68), we obtain:

$$\begin{aligned}
((a_{11})_n F_n + b_1) &\geq 0 \\
F_n &\geq 0 \\
F_n \left((a_{11})_n F_n + b_1\right) &= 0 \;.
\end{aligned} \tag{3.76}$$

Thus, the computation of the contact force for the frictionless case involves solving the system of equations defined in (3.76), which is quadratic on F_n. One way of solving this system is to use quadratic programming. However, such techniques are difficult to implement, often requiring the use of sophisticated numerical software packages.

Fortunately, the system of equations defined in (3.76) is also of the same form of a renowned numerical programming technique called *linear complementarity*. The implementation using linear-complementarity techniques is significantly easier than the implementation of a quadratic program, and is discussed in detail in Appendix G. There, we start presenting solution methods for the frictionless case, and show how to modify them to deal with static and dynamic friction at the contacts. These modifications on the solution method require that equation (3.75) be extended to also consider the relation between the relative-acceleration and contact-force components on the tangent plane of the contact.

Where friction is taken into account, the system of equations become

$$\begin{aligned}
\begin{pmatrix} a_n(t_c) \\ a_t(t_c) \\ a_k(t_c) \end{pmatrix} &= \begin{pmatrix} (a_{11})_n & (a_{12})_t & (a_{13})_k \\ (a_{21})_n & (a_{22})_t & (a_{23})_k \\ (a_{31})_n & (a_{32})_t & (a_{33})_k \end{pmatrix} \begin{pmatrix} F_n \\ F_t \\ F_k \end{pmatrix} + \begin{pmatrix} (b_1)_n \\ (b_1)_t \\ (b_1)_k \end{pmatrix} \\
&= \mathbf{A} \vec{F} + \vec{b} \;,
\end{aligned} \tag{3.77}$$

where

---

[8] Later in this section, we shall relax this assumption to show how the system of equations used in the frictionless case can be expanded to handle friction.

$$a_t(t_c) = (\vec{a}_1(t_c) - \vec{a}_2(t_c)) \cdot \vec{t}(t_c) + 2(\vec{v}_1(t_c) - \vec{v}_2(t_c)) \cdot \frac{d\vec{t}(t_c)}{dt} \quad (3.78)$$

$$a_k(t_c) = (\vec{a}_1(t_c) - \vec{a}_2(t_c)) \cdot \vec{k}(t_c) + 2(\vec{v}_1(t_c) - \vec{v}_2(t_c)) \cdot \frac{d\vec{k}(t_c)}{dt} \quad (3.79)$$

The solution method presented in Appendix G assumes both matrix **A** and vector $\vec{b}$ to be known constants computed from the geometric displacement and dynamic state of the particles at the instant of contact. Therefore, we need to determine the coefficients of matrix **A** and vector $\vec{b}$ before we can safely apply the linear-complementarity techniques of Appendix G.

The first row of matrix **A** and vector $\vec{b}$ is obtained by expressing the normal relative acceleration $a_n(t_c)$ at the instant of contact as a function of the contact-force components F_n, F_t and F_k. This can be done using equation (3.74).

Let's start by examining its first term of equation (3.74), namely the term

$$(\vec{a}_1(t_c) - \vec{a}_2(t_c)) \cdot \vec{n}(t_c) .$$

The acceleration $\vec{a}_1$ of point $\vec{p}_1$ is obtained directly from equation (3.4) as

$$\vec{a}_1 = \frac{\vec{F} + (\vec{F}_1)_{ext}}{m_1}, \quad (3.80)$$

where $(\vec{F}_1)_{ext}$ is the net external force (such as gravity, spring forces, spatially dependent forces, etc.) acting on particle O_1 and $\vec{F}$ is the contact force to be determined. Similarly, the acceleration $\vec{a}_2$ of point $\vec{p}_2$ is given by

$$\vec{a}_2 = \frac{-\vec{F} + (\vec{F}_2)_{ext}}{m_2} . \quad (3.81)$$

Substituting equations (3.80) and (3.81) into the first term of equation (3.74) gives

$$\begin{aligned}
(\vec{a}_1(t_c) - \vec{a}_2(t_c)) \cdot \vec{n}(t_c) &= \left(\frac{1}{m_1} + \frac{1}{m_2}\right) \vec{F} \cdot \vec{n}(t_c) \\
&+ \left(\frac{(\vec{F}_1)_{ext}}{m_1} + \frac{(\vec{F}_2)_{ext}}{m_2}\right) \cdot \vec{n}(t_c) \\
&= \left(\frac{1}{m_1} + \frac{1}{m_2}\right) F_n \\
&+ \left(\frac{(\vec{F}_1)_{ext}}{m_1} + \frac{(\vec{F}_2)_{ext}}{m_2}\right) \cdot \vec{n}(t_c) .
\end{aligned}$$

So, the total contribution to the coefficients of the first row of matrix $\mathbf{A}$ and vector $\vec{b}$ from the first term of equation (3.74) is

$$(a_{11})_n = \left(\frac{1}{m_1} + \frac{1}{m_2}\right)$$
$$(a_{12})_t = 0$$
$$(a_{13})_k = 0 \qquad (3.82)$$
$$(b_1)_n = \left(\frac{(\vec{F}_1)_{ext}}{m_1} + \frac{(\vec{F}_2)_{ext}}{m_2}\right) \cdot \vec{n}(t_c) .$$

Notice that $(a_{12})_t$ and $(a_{13})_k$ are zero because the first term does not depend on F_t and F_k, respectively. Now, let's examine the second term of equation (3.74), namely

$$2\left(\vec{v}_1(t_c) - \vec{v}_2(t_c)\right) \cdot \frac{d\,\vec{n}(t_c)}{dt} .$$

The velocities of points $\vec{p}_1$ and $\vec{p}_2$ are known quantities independent of the contact force. So, the contribution to the first row of matrix $\mathbf{A}$ from the velocity components is zero. However, we still need to compute the rate of change in direction of the contact normal as a function of time and check whether it depends on the contact force.

Section E.3.1 presents a detailed description of how the time derivative of the contact normal for the particle-particle case can be computed. For convenience, the result is reproduced here:

$$\frac{d\,\vec{n}(t)}{dt} = \frac{(\vec{v}_1 - \vec{v}_2)}{|\vec{v}_1 - \vec{v}_2|} ,$$

which is independent of the contact force as well. So, the contribution of the second term of equation (3.74) to the coefficients of matrix $\mathbf{A}$ and vector $\vec{b}$ is

$$(a_{11})_n = 0$$
$$(a_{12})_t = 0$$
$$(a_{13})_k = 0 \qquad (3.83)$$
$$(b_1)_n = 2\left(\vec{v}_1 - \vec{v}_2\right) \cdot \frac{(\vec{v}_1 - \vec{v}_2)}{|\vec{v}_1 - \vec{v}_2|} = 2\,|\vec{v}_1 - \vec{v}_2| .$$

Combining equations (3.82) and (3.83), we obtain the coefficients associated with the first row of matrix $\mathbf{A}$ and vector $\vec{b}$ as

$$(a_{11})_n = \left(\frac{1}{m_1} + \frac{1}{m_2}\right)$$
$$(a_{12})_t = 0$$
$$(a_{13})_k = 0$$
$$(b_1)_n = \left(\frac{(\vec{F}_1)_{ext}}{m_1} + \frac{(\vec{F}_2)_{ext}}{m_2}\right) \cdot \vec{n}$$
$$+ 2|\vec{v}_1 - \vec{v}_2|.$$

The computation of the coefficients associated with the second and third rows[9] of matrix $\mathbf{A}$ and vector $\vec{b}$ is similar to the derivations already obtained for the coefficients of the first row. The main difference is that, instead of computing the dot product with $\vec{n}$, we compute it with $\vec{t}$ for the second row, and with $\vec{k}$ for the third row, as shown in equations (3.78). After some manipulation, we obtain

$$(a_{21})_n = 0$$
$$(a_{22})_t = \left(\frac{1}{m_1} + \frac{1}{m_2}\right)$$
$$(a_{23})_k = 0$$

and

$$(b_1)_t = \left(\frac{(\vec{F}_1)_{ext}}{m_1} + \frac{(\vec{F}_2)_{ext}}{m_2}\right) \cdot \vec{t}(t_c) +$$
$$2\left(\vec{v}_1(t_c) - \vec{v}_2(t_c)\right) \cdot \frac{d\vec{t}(t_c)}{dt} \quad (3.84)$$
$$(3.85)$$

for the second row, and

$$(a_{31})_n = 0$$
$$(a_{32})_t = 0$$
$$(a_{33})_k = \left(\frac{1}{m_1} + \frac{1}{m_2}\right)$$

and

---

[9] These coefficients need only be computed if friction is taken into account. In the frictionless case, both F_t and F_k are zero.

$$(b_1)_k = \left(\frac{(\vec{F}_1)_{ext}}{m_1} + \frac{(\vec{F}_2)_{ext}}{m_2}\right) \cdot \vec{k}(t_c) + $$
$$2\left(\vec{v}_1(t_c) - \vec{v}_2(t_c)\right) \cdot \frac{d\vec{k}(t_c)}{dt} \qquad (3.86)$$

for the third row of matrix $\mathbf{A}$ and vector $\vec{b}$.

The actual determination of the coefficients $(b_1)_t$ and $(b_1)_k$ as shown in equations (3.84) and (3.86) is more involved than that for $(b_1)_n$ because they require computing the rate of change in direction of the vectors $\vec{t}(t)$ and $\vec{k}(t)$ on the tangent plane of the contact. The computation of the time derivative of the tangent-plane vectors is discussed in detail in Section E.4 of Appendix E.

Lastly, having determined the elements of matrix $\mathbf{A}$ and vector $\vec{b}$, we can apply the LCP techniques of Appendix G to compute the contact-force components. We then update the dynamic state of each particle by applying $+\vec{F}$ on particle O_1 and $-\vec{F}$ on particle O_2.

3.6.4 Computing Contact Forces for Multiple Contacts

The principle behind the computation of multiple particle-particle contact forces is the same as the computation of multiple particle-particle collision impulses. Again, the simulation engine needs to group the particles into clusters that share at least one contact. The contacts within each cluster can then be simultaneously resolved independent of all other clusters (see Figure 3.41).

When a particle is involved in multiple contacts, it is possible to have it assigned to different indexes for each contact. For the particular situation of cluster G_2 in Figure 3.41, particle O_2 has index 2 with respect to its contact with particle O_1, and index 1 with respect to its contact with particle O_3. This in turn affects the choice of sign when combining the multiple-contact forces in the system equations. Moreover, the contact normal and tangent plane are different for each contact. So, we also need to carry out a change of base between the contact forces before combining them.

In the single particle-particle contact, the contact-force computation taking friction into account was done using linear-complementarity techniques to solve a system of equations of the form

$$a_n(t_c) \geq 0$$
$$F_n \geq 0$$
$$\vec{F}^t(\mathbf{A}\vec{F} + \vec{b}) = 0 ,$$

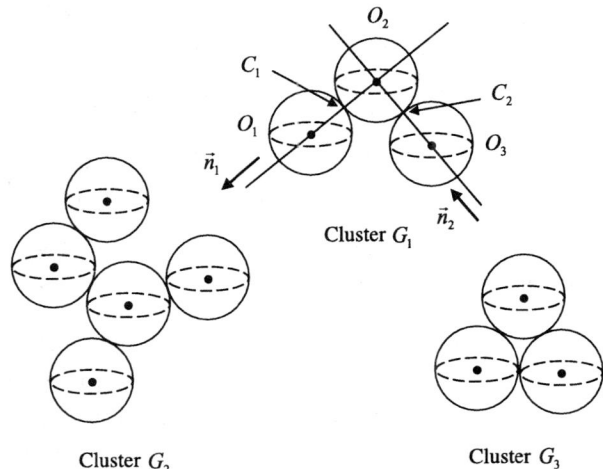

FIGURE 3.41. A multiple particle-particle contact-force computation. In this case, the particles are grouped into three clusters that can be solved in parallel.

where[10]

$$\mathbf{A} = \begin{pmatrix} (a_{11})_n & 0 & 0 \\ 0 & (a_{22})_t & 0 \\ 0 & 0 & (a_{33})_k \end{pmatrix}$$
$$\vec{F} = (F_n, F_t, F_k)^t$$
$$\vec{b} = ((b_1)_n, (b_1)_t, (b_1)_k)^t \ .$$

This solution method can be extended to the case of multiple-contact-force computations. The main difference between multiple- and single-contact-force computations involving a given particle is that the contact force at contact C_i can affect the computation of the contact force at contact C_j. So, instead of solving one contact at a time, we need to simultaneously solve all contacts having a particle in common. This in turn has the same effect of merging the several individual system of equations for each contact into one larger system, and applying the linear-complementarity techniques to merged system.

For example, suppose we have a cluster with m simultaneous contacts. Each contact C_i is defined by its contact-normal $(\vec{n})_i$ and tangent-plane vectors $(\vec{t})_i$ and $(\vec{k})_i$. The contact force at contact C_i is then expressed as

$$\vec{F}_i = ((F_i)_{n_i}, (F_i)_{t_i}, (F_i)_{k_i})^t \ .$$

---

[10] Here, we are already using the result of Section 3.6.3 that the matrix **A** is all zero, save for its diagonal elements.

3.6 Particle-Particle Collision Response

The contact-force vector for the multiple-collision system is obtained by concatenating the contact-force vectors of each of the m contacts, that is

$$\vec{F} = ((F_1)_{n_1}, (F_1)_{t_1}, (F_1)_{k_1}, \ldots, (F_m)_{n_m}, (F_m)_{t_m}, (F_m)_{k_m})^t \ .$$

The vector $\vec{b}$ becomes

$$\vec{b} = ((b_1)_n, (b_1)_t, (b_1)_k, \ldots, (b_m)_n, (b_m)_t, (b_m)_k)^t$$

and the matrix $\mathbf{A}$ is enlarged to accommodate all contact forces. Its partitioned representation is given by

$$\mathbf{A} = \begin{pmatrix} \mathbf{A_{11}} & \mathbf{A_{12}} & \ldots & \mathbf{A_{1m}} \\ \mathbf{A_{21}} & \mathbf{A_{22}} & \ldots & \mathbf{A_{2m}} \\ & \ldots & & \ldots \\ \mathbf{A_{m1}} & \mathbf{A_{m2}} & \ldots & \mathbf{A_{mm}} \end{pmatrix},$$

where each sub-matrix is given by

$$\mathbf{A_{ij}} = \begin{pmatrix} (a_{ij})_{n_i} & (a_{i(j+1)})_{t_i} & (a_{i(j+2)})_{k_i} \\ (a_{(i+1)j})_{n_i} & (a_{(i+1)(j+1)})_{t_i} & (a_{(i+1)(j+2)})_{k_i} \\ (a_{(i+2)j})_{n_i} & (a_{(i+2)(j+1)})_{t_i} & (a_{(i+2)(j+2)})_{k_i} \end{pmatrix}.$$

If contacts C_i and C_j have no particles in common, then the sub-matrix $\mathbf{A_{ij}}$ is set to zero, indicating that their contact forces do not affect each other. However, if contacts C_i and C_j do have a particle in common, then the coefficients a_{ij} are the contribution of the contact force of contact C_j to the relative acceleration at contact C_i. More specifically, the coefficient $(a_{ij})_{n_i}$ is the contribution of the contact-force component $(F_j)_{n_j}$ to the relative normal acceleration at the contact C_i. Analogously, the coefficients $(a_{ij})_{t_i}$ and $(a_{ij})_{k_i}$ are the contribution of the contact-force components $(F_j)_{t_j}$ and $(F_j)_{k_j}$ to the relative normal acceleration at the contact C_i.

Also, notice that the contact force $\vec{F}_j$ is given with respect to the contact frame of C_j, whereas the relative acceleration $\vec{a}_i$ is given with respect to the contact frame of C_i. Therefore, a change of base is required when computing the coefficients of matrix $\mathbf{A_{ij}}$ and vector $\vec{b}_i$.

Suppose contact C_i involves particles O_1 and O_2, and contact C_j involves particles O_2 and O_3, that is, they have particle O_2 in common. We want to determine the contribution of the contact force $\vec{F}_j$ acting on particle O_2 of contact C_j to the relative acceleration of contact C_i. This in turn involves determining the coefficients of the sub-matrix $\mathbf{A_{ij}}$ and the components $(b_i)_{n_i}$, $(b_i)_{t_i}$ and $(b_i)_{k_i}$ of vector $\vec{b}$. The relative acceleration at contact C_i between particles O_1 and O_2 is given by

$$(a_i)_{n_i} = (\vec{a}_1 - \vec{a}_2) \cdot \vec{n}_i + 2(\vec{v}_1 - \vec{v}_2) \cdot \frac{d\vec{n}_i}{dt}$$

$$(a_i)_{t_i} = (\vec{a}_1 - \vec{a}_2) \cdot \vec{t}_i + 2(\vec{v}_1 - \vec{v}_2) \cdot \frac{d\vec{t}_i}{dt} \qquad (3.87)$$

$$(a_i)_{k_i} = (\vec{a}_1 - \vec{a}_2) \cdot \vec{k}_i + 2(\vec{v}_1 - \vec{v}_2) \cdot \frac{d\vec{k}_i}{dt}.$$

As explained in the single-contact case, only the first term of equations (3.87) depends on the forces exerted at contact C_i. The second term depends on the velocities and is added to $(b_i)_{n_i}$, $(b_i)_{t_i}$ and $(b_i)_{k_i}$, as appropriate. Thus, the contribution of the contact force $\vec{F}_j$ acting on particle O_2 of contact C_j does not affect the components of vector $\vec{b}$. In other words, the expressions used to compute the vector $\vec{b}$ for the single-contact case are still valid for the multiple-contact case, that is, the components $(b_i)_{n_i}$, $(b_i)_{t_i}$ and $(b_i)_{k_i}$ of vector $\vec{b}$ are given by

$$\begin{aligned}
(b_i)_{n_i} &= \left(\frac{(\vec{F}_1)_{ext}}{m_1} + \frac{(\vec{F}_2)_{ext}}{m_2}\right) \cdot \vec{n}_i \\
&\quad + 2(\vec{v}_1 - \vec{v}_2) \cdot (\vec{v}_1 - \vec{v}_2) \\
(b_i)_{t_i} &= \left(\frac{(\vec{F}_1)_{ext}}{m_1} + \frac{(\vec{F}_2)_{ext}}{m_2}\right) \cdot \vec{t}_i + \\
&\quad 2(\vec{v}_1 - \vec{v}_2) \cdot \frac{d\vec{t}_i}{dt} \\
(b_i)_{k_i} &= \left(\frac{(\vec{F}_1)_{ext}}{m_1} + \frac{(\vec{F}_2)_{ext}}{m_2}\right) \cdot \vec{k}_i + \\
&\quad 2(\vec{v}_1 - \vec{v}_2) \cdot \frac{d\vec{k}_i}{dt},
\end{aligned}$$

where $(\vec{F}_1)_{ext}$ and $(\vec{F}_2)_{ext}$ are the net external forces (such as gravity, spring forces, spatially dependent forces, etc.) acting on particles O_1 and O_2, respectively.

Using equation (3.4), the contribution of the contact force $\vec{F}_j$ acting on particle O_2 of contact C_j to the acceleration $\vec{a}_1$ of particle O_1 involved in collision C_i is[11]

$$\frac{\vec{F}_j}{m_1}.$$

Conversely, the contribution of $\vec{F}_j$ to $\vec{a}_2$ is

---

[11] Notice that the contact force acting on particle O_2 because of contact C_j can be $+\vec{F}_j$ or $-\vec{F}_j$, depending on particle O_2 having index 1 or 2 with respect to contact C_j. The following derivations assume the contact force is $+\vec{F}_j$.

$$-\frac{\vec{F}_j}{m_2} .$$

The net contribution of $\vec{F}_j$ to the relative acceleration at the contact C_i is then

$$\left(\frac{1}{m_1} + \frac{1}{m_2}\right) \vec{F}_j .$$

Substituting this into the first terms of equations (3.87), we obtain the contributions of $\vec{F}_j$ to each relative-acceleration component at contact C_i:

$$\text{contribution to } (a_i)_{n_i} = \left(\frac{1}{m_1} + \frac{1}{m_2}\right) \vec{F}_j \cdot \vec{n}_i$$

$$\text{contribution to } (a_i)_{t_i} = \left(\frac{1}{m_1} + \frac{1}{m_2}\right) \vec{F}_j \cdot \vec{t}_i$$

$$\text{contribution to } (a_i)_{k_i} = \left(\frac{1}{m_1} + \frac{1}{m_2}\right) \vec{F}_j \cdot \vec{k}_i .$$

Using the fact that

$$\vec{F}_j = (F_j)_{n_j} \vec{n}_j + (F_j)_{t_j} \vec{t}_j + (F_j)_{k_j} \vec{k}_j ,$$

we immediately obtain the coefficients of the sub-matrix $\mathbf{A_{ij}}$ as

$$(a_{ij})_{n_i} = \left(\frac{1}{m_1} + \frac{1}{m_2}\right) \vec{n}_j \cdot \vec{n}_i \tag{3.88}$$

$$(a_{i(j+1)})_{n_i} = \left(\frac{1}{m_1} + \frac{1}{m_2}\right) \vec{t}_j \cdot \vec{n}_i \tag{3.89}$$

$$(a_{i(j+2)})_{n_i} = \left(\frac{1}{m_1} + \frac{1}{m_2}\right) \vec{k}_j \cdot \vec{n}_i \tag{3.90}$$

$$(a_{(i+1)j})_{t_i} = \left(\frac{1}{m_1} + \frac{1}{m_2}\right) \vec{n}_j \cdot \vec{t}_i \tag{3.91}$$

$$(a_{(i+1)(j+1)})_{t_i} = \left(\frac{1}{m_1} + \frac{1}{m_2}\right) \vec{t}_j \cdot \vec{t}_i \tag{3.92}$$

$$(a_{(i+1)(j+2)})_{t_i} = \left(\frac{1}{m_1} + \frac{1}{m_2}\right) \vec{k}_j \cdot \vec{t}_i \tag{3.93}$$

$$(a_{(i+2)j})_{k_i} = \left(\frac{1}{m_1} + \frac{1}{m_2}\right) \vec{n}_j \cdot \vec{k}_i \tag{3.94}$$

$$(a_{(i+2)(j+1)})_{k_i} = \left(\frac{1}{m_1} + \frac{1}{m_2}\right) \vec{t}_j \cdot \vec{k}_i \tag{3.95}$$

$$(a_{(i+2)(j+2)})_{k_i} = \left(\frac{1}{m_1} + \frac{1}{m_2}\right) \vec{k}_j \cdot \vec{k}_i . \tag{3.96}$$

Notice that, if $i = j$, then the sub-matrix $\mathbf{A_{ii}}$ is reduced to

$$\mathbf{A_{ii}} = \begin{pmatrix} (a_{ii})_{n_i} & 0 & 0 \\ 0 & (a_{(i+1)(i+1)})_{t_i} & 0 \\ 0 & 0 & (a_{(i+2)(i+2)})_{k_i} \end{pmatrix},$$

which is the same expression obtained in equation (3.77) for the single-contact case.

When friction is not taken into account, the sub-matrix $\mathbf{A_{ij}}$ is reduced to

$$\mathbf{A_{ij}} = (a_{ij})_{n_i},$$

since the contact-force components $(F_j)_{t_j}$ and $(F_j)_{k_j}$ are zero in the frictionless case. This result is also compatible with that obtained for the frictionless single-contact-force computation explained in Section 3.6.3.

Having computed the contact force $\vec{F}_i$ for each contact C_i, $1 \leq i \leq m$, we update the dynamic state of each particle involved on contact C_i by applying $+\vec{F}_i$ to the particle O_1 (i.e., the particle with index 1) and $-\vec{F}$ to the particle O_2 (i.e., the particle with index 2).

When a particle is involved in multiple contacts, it is possible to have it assigned to different indexes for each contact. For the particular situation of cluster G_2 in Figure 3.41, particle O_2 has index 2 with respect to its contact C_1 with particle O_1, and index 1 with respect to its contact C_2 with particle O_3. So, the net contact force actually applied to particle O_2 after all contact forces have been computed is

$$(\vec{F}_2 - \vec{F}_1),$$

where $\vec{F}_1$ and $\vec{F}_2$ are the contact forces associated with contacts C_1 and C_2.

3.7 Particle-Rigid Body Collision Response

Whenever a particle-rigid body collision is detected, the collision-response module is invoked to compute the appropriate collision impulses and contact forces that will prevent interpenetration. As explained in Section 3.4, the trajectories of the colliding particles are traced back in time to the moment before the collision. The rigid body is not moved from its current position, which happens to be its position at the end of the current time step after any and all rigid body-rigid body collisions were resolved.

In this book, we model the particle-rigid body collision as the particle colliding with another particle on the rigid body's surface. The advantage of doing so is that we can reuse most of the results obtained for the particle-particle single or multiple collisions or contacts developed in Section 3.6.

More specifically, there are only three main modifications that we need to make before we can reuse the equations already derived for the particle-particle collision and contact cases.

The first modification consists of making the mass of the particle on the rigid body's surface be the same as the rigid body's mass. For example, if particle O_1 is colliding, or comes in contact with, particle O_2 on the rigid body's surface, then we make m_2 be the rigid body's mass. Clearly, this modification induces no changes on the equations already derived for the particle-particle cases.

The second modification requires that both velocity and acceleration of the particle on the rigid body's surface be computed using the rigid body's dynamic equations covered in Section 4.2 of Chapter 4. Unlike the particle's dynamics, the rigid-body motion has to take into account rotational motion, making its dynamic equations much more complex than those derived for particles in Section 3.2. Therefore, this modification will cause some changes on the particle-particle collision and contact equations, as will be explained in Sections 3.7.1 and 3.7.2.

The third and last modification involves the way the collision or contact normal is computed. Even though the particle is colliding with another particle on the rigid body's surface, the normal is defined by the rigid body's geometry, as opposed to being the line connecting both particles. This is done as follows.

Let O_1 be the particle colliding or in contact with rigid body B_1. The particle O_2 on the rigid body's surface can be an interior point, a point on an edge, or a vertex of the rigid body's triangle primitive being intersected by O_1's trajectory. So, if O_2 is in the interior of the triangle, then the triangle's normal is taken as the collision or contact normal. If O_2 lies on an edge of the triangle, then the edge's normal is taken as the collision normal. Notice that the edge normal is computed as the average of the normals of the triangle faces that share the edge. Lastly, if O_2 lies on a vertex of the triangle, then the vertex's normal is selected as the collision or contact normal. The vertex's normal is computed as the average of the normals of all triangles that contain the vertex.

If the particles are colliding, then the normal direction is selected such that

$$(\vec{v}_1 - \vec{v}_2) \cdot \vec{n} < 0 \; ,$$

that is, the particles are moving towards each other before the collision. Notice that this may require changing the direction of the normal computed from the triangle normal, edge normal or vertex normal if $\vec{v}_2 > \vec{v}_1$, that is, if the rigid body is moving faster towards the particle, as opposed to the particle moving faster towards the rigid body. On the other hand, if the particles are in contact, we have

$$(\vec{v}_1 - \vec{v}_2) \cdot \vec{n} = 0$$

136   3. Particle Systems

and the normal direction remains unchanged (i.e., pointing outwards the rigid body's surface).

3.7.1 Computing Impulsive Forces

The computation of impulsive forces for the particle-rigid body collision is very nearly the same as that described for the single or multiple particle-particle collision. As mentioned, the main modifications are on the computation of the velocity of particle O_2 located on the rigid body's surface, and the collision normal.

Consider the situation illustrated in Figure 3.42. Particle O_1 is about to collide with particle O_2 on the rigid body's surface.

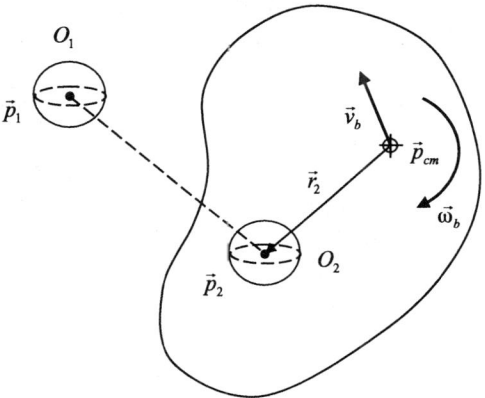

FIGURE 3.42. Particle O_1 is colliding with rigid body B_1 at O_2. The velocity $\vec{v}_2$ and acceleration $\vec{a}_2$ of point $\vec{p}_2$ are computed using the rigid body's dynamic equations.

Let M_b be the mass of the rigid body, and $\vec{v}_b$ and $\vec{w}_b$ be the linear and angular velocities of the rigid body just before the collision. The mass m_2 of the particle O_2 is then set to

$$m_2 = M_b \ . \tag{3.97}$$

Its velocity before the collision is computed as the velocity of a point on the rigid body (see Section 4.2 for more details), that is

$$\vec{v}_2 = \vec{v}_b + \vec{w}_b \times \vec{r}_2 \ , \tag{3.98}$$

where $\vec{r}_2$ is the distance between particle O_2 and the rigid body's center of mass given by

$$\vec{r}_2 = \vec{p}_2 - \vec{p}_{cm} \ .$$

So, using equations (3.97) and (3.98) as the mass and velocity of particle O_2 just before the collision, we can directly apply the single- or multiple-collision equations and compute the collision impulse $\vec{P}_{1,2}$ associated with the collision of particles O_1 and O_2.

The dynamic state of particle O_1 is then updated with the application of the collision impulse $\vec{P}_{1,2}$. However, recall that the update on the dynamic state of the rigid body owing to the impulse $\vec{P}_{1,2}$ is postponed to the next simulation time step. Therefore, for each simulation time step, we need to sum all collision impulses acting on rigid body B_1 owing to one or more particle-rigid body collisions, and apply the resultant impulse at the beginning of the next simulation time step, as if all collisions happened at that time. As explained in Section 3.4, this is an approximation used to improve the overall simulation efficacy.

3.7.2 Computing Contact Forces

The computation of the contact force between a particle and a rigid body follows the same principles explained in the computation of a particle-particle contact. More specifically, the particles should not interpenetrate, the contact force should not prevent contact from breaking apart, and the contact force should be set to zero if contact is about to break. Translating these conditions into equations, we have that the relative acceleration of the particles at the contact point should be greater than or equal to zero, the contact force should be greater than or equal to zero, and the contact force should be zero if the relative acceleration is positive, or it should be positive if the relative acceleration is zero.

Again, we use the vector $\vec{q}(t) = (\vec{p}_1 - \vec{p}_2) \cdot \vec{n}$ to compute the relative acceleration of the particles at the contact point, given by

$$a_n(t_c) = (\vec{a}_1(t_c) - \vec{a}_2(t_c)) \cdot \vec{n}(t_c) + 2(\vec{v}_1(t_c) - \vec{v}_2(t_c)) \cdot \frac{d\vec{n}(t_c)}{dt}, \quad (3.99)$$

where t_c is the instant when contact was established. Equation (3.99) is the same as equation (3.74) rewritten here for convenience.

Clearly, the velocity $\vec{v}_2$ and acceleration $\vec{a}_2$ of particle O_2 should be computed using the rigid-body dynamics equations covered in Section 4.2. Moreover, since the contact normal is either a triangle normal, an edge normal or a vertex normal, the time derivative of the normal vector should also take into account the dynamics of the rigid-body motion. In other words, the rate of change in direction of the normal vector is a function of the linear and angular velocities of the rigid body.

Unfortunately, the derivation of these relations requires a significant knowledge of rigid-body dynamics, which is only covered in Chapter 4. Therefore, for the sake of clarity, we shall postpone the computation of the contact forces between a particle and a rigid body until Section 4.9. By so doing, the reader will have a chance to assimilate the concepts of rigid-body dynamics necessary to understand the contact-force computations.

3.8 Particle Emitter

The particle emitter is used to create and release particles in the simulated environment. Usually, particle emitters are implemented as an invisible planar quadrangular surface that can be attached to other objects in the simulation, or be used as a stand-alone object, in which case it can be displayed attached to a default cube (see Figure 3.43).

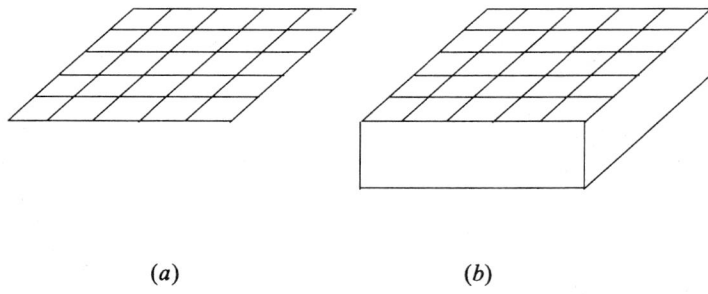

(a)   (b)

FIGURE 3.43. (a) In most cases, particle emitters are represented by a planar quandrangular face. The grid is created from the size of each particle and the total number of particles to be released at a time; (b) In the stand-alone version, a default cubic object can be created, and the particle emitter is attached to its top face.

The actual emission of particles in the simulated environment is not as straightforward as it may seem. Some precautionary measures should be taken to avoid unnecessary implementation problems. For example, the particle-particle collision-detection module checks for geometric intersections of the particles' trajectories as the system evolves. It relies heavily on the fact that the particles are not intersecting at the beginning of the current time step, that is, at the starting point of their trajectories, so that the trace back in time can always detect a state wherein the particles are not intersecting. So, if the particles are created in such a manner that they overlap each other right from the start, then any particle-particle collisions detected immediately after they are released cannot be resolved using the collision-detection techniques presented in this book. The approach fails because the trace back in time is unable to reach a state wherein the particles are not intersecting. Therefore, we do need to assure that particles not be overlapping each other at the time they are released.

Also, it is a good idea to establish a clearance region around the emitter surface to prevent situations in which an object is placed on top of the emitter and the particles either jam immediately after they are released, or simply cannot be released at all. This is illustrated in Figure 3.44.

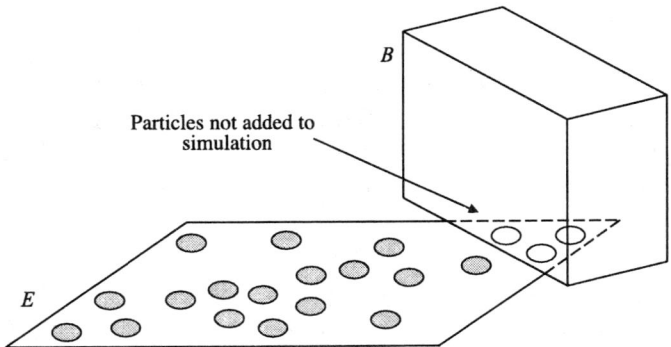

FIGURE 3.44. Object B is partially obstructing the emitter surface E. Particles created on the obstructed area are not added to the simulation.

Our approach to these practical problems is as follows. Every time a new round of particles is released, the particles just created are checked for intersections with other particles already released, as well as rigid bodies that are in the same world-cell decomposition as the emitter. Whenever an intersection is detected, the particles just created are immediately removed from the simulation. In other words, the emitter will only add new particles to the system if there is enough room for them.

Clearly, this approach requires that we know the position on the emitter surface at which each particle is released. In order to do so, we consider a grid decomposition of the emitter surface such that a single particle is released at the center of each cell of the decomposition.

Let d be the desired diameter of the spheres representing the particles being emitted, and let n_t be the number of particles simultaneously released at a time. The dimension of each cell in the decomposition is then

$$f_s = 2d\sqrt{n_t} \qquad (3.100)$$

for a square emitter surface, where the factor of 2 is used to allow for some extra space between each particle positioned at the center of its associated cell.

Another practical problem that should be considered when emitting particles is that in most cases the particles should not behave exactly like each other. For instance, the introduction of some variations on size, speed and direction of movement of each particle often makes the particle system look and feel more real. In our approach, we define a set of user-adjustable parameters to control the dynamic behavior of the particles being emitted. For each parameter, we define its mean value and deviation. So, for example, a particle system with particles of size four and deviation two will have

140   3. Particle Systems

particles with size ranging from two to six, and average size of four. The actual size of each particle can be dynamically computed assuming a normal probability distribution for each parameter. Equation (3.100) should then be modified to accommodate this deviation on size.

Each time new particles are released, we need to compute their maximum diameter d_{max}. Notice that the value of d_{max} will probably vary from consecutive releases, that is, the emitter grid dynamically grows and shrinks to the appropriate size before each release. The dimensions of each cell in the decomposition is then dynamically obtained from

$$f_s = 2\, d_{max}\, \sqrt{n_t}\,. \tag{3.101}$$

This in turn minimizes the risk of having particles just released intersecting each other.

3.8.1 User-Definable Parameters

The desired dynamic behavior of the particle system can usually be adjusted through a set of particle parameters. Each parameter is defined by its nominal (i.e., mean) value and an optional deviation value. Here, we provide a list of default parameters commonly found in animation packages to set up particle-emitter properties. Other parameters used to implement specialized particle systems we cover in Section 3.9.1.

1. **Particle size**

    Defines the diameter of the spheres representing the emitted particles.

2. **Particle mass**

    Defines the mass of each particle.

3. **Particle age**

    Defines the age of a particle. This parameter is used to determine the maximum time interval the particle will exist in the simulation, in case the *particle resilience* parameter is set to true.

4. **Particle resilience**

    Regulates whether particles that exist for a time span longer than their age should be removed from the simulation.

5. **Emitter size**

    Limits the maximum size of the emitter surface. If the size computed using equation (3.101) turns out to be greater than this value, then the number of particles simultaneously released is reduced to

    $$n_r = \left\lfloor \left(\frac{(f_s)_{max}}{2 d_{max}} \right)^2 \right\rfloor,$$

where $(f_s)_{max}$ is the maximum size of the emitter surface and n_r is the actual number of particles being released.

6. **Emission velocity**

    Defines each particle's velocity at time of emission.

7. **Emission direction**

    Defines the direction in which the particles will move as soon as they are released.

8. **Emission delay**

    Controls the time interval between two consecutive emissions, as a multiple of the simulation time step being used. For example, if the delay is set to three, then the emitter will skip two simulation time steps, and will emit particles again only on the third time step after the last emission.

9. **Emission rate**

    Defines the number of particles emitted at each emission. Notice that the maximum possible emission rate is constrained by the emitter size.

10. **Maximum emission**

    Defines the maximum number of particles emitted by this source. This is required not only to limit the overall number of particles being used in the simulation, but also to simulate effects such as smoke resulting from a fire being dissipated after the fire is out.

11. **Particle spawn**

    Regulates whether particles that were removed from the simulation should be reused by the emitter. If set to true, then the interpretation of the *maximum emission* parameter will be slightly different than its default interpenetration. In this case, the *maximum emission* parameter will regulate the total number of particles emitted by this source that are *active* in the simulation. (Active here means that the particle's lifespan has not yet reached its age.)

12. **Collision detection**

    This flag is used to indicate whether particle collisions should be taken into account during the simulation. The flag can turn on and off internal, external and complex collisions. Recall from Section 3.4 that internal collisions refer to collisions between particles emitted by the same particle system. External collisions indicate that collisions between particles emitted by different particle systems should be taken into account, whereas complex collisions refers to collisions between particles and rigid bodies. Notice that there is no deviation value associated with this parameter.

13. **Particle static friction**

    Defines the static-friction coefficient of the particle when in contact with other particles and rigid bodies. The actual static-friction coefficient used in the computations is the average value of the static friction assigned to each of the contacting particles and rigid bodies.

14. **Particle dynamic friction**

    Defines the dynamic-friction coefficient of the particle when in contact with other particles and rigid bodies. The actual dynamic-friction coefficient used in the computations is the average value of the dynamic friction assigned to each of the contacting particles and rigid bodies.

15. **External forces**

    Defines a list of external forces globally applied to all (and only) particles emitted from this source. For example, this listing can be used to define a viscous drag force to simulate air resistance, add gravity to pull the particles towards the ground, or to force all particles to have a default movement along a specific direction, as happens when modeling rain falling diagonally to the ground.

3.9 Specialized Particle Systems

In this Section, we shall discuss some specialized particle systems commonly found in animation packages. The idea behind such systems is to swap accuracy of the physical modeling of the particles' interactions for overall system efficiency without sacrificing the look of the simulation. For example, instead of using the Navier-Stokes equations to simulate vapor leaving a cup of hot water as a turbulent gas, we obtain the same effect by using a set of user-definable parameters to adjust some properties of the particle system representing the vapor. The result is a simulation that has a behavior similar to the hot turbulent gas modeled by the actual physics-based equations, but using a simpler mathematical framework.

There are two main reasons to include this and the previous sections in this book, despite the methods used *not* being based on the accurate physical modeling of the particles' interactions. First of all, these specialized systems are very popular among computer graphics practitioners and animators, and we feel that this chapter would be incomplete without mentioning such systems. Besides, discussing in detail the mathematical framework needed to precisely model these systems, which require specialized forms of equations of motion and force interactions, is beyond the scope of this book.

3.9 Specialized Particle Systems

Secondly, we want to make the point that the physics-based particle-dynamics and collision-detection and response modules developed here can be used in these systems as a fast and reliable way to evolve the particles' trajectories over time. The benefit of doing so from the object-oriented point of view is that particles and rigid bodies can use the same underlying simulation engine. This includes using the same numerical methods, collision-detection algorithms and collision-response-force computations. Particles and rigid-body objects can then be derived from a common parent class, and the shared functionality can be implemented as virtual methods of the parent class. So, as far as the simulation engine is concerned, virtual methods on the parent class are invoked and the appropriate particle or rigid-body behavior is obtained. In other words, the simulation engine makes no distinction between particles and rigid bodies when advancing the system by one more time step because it only manipulates their parent class's objects.

3.9.1 User-Adjustable Parameters

The set of particle system parameters presented here is intended to complement the set already presented in Section 3.8.1 with new functionality needed to create visual effects like smoke, rain, jet flows of water, and explosions. Notice that these parameters are used to mimic the coarse behavior of such systems without having to solve the sophisticated equations of motion associated with the simulation of gases, liquids and explosions[12]. As mentioned in Section 3.1, the detailed explanation of the theoretical framework needed to capture the precise behavior of such specialized systems is beyond the scope of this book. The interested reader is referred to Section 3.10 for pointers to the literature wherein in-depth explanations of such techniques can be found.

1. **Particle split**

   Indicates whether particles can generate new child particles. This parameter, combined with the age parameter, lets users create several interesting effects, such as air being burned by a rocket launcher and the smoke getting first more dense, then dispersed in air a few seconds later.

2. **Particle-split number**

   Defines the total number of child particles that will be created by their parent after each split.

3. **Particle-split age**

---

[12]Our approach is the same as that commonly used in most animation environments where parameters are fine tuned to achieve the desired visual effect.

Defines the age of the particle at which it will start to create its children. This age should be less than the particle age.

4. **Particle-split delay**

   Defines the time interval between two consecutive splits, as a multiple of the simulation time step. For example, a split delay of two indicates that, after the particle-split age has been reached, the particle will create new child particles every other simulation time step.

5. **Particle-split depth**

   Defines the maximum number of splits a particle may have.

6. **Particle-split velocity**

   Defines the percentage value of the parent's velocity inherited by its children at the time of the split.

7. **Particle-split direction**

   Defines a maximum deviation from the parent's direction inherited by its children at the time of the split. The deviation is used to spread the children round their parents, along the direction of movement.

8. **Particle-split size**

   Defines the percentage value of the parent's size inherited by its children at the time of the split.

9. **Particle-split lifespan**

   Defines a maximum deviation from the parent's age inherited by its children at the time of the split.

10. **Motion trail**

    Regulates whether particles should leave a motion trail behind them.

11. **Motion-trail age**

    Defines the amount of time the motion trail remains visible. The "motion trail" particles are then removed from the simulation as soon as this time expires.

12. **Color evolution**

    Defines a sequence of RGB color intensities of the particle over its lifetime. A time value is associated with each element of the color sequence, such that the color changes as soon as its corresponding time has been reached. For example, a color evolution of the form (R, G, B, t) with the values

$$\{(1.0, 1.0, 1.0, 0.0), (0.6, 0.6, 0.6, 2.0),$$
$$(0.3, 0.3, 0.3, 4.0), (0.0, 0.0, 0.0, 7.0)\}$$

assigns an initial white to the particle, then changes it to gray after two seconds, changes it again to light gray at four seconds from the previous change, and lastly sets it to black when seven seconds have passed from the last change.

13. **Particle electrical charge**

    Defines the electrical charge of the particle; useful when electrical force fields are used as external forces acting on the particle system.

There are several ways to create the visual effect of smoke using the particle system parameters described so far. We can define a large number of particles with small mass and size, and set up an external force that makes the particles flow from the emitter in the desired direction to form the column of smoke. Notice that there is no need to define gravity in these systems; only a viscous-drag force to compensate for the acceleration increase owing to the external force being applied.

Particles should be assigned a large value for their age, such that the column of smoke can rise slowly over time without disappearing too soon as it rises. On the other hand, the split age should be set fairly soon compared with the particle's age so that the smoke starts dense near the emitter. The dissipation is regulated by the deviations in size, velocity and direction of the split. These deviations should be set to small values such that the column of smoke remains concentrated along the direction of movement near the emitter and opens up like a cone as the particles move away from it, that is, as their deviation values accumulate to significant values after a certain number of splits.

When we are creating the visual effect of smoke from a fire, the particles can be set to an initial RGB color intensity of yellow, then change to orange, then to dark gray and finally to black. The time span used for each color depends on the desired visual effect for the intensity of the fire and the material being burned. The material generally determines how quickly the smoke turns black or dark gray.

The type of visual effects involving liquids that can be created using the specialized particle-system framework described in this book is limited to liquids that can be effectively represented by their drops. This includes rain, sprinklers and jet flows of water.

In the case of rain, the emitter surface should be set as large as necessary to cover the area where the rain will fall. The size of the particles (i.e., rain drops) should be kept small, unless you want to create the effect of a summer thunderstorm or hail-storm. The number of particles used combined with the particle's velocities defines the intensity of the rain. A small number falling slowly characterizes a light rain, whereas a large number falling fast resembles a tropical rain. The direction of the rain can be set using an appropriate external force to pull the particles down, as opposed to using gravity. Again, viscous drag should be used to counter balance the acceleration increase owing to the external force and give the impression the rain is falling at constant speed. Also, there is no need to split the particles or change their color.

In the case of sprinklers, the mass of the particles and their initial velocity defines the maximum height the particles will reach along their entire trajectories. Notice that the trajectory will be parabolic if we consider only gravity as the external force. The deviation is used to control the "aperture" of the sprinkler. Small deviations force the particles to be concentrated near the emitter, whereas large deviations make the particles disperse all around the emitter. Particles may or may not be split, depending on the desired volume of liquid being simulated.

Jet flows are similar to sprinklers. The main difference is that the initial velocity on jet flows is usually considerably higher than the velocity used to model sprinklers. Also, the deviation is maintained as small as possible to force the particles to stay close together during their motion. When we use jet flows to push rigid bodies through the simulated environment, we should use a moderate number of particles with increased mass values so that the net reaction force to the impact of the particles on the rigid body is large enough to move it.

The visual effect of explosions can use either a planar or spherical emitter surface, depending on what is being exploded. For instance, if we are simulating the explosion of a grenade, then we can safely use the planar emitter with a large number of particles of small size, short lifespan and large deviation. The splitting of particles is usually over used in these explosions, since we want to form a dense cloud immediately after the explosion. The particle's age is then used to control how fast the cloud will dissipate. There is no need to use external forces in these effects.

The spherical emitter is useful if we are creating the visual effect of fireworks, or the explosion of an artifact in a game environment. The idea is the same as that presented for planar emitters. The sphere's surface is decomposed into a grid where each cell releases one particle at a time along its radial direction. The direction deviation is then used to vary the emission direction with respect to the nominal radial direction. Also, the initial velocity deviation can be used to obtain an uneven cloud of particles.

3.9.2 Overview of Cloth Simulation

One possible way of simulating objects made of cloth is to use a combination of particles and springs to model the local deformations of the cloth as it moves. In this context, the cloth surface is represented by a uniform rectangular grid with each vertex being assigned to a particle. Neighboring particles are then strategically connected by springs to represent the stretch, shear and bend forces of the cloth. Figure 3.45(a) shows a small cloth grid containing sixteen vertices. The stretch force is modeled by springs connecting adjacent vertices of the grid, as illustrated in Figure 3.45(b). The shear force is modeled by another set of springs connecting diagonally opposed vertices of the grid (see Figure 3.45(c)). Finally, the bend force is modeled by a third set of springs connecting every other adjacent vertex of the grid, as indicated in Figure 3.45(d). By changing the stiffness of the three sets of springs, we can obtain the desired surface property of the cloth being simulated.

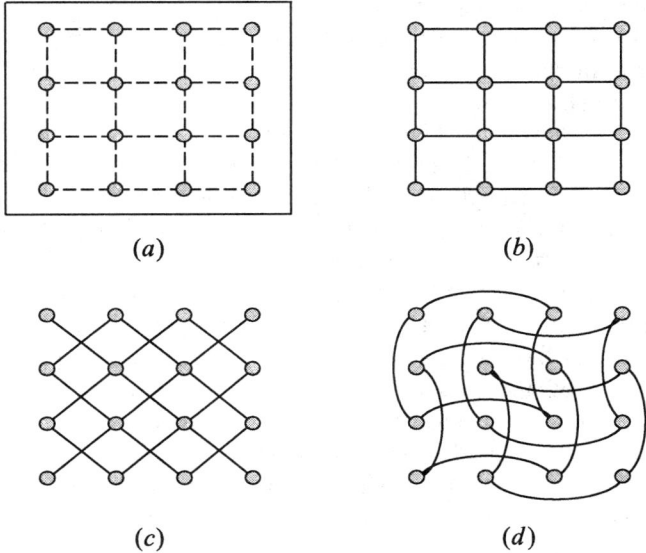

FIGURE 3.45. (a) Rectangular grid representing a piece of cloth. The broken lines connect adjacent vertices of the grid; (b) Stretch forces are represented by springs connecting particle (i,j) to particles $(i+1,j)$ and $(i,j+1)$; (c) Shear forces are represented by springs connecting particles (i,j) and $(i+1,j+1)$, and particles $(i+1,j)$ and $(i,j+1)$; (d) Bend forces are represented by springs connecting particles (i,j) and $(i+2,j)$, and particles (i,j) and $(i,j+2)$.

The collision detection between cloth objects and other particle or rigid-body objects in the simulation follows the same principles already explained in Section 3.4. The rectangular grid is internally decomposed into a triangular mesh by simply connecting one of the diagonally opposing vertices of each grid cell, say vertex (i,j) to $(i+1,j+1)$. Particle-cloth and rigid body-cloth collisions are detected by checking for geometric intersections between the particle or rigid body with the triangular mesh. This intersection test can be efficiently implemented if we consider a hierarchical tree representation of the triangular mesh to prune out unnecessary primitive-primitive intersection tests, as explained in detail in Chapter 2.

Self-intersections are also handled using the same technique. We first check for intersections between the intermediate nodes at the first level of the hierarchical tree, that is, between the two children of the root node (assuming a binary tree). If they do not intersect, then the cloth is guaranteed to be not self-intersecting. Otherwise, we check for intersections between the intermediate nodes at the second level of the three. We recursively proceed checking for intersections between the intermediate nodes at the ith level that have their parents intersecting at the $(i-1)$-th level until we reach the leaf nodes, or until no intermediate intersections are found. Self-intersections exist only if there are primitive-primitive intersections between the primitives associated with overlapping leaves of the tree. In the case of rigid body-cloth intersections, the primitives are both triangles, whereas in the case of particle-cloth intersections the primitives are triangles and cylinders with a spherical cap, respectively.

Whenever a collision is detected, the appropriate impact or contact forces are computed and applied to the cloth particles associated with the vertices of each intersected triangle according to the following rules. If the contact point is a vertex of the triangle, then the force is applied only to the particle associated with that vertex. If the contact point is on an edge of the triangle, then half the force is applied to each particle forming the edge. Otherwise, if the contact point lies inside the triangle, then one third of the force is applied to each of the three particles of the intersected triangle. Also, notice that, since all particles in the cloth are interconnected with their neighbors, we shall need to solve a linear system of the same order of magnitude as the total number of particles defining the cloth surface for each simulation time step. So, as we attempt to obtain more realistic simulation results by increasing the number of particles used to represent the cloth object, computational time per simulation time step increases considerably, thus reducing the overall efficacy of the simulation.

Another common problem observed when modeling cloth with a spring-particle system is that the system of equations of motion can become stiff, especially when one of the spring constants used is orders of magnitude different from the others. One way of minimizing this stiffness is to use a numerical-integration method appropriate to such situations; see the implicit Euler method discussed in Appendix B.

There is also one more problem worth mentioning when modeling cloth as a spring-particle system. In some situations, one or more springs can end up being elongated to extremely high values, sometimes as large as twice their resting displacement. This might be acceptable if we were simulating an actual spring-mass system, but such *super elastic* behavior is unrealistic in the case of cloth. The problem is that the stretch and shear forces of real woven fabrics vary non-linearly with the distance between their fibers, such that the resistance increases very rapidly as a function of the relative displacement of the fibers until either the movement stops, or the fabric fails to withstand the force, and ruptures.

One way of preventing this super-elastic behavior would be to dynamically increase the stiffness of the springs being stretched or sheared above a threshold limit, such that the spring resistance dramatically increases in these cases to prevent further elongation. Unfortunately, this is not practical because the dynamic increase of the spring resistance can introduce stiffness into the numerical integration, forcing it to reduce the size of the simulation time steps being taken, and thus adversely affect the overall simulation. As a matter of fact, there are situations in which the time step can be reduced to such a small value that the simulation seems to be halted forever.

Another way to deal with the super-elastic effect without having to change the stiffness of the springs is to set up a maximum spring deformation and dynamically enforce it during the simulation. The maximum deformation should be set as a percentage of the resting size of the spring. We can assign a maximum deformation for the stretch, shear and blend strings, although the latter is usually not considered because real cloth objects have no significant limits on how much they can be bended. Also, because the particles are laid out as vertices of a uniform rectangular mesh, we set the shear and stretch maximum deformations be the same. This last assumption can considerably simplify the enforcement of the maximum-deformation condition because we shall always be able to ignore the elongation of the stretch springs, and only consider the elongation of the shear springs diagonally placed on each rectangular cell of the grid, as shown in Figure 3.45(c). In other words, the stretch deformation is automatically enforced whenever we enforce the shear deformation.

This approach of enforcing maximum deformations can be implemented as follows. On a first pass, the spring forces are computed as usual, that is, from the initial displacement of their associated particles. The net force acting on each particle is then computed by summing the spring forces acting on the particle with any other external forces, as appropriate. The particles are moved according to the results obtained from the numerical integration of their equations of motion, and their associated springs are stretched and compressed accordingly. Then, on a second pass, the displacement of each shear spring is checked against the maximum allowable spring deformation. If the shear spring is elongated more than the maximum allowable deformation, then its associated particles are brought up close together to the point where the spring displacement is equal to the maximum deformation allowed. Figure 3.46 shows how the particles are moved together without violating the direction of motion obtained from the numerical integration.

Consider a shear spring connecting particles P and Q located at positions $\vec{p}_1$ and $\vec{q}_1$ at time $t = t_0$. The current spring elongation is $d_1 < d_{max}$, where d_{max} is the maximum deformation allowed. The particles are then moved to positions $\vec{p}_2$ and $\vec{q}_2$ at time $t = (t_0 + h)$, and the spring is stretched to

150   3. Particle Systems

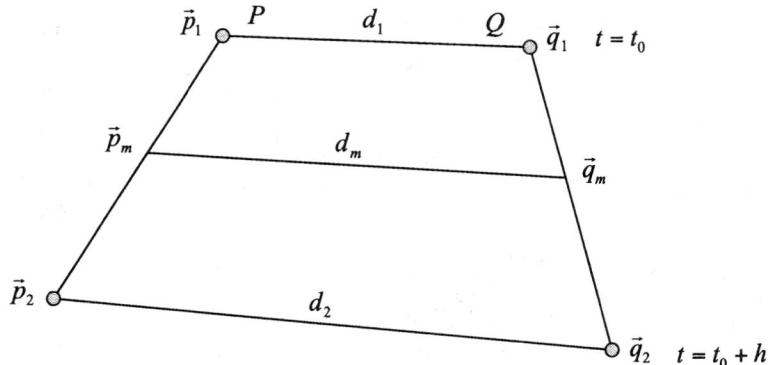

FIGURE 3.46. The shear spring connecting particles P and Q is elongated above its maximum deformation value from time $t = t_0$ to $t = (t_0 + h)$. We need to determine positions $\vec{p}_m$ and $\vec{q}_m$ at which the elongation d_m equals its maximum value d_{max}.

a size $d_2 > d_{max}$. The problem is then to find the intermediate particle positions $\vec{p}_m$ and $\vec{q}_m$ along their direction of motion, at which the spring elongation is d_{max}.

Since the particles' path is approximated by a straight line connecting their initial and final positions, the intermediate point $\vec{p}_m$ belongs to the line segment connecting $\vec{p}_1$ and $\vec{p}_2$, that is

$$\vec{p}_m = \vec{p}_1 + t_p (\vec{p}_2 - \vec{p}_1), \quad (3.102)$$

where $0 \le t_p \le 1$ is a scalar variable to be determined. Analogously we have

$$\vec{q}_m = \vec{q}_1 + t_q (\vec{q}_2 - \vec{q}_1), \quad (3.103)$$

with $0 \le t_q \le 1$. Notice that, since the particles are moved from their initial to their final positions within the same simulation time step, we have

$$t_p = t_q. \quad (3.104)$$

The distance d_m between $\vec{p}_m$ and $\vec{q}_m$ is computed as

$$d_m^2 = |\vec{p}_m - \vec{q}_m|^2. \quad (3.105)$$

Substituting equations (3.102) and (3.103) into (3.105), and using equation (3.104), we have

$$\begin{aligned} d_m^2 &= |(\vec{p}_1 + t_p (\vec{p}_2 - \vec{p}_1)) - (\vec{q}_1 + t_q (\vec{q}_2 - \vec{q}_1))|^2 \\ &= |(\vec{p}_1 - \vec{q}_1) + t_p ((\vec{p}_2 - \vec{q}_2) - (\vec{p}_1 - \vec{q}_1))|^2 \\ &= |\vec{k}_1 + t \vec{k}_2|^2 \end{aligned}$$

$$= ((k_1)_x + t(k_2)_x)^2 + ((k_1)_y + t(k_2)_y)^2$$
$$+ ((k_1)_z + t(k_2)_z)^2 ,$$

where $\vec{k}_1 = (\vec{p}_1 - \vec{q}_1)$ and $\vec{k}_2 = ((\vec{p}_2 - \vec{q}_2) - (\vec{p}_1 - \vec{q}_1))$ are known vectors. We want $d_m = d_{max}$, that is, we need to solve the quadratic equation

$$(k_1)_x^2 + (k_1)_y^2 + (k_1)_z^2 - d_{max}^2 +$$
$$2((k_1)_x(k_2)_x + (k_1)_y(k_2)_y + (k_1)_z(k_2)_z)t_p +$$
$$((k_2)_x^2 + (k_2)_y^2 + (k_2)_z^2)t_p^2 = 0 \qquad (3.106)$$

for t_p. Let $(t_p)_1$ and $(t_p)_2$ be the two possible solutions for equation (3.106). If both solutions are less than zero, then the initial elongation d_1 is greater than d_{max}, which contradicts the fact that the maximum deformation is enforced on every time step (i.e., if $d_1 > d_{max}$, then the maximum deformation enforcement would have been applied to the particles during the previous simulation time step). Also, if both solutions are greater than one, then the final elongation d_2 is less than d_{max}, which contradicts the fact that the maximum-deformation condition has been violated. If just one of the solutions, say $(t_p)_1$, is between zero and one, then we set $t_p = (t_p)_1$. Otherwise, if both solutions are between zero and one, say $0 \leq (t_p)_1 \leq (t_p)_2 \leq 1$, then we select the largest solution, that is, we set $t_p = (t_p)_2$. Having determined the parameter t_p, we substitute its value on equations (3.102) and (3.103) to obtain $\vec{p}_m$ and $\vec{q}_m$, respectively.

The worst-case scenario for this procedure occurs when all shear springs are already elongated at their limits at the beginning of the current time step. In such situations, the particles are initially moved to their new positions at the end of the current time step, and some shear springs may have been further elongated, and others compressed. The particles associated with the springs that were elongated are then moved back to their initial positions. So, the computational time spent processing these particles was wasted. However, in spite of this apparent inefficiency, the algorithm is guaranteed to assure the maximum-deformation condition in just one pass through all particles. Moreover, because we are moving the particles backwards along their computed paths, any particle can be chosen as the starting point.

One final remark: when we have multiple updates in the position of a particle owing to several shear springs attached to it, the selected parameter t_p will be the minimum of all values obtained to enforce the maximum-deformation condition at each shear spring. Again, this is possible because we are moving the particles backwards along their computed paths, so that each parameter t_p obtained for different shear springs refers to the same path. Therefore, they can be compared with each other, and selecting the minimum value guarantees that all other conditions are satisfied.

3.10 Notes and Comments

The idea of using particle systems as a modeling tool in computer graphics was originally introduced by Reeves [Ree83]. Nowadays, particle systems are so widely used to model diverse sets of systems that the large number of different techniques, implementations and parameter settings available makes it difficult to address this topic in depth.

From our experience, the selection of the particle system depends on the application at hand. For instance, if you are working on molecular dynamics, then you will probably be willing to trade efficiency for accuracy in your models, and so you need to use a particle system that models the details of particle interactions to a great extent. On the other hand, if your concern is to get a visually appealing simulation of smoke coming out of a fire, then you will probably need to use a flexible particle system that lets you adjust the look of the simulation without worrying too much about the underlying accuracy of the model.

Our approach in this chapter was to first provide the foundation needed to implement accurate models of particle interactions based on classical mechanics, then extend our model with some user-definable parameters commonly found in animation packages to support a diverse set of specialized particle systems. Frenkel *et al.* [FS96] also used the mathematical foundation of classical mechanics to model several molecular-dynamics systems. The types of force interactions described in Section 3.3 were inspired from the work of Frenkel and Baraff *et al* [BW98a]. A detailed discussion on under, over and critically damped spring systems can be found in Beer *et al.* [BJ77b].

The particle-particle collision detection representing the particle's trajectory as a cylinder with spherical cap was originally proposed by Karabassi *et al.* [KPTB99]. The cylinder-cylinder, cylinder-sphere and cylinder-line segment intersection tests were obtained from Held [Hel97]. The cylinder-box intersection test presented in Section 3.5.2 was adapted from Held's cylinder-triangle intersection test. However, the actual cylinder-triangle test used in Section 3.5.4 was suggested by Karabassi.

The use of a critical-coefficient-of-friction value to determine whether the particles are sliding at their collision point was obtained from Brach [Bra91]. This in turn allows us to express the multiple particle-particle collision case in a concise block-partitioned matrix representation, which can be efficiently solved by Gaussian elimination, or by a specialized sparse-matrix solver routine. The mathematical foundation used to derive the single and multiple contact equations was adapted from Baraff *et al* [BW98a], which dealt with rigid body-rigid body contact-force computation.

Several parameters used in Section 3.9 were inspired from commercial software packages, such as Visviva's Animation Engine and SoftImage's Particle System. The suggested parameter settings for simulating smoke, liquids and explosions were obtained by trial and error.

The cloth model was by far the most demanding. The model presented in Section 3.9.2 is very similar to that proposed by Provot [Pro95, Pro97]. The main differences are in the collision-detection and response techniques used, and in the way the particles are moved back to enforce the spring's maximum deformation condition. There are, however, other models that could be used to simulate cloth. These derive from the work of Baraff *et al.* [BW99], Breen *et al.* [BHW94], Demetri *et al.* [DJAK87] and Michel *et al.* [CYND92], to name a few.

Lastly, the mathematical theory of the Navier-Stokes equations and the physics of the conventional theory of turbulence can be found in Wilkins [Wil99], Stam *et al.* [SF95, Sta99] and Foster *et al.* [FM96, FM97]. An introduction to the Chapman-Jouget theory used to model detonation of explosives can also be found in Wilkins [Wil99].

4
Rigid-Body Systems

4.1 Introduction

Rigid-body dynamic simulations are by far the most interesting ones, with applications ranging from mechanical-systems design and prototyping, to robotic motion, to physics-based computer-graphics animations. A rigid body is modeled as a collection of particles that make up its geometric shape. During motion, the relative position of each particle forming the rigid body must remain constant so that its shape remains unchanged throughout the motion. This requires taking into account the rotational motion of the rigid body, which in turn considerably complicates the derivation of the equations of motion, the collision-detection techniques to be used, and the computation of all impulsive and contact forces that prevent their interpenetration during a simulation.

Here, we shall focus on both unconstrained and constrained motion of rigid bodies. The unconstrained motion deals with the dynamic equations that govern the free motion of a rigid body as a function of the net force and net torque acting on it. By free motion we mean that the motion is determined without worrying about collision detection and response. The constrained motion, on the other hand, deals with the computation of all impulsive and contact forces resulting from single or multiple collisions, or contacts between two or more rigid bodies during a simulation. Notice that the constraints considered in this chapter are imposed solely to prevent interpenetration of the rigid bodies. In the next chapter, we shall study other types of constrained motion where two or more rigid bodies are interconnected through joints that limit their relative degree of freedom, forcing them to stay "connected" throughout the entire simulation.

All algorithms presented in this book assume that a rigid body is given by its boundary representation, that is, by the faces defining its geometric shape. From the boundary representation, we can compute the rigid body's mass properties and convex decomposition, as explained in Appendices D and F, respectively. The mass properties of a rigid body are used in the dynamic-equations formulation presented in Section 4.2. The convex decomposition can be used to speed the collision-detection phase using specialized algorithms tailored for convex objects, as discussed in Section 4.4. Also, we assume that the faces describing the rigid body geometry are all triangular. This is not a limitation, since the faces can be efficiently triangulated during the convex-decomposition computation (see Appendix F for details).

4.2 Rigid-Body Dynamics

The dynamic equations that govern the motion of a rigid body need to capture both translational and rotational effects owing to external forces acting on it. Moreover, the rigid body's motion is influenced not only by the external forces, but also by its shape and mass distribution. The former defines a set of variables known as the rigid body's *mass properties*.

The mass properties of a rigid body are its volume, total mass, center of mass and inertia tensor. The inertia tensor is the equivalent of the total mass for rotational motion. That is, in the same way that the mass is used to relate the linear acceleration with the net external force, the inertia tensor is used to relate the angular acceleration with the net external torque acting on the rigid body. The mass properties can be directly computed from the rigid body's boundary representation, as explained in detail in Appendix D.

The boundary representation of a rigid body is usually given with respect to a local-reference frame known as the *body frame*. The mass properties of the rigid body are therefore computed with respect to this local frame. Scalar quantities, such as the rigid body's mass and volume, are independent of the reference frame being used. However, the position of the center of mass and the inertia tensor are affected by the choice of reference frame.

Let $\mathcal{F}_1$ be the body frame associated with the rigid body B_1, and let $\mathcal{F}$ be the world frame used in the simulation. Also, let the body frame be rotated by $\mathbf{R}(t)$ with respect to the world frame at time t. This is illustrated in Figure 4.1.

As mentioned in Appendix D, the inertia tensor is computed using a body frame that is parallel to the world frame, but has its origin at the rigid body's center of mass. The main advantage of using such body frame is that

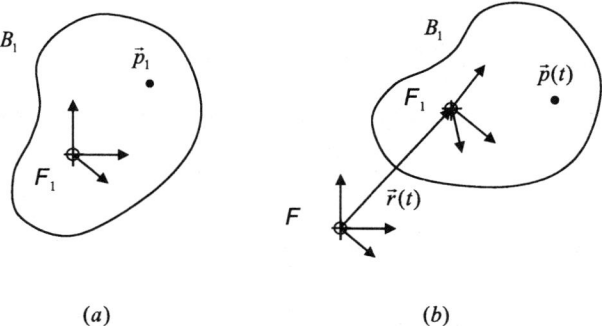

(a) (b)

FIGURE 4.1. (a) Rigid body positioned and oriented with respect to its body frame; (b) Same rigid body described with respect to the world frame.

the transformation between the inertia tensor from body-frame coordinates to world-frame coordinates is reduced to

$$\mathbf{I}(t) = \mathbf{R}(t)\,\mathbf{I_{body}}(t)\,\mathbf{R}^t(t)\,. \tag{4.1}$$

Given the position of a point $\vec{p}_1$ on the rigid body expressed in body-frame coordinates, its corresponding position $\vec{p}(t)$ with respect to the world frame is then computed as:

$$\vec{p}(t) = \mathbf{R}(t)\,\vec{p}_1 + \vec{r}(t) \tag{4.2}$$

where $\vec{r}(t)$ is the position of the center of mass of B_1 with respect to $\mathcal{F}$ at time instant t. Notice that $\vec{p}_1$ is computed with respect to the body frame $\mathcal{F}_1$, which moves with the body. Thus $\vec{p}_1$ is a constant vector over time. The actual motion of the body with respect to $\mathcal{F}$ is encoded in the rotation matrix $\mathbf{R}(t)$ and the translation vector $\vec{r}(t)$.

The velocity of $\vec{p}(t)$ is obtained by computing the time derivative of its position vector, namely

$$\frac{d\,\vec{p}(t)}{dt} = \frac{d\,\mathbf{R}(t)}{dt}\,\vec{p}_1 + \mathbf{R}(t)\,\frac{d\,\vec{p}_1}{dt} + \frac{d\,\vec{r}(t)}{dt}\,.$$

Since $\vec{p}_1$ is a constant vector, we have

$$\frac{d\,\vec{p}(t)}{dt} = \frac{d\,\mathbf{R}(t)}{dt}\,\vec{p}_1 + \frac{d\,\vec{r}(t)}{dt}\,. \tag{4.3}$$

Let $\vec{v}(t)$ be the linear velocity of the center of mass of body B_1 with respect to $\mathcal{F}$. Thus,

$$\frac{d\,\vec{r}(t)}{dt} = \vec{v}(t)\,. \tag{4.4}$$

Also, let $\vec{\omega}(t)$ be the angular velocity of body B_1 with respect to $\mathcal{F}$. According to Section E.5 of Appendix E, the time derivative of the rotation matrix is given by

158   4. Rigid-Body Systems

$$\frac{d\mathbf{R}(t)}{dt} = \vec{\omega} \times \mathbf{R}(t) = \tilde{\omega}(t)\,\mathbf{R}(t)\,, \qquad (4.5)$$

where $\tilde{\omega}(t)$ is the matrix-vector representation of a cross-product, described in Section A.7 of Appendix A. Substituting equations (4.4) and (4.5) into (4.3), we obtain

$$\frac{d\vec{p}(t)}{dt} = \tilde{\omega}(t)\,\mathbf{R}(t)\,\vec{p}_1 + \vec{v}(t)\,.$$

Using equation (4.2) we get

$$\begin{aligned}\frac{d\vec{p}(t)}{dt} &= \tilde{\omega}(t)\,\mathbf{R}(t)\,\mathbf{R}(t)^{-1}(\vec{p}(t) - \vec{r}(t)) + \vec{v}(t) \\ &= \tilde{\omega}(t)\,(\vec{p}(t) - \vec{r}(t)) + \vec{v}(t) \\ &= \vec{\omega}(t) \times (\vec{p}(t) - \vec{r}(t)) + \vec{v}(t)\,. \end{aligned} \qquad (4.6)$$

Equation (4.6) gives us a way of computing the velocity of any point $\vec{p}(t)$ in the rigid body. Finally, the acceleration of point $\vec{p}(t)$ is obtained by computing the time derivative of its velocity vector, namely

$$\frac{d^2\vec{p}(t)}{dt^2} = \frac{d\vec{\omega}(t)}{dt} \times (\vec{p}(t) - \vec{r}(t)) + \vec{\omega}(t) \times \left(\frac{d\vec{p}(t)}{dt} - \frac{d\vec{r}(t)}{dt}\right) + \frac{d\vec{v}(t)}{dt}\,. \qquad (4.7)$$

Let $\vec{a}(t)$ be the linear acceleration of the rigid body's center of mass, that is

$$\vec{a}(t) = \frac{d\vec{v}(t)}{dt}$$

and let $\vec{\alpha}(t)$ be the angular acceleration of the rigid body defined as

$$\vec{\alpha}(t) = \frac{d\vec{\omega}(t)}{dt}\,.$$

Notice that both linear and angular accelerations are expressed in world-frame coordinates. Substituting these into equation (4.7), we obtain

$$\frac{d^2\vec{p}(t)}{dt^2} = \vec{\alpha}(t) \times (\vec{p}(t) - \vec{r}(t)) + \vec{\omega}(t) \times (\vec{\omega}(t) \times (\vec{p}(t) - \vec{r}(t))) + \vec{a}(t)\,. \qquad (4.8)$$

Let $\vec{F}(t)$ be the net external force acting on the rigid body's center of mass at time instant t. Using Newton's law, we have

$$\vec{F}(t) = \frac{d\vec{L}(t)}{dt}\,, \qquad (4.9)$$

where $\vec{L}(t)$ is the linear momentum of the rigid body computed from

$$\vec{L}(t) = m\,\vec{v}(t)\,, \qquad (4.10)$$

with m and $\vec{v}(t)$ being the rigid body's mass and linear velocity of its center of mass. Substituting equation (4.10) into (4.9), we have

$$\vec{F}(t) = \frac{d\,(m\,\vec{v}(t))}{dt} = m\,\frac{d\,\vec{v}(t)}{dt} = m\,\vec{a}(t)\,. \qquad (4.11)$$

A similar equation can be obtained relating the net torque $\vec{\tau}(t)$ acting on the rigid body's center of mass, with the angular acceleration $\vec{\alpha}(t)$. It is known as the Euler equation and is given by

$$\vec{\tau}(t) = \frac{d\,\vec{H}(t)}{dt}\,, \qquad (4.12)$$

where $\vec{H}(t)$ is the rigid body's angular momentum computed as

$$\vec{H}(t) = \mathbf{I}(t)\,\vec{\omega}(t)\,, \qquad (4.13)$$

with $\mathbf{I}(t)$ being the inertia tensor expressed in *world frame* coordinates, obtained from equation (4.1).

The angular acceleration $\vec{\alpha}(t)$ is related to the angular momentum $\vec{H}(t)$, as follows. We already know that the angular acceleration is computed as the time derivative of the angular velocity. Since equation (4.13) relates the angular velocity and the angular momentum, we have

$$\vec{\alpha}(t) = \frac{d\,\vec{\omega}(t)}{dt} = \frac{d\,(\mathbf{I}^{-1}(t)\,\vec{H}(t))}{dt} = \frac{d\,\mathbf{I}^{-1}(t)}{dt}\,\vec{H}(t) + \mathbf{I}^{-1}(t)\,\frac{d\,\vec{H}(t)}{dt}\,.$$

Using equation (4.12),

$$\vec{\alpha}(t) = \frac{d\,\mathbf{I}^{-1}(t)}{dt}\,\vec{H}(t) + \mathbf{I}^{-1}(t)\,\vec{\tau}(t)\,. \qquad (4.14)$$

From equation (4.1), the inverse of the inertia tensor is given by

$$\mathbf{I}^{-1} = \mathbf{R}(t)\,(\mathbf{I_{body}})^{-1}(t)\,\mathbf{R}^t(t)\,. \qquad (4.15)$$

The time derivative of the inverse of the inertia tensor is then

$$d\,\mathbf{I}^{-1}(t)/dt = \frac{d\,\mathbf{R}(t)}{dt}\,(\mathbf{I_{body}})^{-1}(t)\,\mathbf{R}^t(t) + \mathbf{R}(t)\,(\mathbf{I_{body}})^{-1}(t)\,\frac{d\,\mathbf{R}^t(t)}{dt}$$
$$= \tilde{\omega}(t)\,\mathbf{R}(t)\,(\mathbf{I_{body}})^{-1}(t)\,\mathbf{R}^t(t) + \mathbf{R}(t)\,(\mathbf{I_{body}})^{-1}(t)\,(\tilde{\omega}(t)\,\mathbf{R}(t))^t$$
$$= \tilde{\omega}(t)\,\mathbf{R}(t)\,(\mathbf{I_{body}})^{-1}(t)\,\mathbf{R}^t(t) + \mathbf{R}(t)\,(\mathbf{I_{body}})^{-1}(t)\,\mathbf{R}^t(t)\,\tilde{\omega}^t(t)$$
$$= \tilde{\omega}(t)\,\mathbf{R}(t)\,(\mathbf{I_{body}})^{-1}(t)\,\mathbf{R}^t(t) - \mathbf{R}(t)\,(\mathbf{I_{body}})^{-1}(t)\,\mathbf{R}^t(t)\,\tilde{\omega}(t)$$

because $\tilde{\omega}^t(t) = -\tilde{\omega}(t)$. Using equation (4.1) again, we can simplify the above expression to

$$\frac{d\,\mathbf{I}^{-1}(t)}{dt} = \tilde{\omega}(t)\,\overbrace{\mathbf{R}(t)\,(\mathbf{I_{body}})^{-1}(t)\,\mathbf{R}^t(t)}^{\mathbf{I}^{-1}(t)} - \overbrace{\mathbf{R}(t)\,(\mathbf{I_{body}})^{-1}(t)\,\mathbf{R}^t(t)}^{\mathbf{I}^{-1}(t)}\,\tilde{\omega}(t)$$
$$= \tilde{\omega}(t)\,\mathbf{I}^{-1}(t) - \mathbf{I}^{-1}(t)\,\tilde{\omega}(t) . \qquad (4.16)$$

Substituting equation (4.16) into (4.14), we obtain

$$\begin{aligned}
\vec{\alpha}(t) &= (\tilde{\omega}(t)\,\mathbf{I}^{-1}(t) - \mathbf{I}^{-1}(t)\,\tilde{\omega}(t))\,\vec{H}(t) + \mathbf{I}^{-1}(t)\,\vec{\tau}(t) \\
&= \tilde{\omega}(t)\,\mathbf{I}^{-1}(t)\,\vec{H}(t) - \mathbf{I}^{-1}(t)\,\tilde{\omega}(t)\,\vec{H}(t) + \mathbf{I}^{-1}(t)\,\vec{\tau}(t) \\
&= \tilde{\omega}(t)\,\overbrace{\mathbf{I}^{-1}(t)\,\vec{H}(t)}^{\vec{\omega}(t)} - \mathbf{I}^{-1}(t)\,\tilde{\omega}(t)\,\vec{H}(t) + \mathbf{I}^{-1}(t)\,\vec{\tau}(t) \\
&= \overbrace{\tilde{\omega}(t)\,\vec{\omega}(t)}^{\vec{\omega}(t)\times\vec{\omega}(t)=\vec{0}} - \mathbf{I}^{-1}(t)\,\tilde{\omega}(t)\,\vec{H}(t) + \mathbf{I}^{-1}(t)\,\vec{\tau}(t) \\
&= -\mathbf{I}^{-1}(t)\,\vec{\omega}(t) \times \vec{H}(t) + \mathbf{I}^{-1}(t)\,\vec{\tau}(t) \\
&= \mathbf{I}^{-1}(t)\,\vec{H}(t) \times \vec{\omega}(t) + \mathbf{I}^{-1}(t)\,\vec{\tau}(t) . \qquad (4.17)
\end{aligned}$$

So, the relation between the angular acceleration and angular momentum is expressed as

$$\vec{\alpha}(t) = \mathbf{I}^{-1}(t)\,(\vec{H}(t) \times \vec{\omega}(t) + \vec{\tau}(t)) , \qquad (4.18)$$

or alternatively

$$\begin{aligned}
\vec{\tau}(t) &= \mathbf{I}(t)\,\vec{\alpha}(t) - \vec{H}(t) \times \vec{\omega}(t) \\
&= \mathbf{I}(t)\,\vec{\alpha}(t) + \vec{\omega}(t) \times \vec{H}(t) . \qquad (4.19)
\end{aligned}$$

Let $\vec{y}(t)$ denote the dynamic state of the rigid body at time t, that is, the vector comprising all variables necessary to define the dynamics of the rigid body at any instant during the simulation. Since the rigid body's motion can be sub-divided into translational and rotational components, we shall pick the center of mass's position, the rigid body's orientation, and its linear and angular momenta to define its dynamic state, namely

$$\vec{y}(t) = \begin{pmatrix} \vec{r}(t) \\ \mathbf{R}(t) \\ \vec{L}(t) \\ \vec{H}(t) \end{pmatrix} .$$

So, the dynamic state of the rigid body at time $t = t_0$ is defined by the center of mass's position $\vec{r}(t_0)$, the rigid body's orientation $\mathbf{R}(t_0)$, its linear momentum $\vec{L}(t_0)$ computed as $m\,\vec{v}(t_0)$, and its angular momentum $\vec{H}(t_0)$ computed as $\mathbf{I}(t)\,\vec{\omega}(t)$.

The time derivative of the dynamic state defines how the dynamic state of the rigid body changes over time, and is given by

$$\frac{d\,\vec{y}(t)}{dt} = \begin{pmatrix} d\,\vec{r}(t)/dt \\ d\,\mathbf{R}(t)/dt \\ d\,\vec{L}(t)/dt \\ d\,\vec{H}(t)/dt \end{pmatrix} = \begin{pmatrix} \vec{v}(t) \\ \tilde{\omega}(t)\,\mathbf{R}(t) \\ \vec{F}(t) \\ \vec{\tau}(t) \end{pmatrix}.$$

So, the time derivative of the dynamic state at time $t = t_0$ is defined by the center of mass's velocity $\vec{v}(t_0)$ computed as $(\vec{L}(t_0)/m)$, the updated orientation $\tilde{\omega}(t_0)\,\mathbf{R}(t_0)$ with $\vec{\omega}(t_0)$ being computed as $\mathbf{I}^{-1}(t_0)\,\vec{H}(t_0)$, and the net force $\vec{F}(t_0)$ and net torque $\vec{\tau}(t_0)$ acting on the center of mass.

For a system with N rigid bodies, we can combine their individual dynamic states into a single system-wide dynamic-state vector

$$\vec{Y}(t) = \left(\vec{r}_1(t), \mathbf{R}_1(t), \vec{L}_1(t), \vec{H}_1(t), \ldots, \vec{r}_N(t), \mathbf{R}_N(t), \vec{L}_N(t), \vec{H}_N(t)\right)^t$$

with its corresponding time derivative being

$$\begin{aligned} \frac{d\vec{Y}(t)}{dt} &= \left(\vec{v}_1(t), \tilde{\omega}_1(t)\mathbf{R}_1(t), \vec{F}_1(t), \vec{\tau}_1(t), \ldots \right. \\ & \left. \vec{v}_N(t), \tilde{\omega}_N(t)\mathbf{R}_N(t), \vec{F}_N(t), \vec{\tau}_N(t)\right)^t. \end{aligned} \quad (4.20)$$

The general description of how the dynamic simulation of rigid-body systems work is very similar to the one already given in Chapter 3 for particle systems. At the beginning of the simulation, we have the dynamic state of each rigid body, namely its position, orientation and linear and angular momenta defined with respect to the world-reference frame. Each simulation time step consists of numerically integrating equation (4.20), using the dynamic state of the rigid bodies at the beginning of the time step as the initial condition for the numerical integration. There are several numerical methods that can be used to integrate equation (4.20), and the most popular ones are discussed in detail in Appendix B.

The computation of the net external force acting on each rigid body at each intermediate step of the numerical integrator is determined by summing all external forces acting on different points of the rigid body. These forces are then substituted for a force-torque pair acting on the rigid body's center of mass before they are used in the equations of motion. A force $\vec{F}_i(t)$ acting at point $\vec{p}_i(t)$ is substituted for a force-torque pair

$$\begin{aligned} \vec{F}_{cm}(t) &= \vec{F}_i(t) \\ \vec{\tau}_{cm}(t) &= (\vec{p}_i(t) - \vec{r}(t)) \times \vec{F}_i(t), \end{aligned}$$

where the (cm) index stands for "center of mass." The types of external forces considered in this book for rigid-body simulations range from simple global forces (such as gravity) to point-to-point forces (such as springs), and are discussed in detail in Section 4.3.

In a first pass, the determination of the dynamic state of each rigid body at the end of the current time step is done without taking into account any possible collisions between rigid bodies and other objects in the simulation environment. The information about the final dynamic state of each rigid body is then used on a second pass to check for collisions between themselves and other particles in the simulation. The collision detection consists of checking for geometric intersections between the bodies positioned at the end of the current simulation time step with all other bodies and particles in the simulation. As explained in Section 3.4.2, the particular case of rigid body-particle collision is handled only after all rigid body-rigid body collisions have been detected and resolved. This is necessary because the rigid body-particle collision model used in this book is such that, when a particle collides with a rigid body, only the particle is traced back in time to the moment before the collision.

Whenever a rigid body-rigid body collision is detected, the colliding rigid bodies have their positions and orientations traced back in time to the moment before their collision. The collision point and collision normal are then computed from the relative displacement of the colliding bodies. Only after this information is obtained does the collision-response module engage to compute the appropriate impulsive or contact forces that will be applied to change the direction of motion of the colliding rigid bodies.

The dynamic equations of all rigid bodies involved in a collision are then numerically integrated for the remaining period of time, that is, from the collision time to the end of the current time step. This new numerical integration will update the current rigid bodies' positions and orientations to account for all collision forces. Notice that this also requires the numerical integration of the dynamic state of all other rigid bodies connected to one or more rigid bodies involved in a collision, since the connection usually implies the existence of a force component between them. For example, consider a simple rigid-body system consisting of four rigid bodies, namely B_1, B_2, B_3 and B_4, and suppose rigid bodies B_1 and B_2 are connected by a spring.

Initially, the dynamic state of the system is numerically integrated from t_i (the beginning of the current time step) to t_f (the end of the current time step). Now, assume the collision-detection module detected a collision between bodies B_2 and B_3 at time t_c such that $t_i < t_c < t_f$ (see Figure 4.2). The colliding bodies are then traced back in time to the moment just before their collision (i.e., traced back to t_c) and the collision impulses are computed so as to prevent their interpenetration. Having applied the collision impulses to both bodies, their dynamic state is numerically integrated again save for the remaining period of time only, that is, from t_c to t_f.

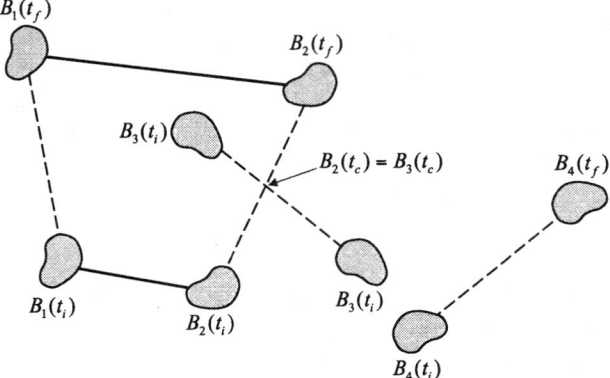

FIGURE 4.2. A simple rigid-body system containing four particles. The dynamic state of the system is numerically integrated from t_i to t_f. A collision between bodies B_2 and B_3 is detected at time t_c.

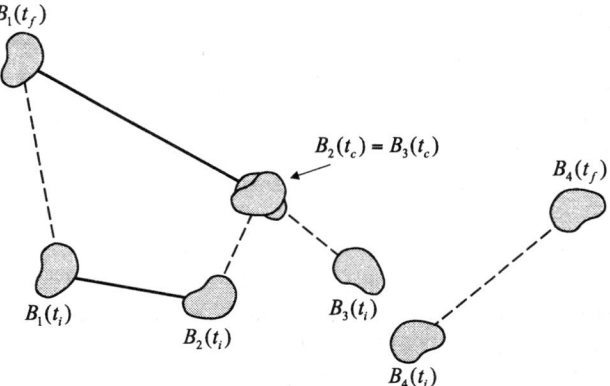

FIGURE 4.3. Bodies B_2 and B_3 are traced back in time to the moment just before their collision.

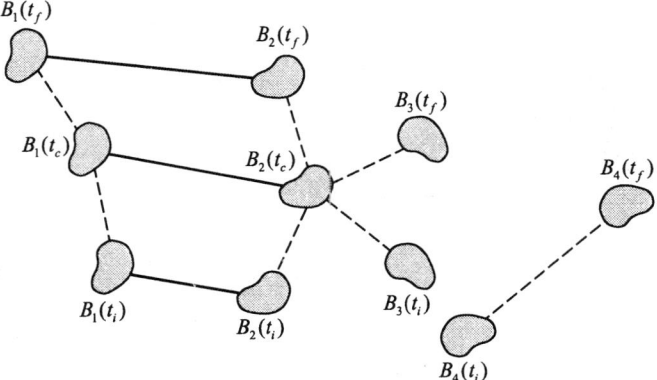

FIGURE 4.4. Body B_1 is also traced back in time to t_c before the numerical integrator is used to recompute the positions and orientations of bodies B_1, B_2 and B_3. Notice that body B_4 was not affected by the collision, and therefore remained unchanged throughout the collision-detection and response phase.

However, if we just trace back in time bodies B_2 and B_3, the spring-force computation between bodies B_1 and B_2 will be incorrect in the numerical integration for the remaining period of time. The problem is that, since body B_1 was not involved in any collision, its dynamic state corresponds to time t_f, whereas the dynamic state of body B_2 corresponds to time t_c at the beginning of the integration. So, if we do not trace back in time B_1 either, the spring-force computation will use B_1's position and orientation at time t_f when it should use them at the same simulation time of B_2, namely time t_c (see Figure 4.3). In other words, the numerical integration of the dynamic state of all interconnected bodies should be synchronized to provide the correct system behavior (see Figure 4.4). On the other hand, bodies that are not involved in a collision can be asynchronously moved within the same simulation time step. This is the case of rigid body B_4, as shown in Figure 4.4, since the numerical integration of B_2 and B_3 for the remaining period of time did not affect its dynamic state already computed in the first pass. As far as implementation is concerned, this approach requires some bookkeeping mechanism to efficiently determine which rigid bodies are connected to others. The payoff is the significant efficiency gain over the alternate solution of tracing back in time all rigid bodies, even those not involved in a collision, to the moment before the most recent collision.

The updated information about the final dynamic state of each rigid body involved in a collision is used to check again for collisions between them and all other rigid bodies in the simulation. This process repeats until all rigid body-rigid body collisions detected within the current time step have been resolved.

Clearly, the collision check is an intense process that can take up a lot of computational time, especially in a naive implementation. We suggest using the cell decomposition of the simulated world already discussed in Section 2.3 to speed the collision-detection checks. As the system evolves, the position of each rigid body relative to the cell decomposition of the simulated world is tracked by assigning the rigid body to the cells its hierarchical representation intersects. By so doing, only rigid bodies assigned to the same cell need be checked for collisions. This dynamic assignment can be efficiently implemented by observing that the cell decomposition defines a uniform subdivision of the simulated world into cubic cells (see Section 2.3 for more details).

4.3 Interaction Forces

The interaction forces used in most rigid-body system simulations can be categorized into two different types of forces. The first type considers global interaction forces, that is, forces independently applied to all rigid bodies in the system. Examples include gravity and viscous drag (used to simulate air resistance). These are the least expensive interaction forces available and their required computational cost is often negligible compared with that of the second type of interaction forces discussed in this book.

The second type considers point-to-point forces between a specific number of rigid bodies. Damped springs are a good example of such interaction forces between two given bodies. Interactive-user manipulation is also modeled as a point-to-point force between the current mouse position and the selected rigid body. The idea of using a fictitious interaction force between the mouse and the selected rigid body is to prevent the introduction of unstable configurations owing to abrupt mouse movements, as explained in Section 4.3.4.

4.3.1 Gravity

The force contribution of gravity acting on each rigid body owing to its attraction to the ground is directly obtained from

$$\vec{F} = m\,\vec{g},$$

where $\vec{g}$ is the gravitational acceleration, m is the rigid body's mass and $\vec{F}$ acts on the body's center of mass (see Figure 3.4). The gravitational acceleration is in most cases assumed to have constant magnitude equal to 9.81 meters per second squared and downwards direction (i.e., towards the "ground" of the simulated environment.)

166   4. Rigid-Body Systems

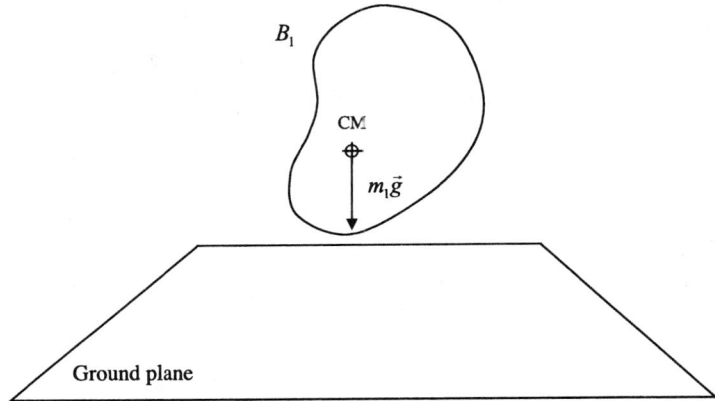

FIGURE 4.5. Gravity pulling rigid body B_1 with mass m_1 towards the ground plane.

4.3.2 Viscous Drag

The most common use of viscous drag in dynamic simulations of rigid-body systems is to model the air resistance to the body's movement. The goal is to ensure that rigid bodies will eventually come to rest if there are no other external forces acting on them. Figure 4.6 illustrates this. The force component of the viscous drag is computed as

$$\vec{F} = -k_v \, \vec{v} \, ,$$

where $\vec{v}$ is the linear velocity of the rigid body's center of mass and k_v is the linear drag coefficient.

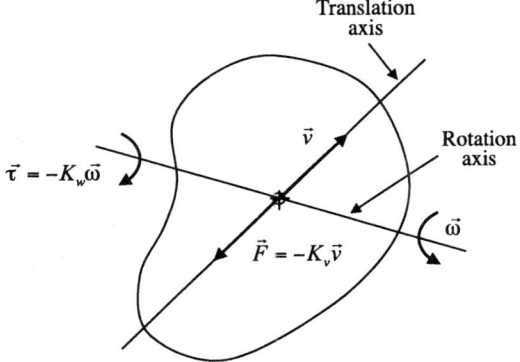

FIGURE 4.6. Rigid body B_1 is moving in a random direction experiencing air resistance modeled as viscous drag with force component $\vec{F}$ and torque component $\vec{\tau}$ acting on B_1's center of mass.

The torque component of the viscous drag is determined using a similar formula given by

$$\vec{\tau} = -k_\omega \vec{\omega},$$

where $\vec{\omega}$ is the angular velocity of the rigid body and k_ω is the angular drag coefficient.

4.3.3 Damped Springs

Springs are mostly used to keep the distance between pairs of rigid bodies at a known value. Whenever the rigid bodies are pushed apart or pulled together, a spring force is applied to both bodies, with same magnitude and opposite direction.

Let B_1 and B_2 be two rigid bodies connected by a spring of resting length r_0. The spring is attached to bodies B_1 and B_2 at points $\vec{p}_1$ and $\vec{p}_2$, respectively. Let $\vec{v}_1$ and $\vec{v}_2$ be the velocities of points $\vec{p}_1$ and $\vec{p}_2$, computed using equation (4.6) on page 158. The spring-force component acting on the center of mass of both rigid bodies is then given by

$$\vec{F}_2 = -\left[k_s \left(|\vec{p}_2 - \vec{p}_1| - r_0\right) + k_d \left(\vec{v}_2 - \vec{v}_1\right) \frac{(\vec{p}_2 - \vec{p}_1)}{|\vec{p}_2 - \vec{p}_1|}\right] \frac{(\vec{p}_2 - \vec{p}_1)}{|\vec{p}_2 - \vec{p}_1|}$$
(4.21)
$$\vec{F}_1 = -\vec{F}_2,$$

with $\vec{F}_i$ being the spring force acting on rigid body B_i for $i \in \{1, 2\}$, k_s being the spring constant and k_d being the damping constant (see Figure 4.7). Since the spring force is applied to points $\vec{p}_1$ and $\vec{p}_2$, it is substituted for a force-torque pair ($\vec{F}_1$ and $\vec{\tau}_1$) acting on body B_1, and $\vec{F}_2$ and $\vec{\tau}_2$ acting on body B_2, with

$$\vec{\tau}_1 = (\vec{p}_1 - \vec{r}_1) \times \vec{F}_1,$$
(4.22)
$$\vec{\tau}_2 = (\vec{p}_2 - \vec{r}_2) \times \vec{F}_2.$$

The damping term of equation (4.21) is used to prevent oscillation, and does not affect the motion of the center of mass of the connected bodies.

4.3.4 User-Interaction Forces

The user-interaction force is modeled as a damped spring connecting the current mouse position to the position of the rigid body's point being dragged. The idea of using this fictitious spring is to avoid the introduction of unrealistically large external forces acting on the selected body owing

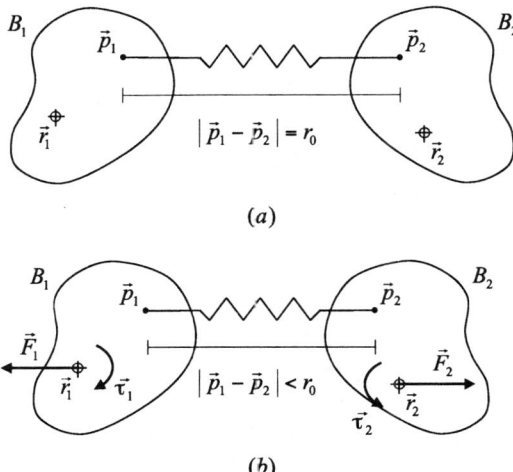

FIGURE 4.7. Rigid bodies B_1 and B_2 are connected by a damped spring; (a) Bodies at resting position; (b) Spring forces exerted on the rigid bodies when they are pushed together.

to abrupt mouse movements. These large external forces can make the dynamic equations describing the rigid body's motion stiff. Stiff systems are more susceptible to round-off errors and usually require the use of more elaborate and time-consuming numerical-integration methods, such as the implicit Euler method described in Appendix B.

The main difference between the damped spring described in Section 4.3.3 and the fictitious spring used here, is that the resting length of the fictitious spring should be set to zero. A zero resting length means that the selected rigid body will only stabilize its motion when its position is coincident with the mouse position. Therefore, as the user drags the body around, the current mouse position is used to update the actual distance between the rigid body and the mouse. This distance is then used in equations (4.21) and (4.22) to compute the appropriate spring force-torque pair to be applied to the rigid body's center of mass.

4.4 Collision Detection

Accurate collision detection on rigid-body dynamic simulations is much more difficult to achieve than in particle systems because of a fundamental difference in modeling these systems. Particles are usually modeled as simple spherical objects that can translate, but not rotate, during their motion. Their trajectories therefore can be analytically computed as the union of a cylinder and a sphere. Collision detection is then carried out by checking for geometric intersections of the cylinders and spheres repre-

4.4 Collision Detection

senting the trajectories of all particles that are potentially colliding. Rigid bodies, on the other hand, have convex and non-convex shapes and are allowed to translate and rotate between two consecutive time steps. This makes it very difficult to determine the volume spanned by their trajectories, and even when we can do it the added complexity and computational time required would probably not meet the real-time interactive efficiency goals of this book.

For example, an alternate approach to detect collisions taking into account the rigid body's trajectory considers time as an additional axis to augment the three-dimensional space defining the simulation environment to a four-dimensional space. By so doing, both numerical-integration and collision-detection phases are carried out with respect to the four-dimensional space. In particular, collision detection is carried out by checking for geometric intersections between the volume swept in this four-dimensional space. This by itself is not an easy task, especially if the rigid bodies have non-convex shapes. Nonetheless, the problem of having geometric intersections between trajectory positions associated with different simulation times (such as we had with particle systems) is no longer relevant, since time itself is used as an extra variable describing the trajectory. Even though approaches like this one may work well for particular classes of rigid-body shapes, or even general shapes moving along known parameterized trajectories, they are not directly applicable to the general rigid-body simulation scenario considered in this book. Here, we want solutions capable of efficiently handling the motion of convex and non-convex rigid bodies following the laws of physics, as opposed to being limited to scripted motions, as is the case of parameterized trajectories.

Because considering the rigid body's trajectory to undertake collision detection is infeasible for real-time simulations, the approach we shall take in this book is to simplify the collision-detection problem by checking for collisions between the bodies only after they are placed at the end of the current simulation time step. In other words, we first numerically integrate the dynamic state of the system without taking into account possible collisions between the rigid bodies. The result of the numerical integration is then the position and orientation of each rigid body in the system at the end of the current simulation time step. Only then, do we check for collisions between the bodies. Again, the collision check is undertaken by ascertaining geometric intersections between the bodies. Whenever an intersection is detected, the bodies are traced back in time to the moment before their collision.

The trace back in time can be efficiently implemented using a root-finding algorithm such as the *bisection* method. The bisection method works by moving each colliding rigid body half way through the current time interval. An intersection test is then conducted to determine whether the bodies intersect at the middle point. If the bodies intersect at the middle point, then the current time interval is substituted for its first half (i.e., from the start to middle points). Otherwise, it is substituted for its second half (i.e., from the middle to end points). The current time interval is then recursively subdivided into halves using the bisection method until it is shrunk to less than a threshold value. In this case, the bodies will be intersecting at one end of the time interval, and will not be intersecting at the other end. The collision time (i.e., time just before the collision) is approximated by the end of the time interval in which the bodies are not colliding.

The rigid bodies' geometric displacement at the collision time is used to compute the collision information, such as the collision point at each rigid body, the collision normal and tangent plane. The collision points are assumed to be the closest points between the bodies at the moment just before their collision. The techniques used to determined such points vary depending on both rigid bodies being convex, or at least on one being non-convex. In Section 4.5 we address the problem of checking for geometric intersections and computing the collision information when at least one of the colliding bodies is non-convex. More specialized techniques to address the case in which both bodies are convex are presented in Section 4.6. Basically, these techniques use the convexity properties of the rigid bodies to boost the efficiency of the collision-detection tests, providing substantial gains over the more general non-convex technique.

The collision normal is computed from the closest features between the rigid bodies at the moment just before their collision. A feature can be a vertex, an edge or a face of the body. We say that the closest feature is a vertex-face if it consists of a vertex of one body and a face of the other. Possible closest-feature combinations are therefore vertex-vertex, vertex-edge, edge-edge, vertex-face and edge-face. The face-face case is substituted for three vertex-face cases, that is, one for each face vertex. By so doing, the collision normal can be computed depending on the closest feature case, as follows

1. *Vertex-Vertex:* unit vector connecting the vertices.

2. *Vertex-Edge:* unit vector parallel to the edge normal.

3. *Edge-Edge:* unit vector perpendicular to both edges. Can be computed as the cross-product between the vectors defining the direction of each edge.

4. *Vertex-Face:* unit vector parallel to the face normal.

5. *Edge-Face:* unit vector perpendicular to the face normal and the edge. Can be computed as the cross-product between the face normal and the vector defining the edge direction.

In all these cases, the actual direction of the collision normal is chosen such that the relative velocity of the rigid bodies at the collision point along the collision normal is negative, indicating that the bodies are moving towards each other. The tangent plane is directly obtained from the collision normal and collision points, as explained in Section A.6. Once the collision information is gathered, a collision impulse or contact force of same magnitude and opposite direction is then applied to each colliding body to prevent their interpenetration.

Unfortunately, there are a couple of drawbacks to ignoring the rigid body's trajectory when checking for rigid body-rigid body collisions. First of all, collisions that occur at the beginning of the current simulation time step have a higher probability of being missed than collisions that occur near the end of the time step. This situation is illustrated in Figure 4.8.

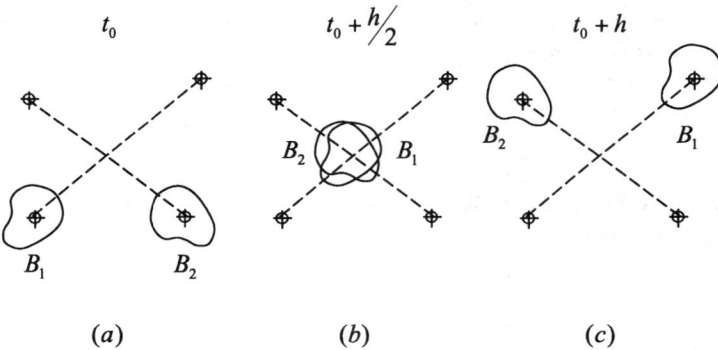

FIGURE 4.8. (a) Rigid bodies B_1 and B_2 positioned and oriented at the beginning of the current time step; (b) The bodies intersect at time $t_0 = \frac{h}{2}$; (c) But no intersections occur at the end of the current simulation time step and the bodies are treated as non-intersecting.

Since the bodies are positioned at the end of the current time step before any collision-checking is undertaken, intermediate collisions may be missed during simulation. Of course, we could reduce the simulation time step to catch these collision misses, but this would considerably slow the simulation, defeating the purpose of using the simplification in the first place.

The second drawback to our collision-detection scheme is that it is possible to have situations in which, even though the rigid bodies do not intersect each other at the end of the current simulation time step, one is completely inside the other (see Figure 4.9).

This is a difficult problem to deal with because using standard point-in-polygon tests for checking whether the vertices of one rigid body are included in the other rigid body's shape can be very time consuming, especially if we don't have a convex decomposition of the rigid body at hand. Fortunately, this difficult problem can be efficiently resolved if we have a convex decomposition of both rigid bodies. In this case, the specialized algorithms covered in Section 4.6 are capable of detecting interpenetration

172   4. Rigid-Body Systems

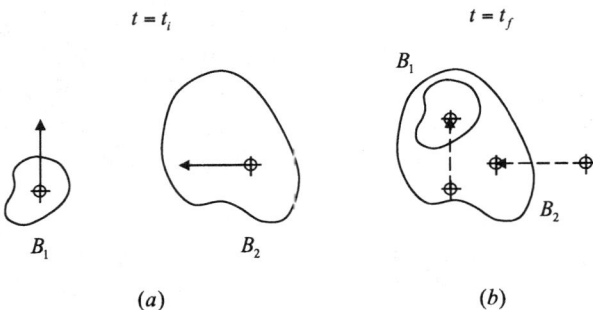

FIGURE 4.9. Drawbacks of considering only the final position and orientation to undertake collision checks; (a) Rigid bodies B_1 and B_2 positioned at the beginning of the current time step. No interpenetration occurs; (b) Bodies positioned at the end of the current time step. Body B_1 is completely included in body B_2.

(partial or complete inclusion) between the convex bodies. However, if at least one of the bodies is non-convex, and its convex decomposition is not available at hand, then accurate inclusion tests are too time consuming to warrant their use.

4.5   Collision Detection between Non-Convex Bodies

Whenever two rigid bodies are checked for collision against each other, the first thing the collision detection module does is verify that the bodies are either convex, or decomposed into convex parts. The convex decomposition can in most cases be computed using the algorithm presented in Appendix F. However, sometimes the convex-decomposition algorithm is incapable of determining the convex decomposition of the rigid body, possibly owing to a minimum dihedral angle or cut-face[1] requirement that cannot be satisfied.

When both bodies are convex, or are represented by their convex decomposition, the collision-detection module uses the more specialized algorithms presented in Section 4.6 that take advantage of the convexity properties to significantly improve overall collision detection. Otherwise, a more costly collision-detection test for non-convex rigid bodies is used.

The collision-detection scheme for non-convex rigid bodies relies on their hierarchical tree representation to detect and pin-point the colliding points,

---

[1] The algorithm presented in Appendix F limits the set of valid cut faces to simple polygons (without holes or double edges). This, in turn, makes the algorithm unsuitable for decomposing complex geometric shapes.

4.5 Collision Detection between Non-Convex Bodies

and the moment before the collision. This in turn has the implicit assumption that the bodies are treated as a polygon soup[2].

The algorithm's purpose is to check for geometric intersections between the non-convex rigid bodies using their hierarchical representations. If their hierarchical representations do not intersect, then the bodies are quickly dismissed and no collisions are reported from the collision-detection module. However, if their hierarchical representations do intersect, then further analysis is required to determine the collision point and normal, and to fix the time just before the collision.

The collision time is computed by tracing back in time both bodies to the moment before their collision. The trace back in time can be efficiently implemented using a root-finding algorithm, such as the *bisection* method. The bisection method works by moving each rigid body half way through the current time interval. A hierarchical tree intersection test is then undertaken to determine whether the bodies intersect at the middle point. If so, the current time interval is substituted for its first half (i.e., from the start point to the middle point). Otherwise, it is substituted for its second half (i.e., from the middle point to the end point). This process is repeated until the time interval is less than a threshold value, in which case the collision time is assumed to be the one corresponding to the starting point of the interval (i.e., the point at which the bodies do not intersect).

The actual collision point and collision normal can be determined as a byproduct of the back-tracing algorithm. Every time the hierarchical representations intersect at the middle point of the current time interval, the information about all pairs of intersecting faces is cached, replacing any previously cached information. When the tracing back in time is completed, we assign the collision time to the starting point of the current interval mainly because the bodies intersect at the end point of the interval. Since we have cached the information about which pairs of faces intersect at the end point, we can use it to compute their distance at the starting point of the interval, that is, at the time just before their intersection. This distance can then compared against a threshold value to select one or more pairs of faces that are close enough to be considered colliding faces. We can then compute the closest points between such pairs of faces and determine the collision point and normal. Since in this book we restrict the polyhedral representation to contain only triangles, most of the triangle-triangle intersection tests presented in Section 2.4.3, and the computation of the closest point between a point and a plane (see Section A.3 of Appendix A), can be used here.

---

[2]By "polygon soup" we mean the information about the face neighbor's connectivity, that is, the information about which faces share an edge is not used.

4.6 Collision Detection between Convex Bodies

The collision detection between rigid bodies can be carried out significantly faster if the bodies are convex, or represented by their convex decomposition. In this case, convexity is used to gather coherence information between two consecutive time steps, such that the result of the previous collision check is used as the starting point of the following collision check. Usually, the result is a pair of intersecting or closest (if no intersections are detected) features[3].

In this book, we present two extremely efficient algorithms to detect collisions between convex rigid bodies. Mirtich's *Voronoi Clip* algorithm is the most efficient feature-based algorithm known. The term feature-based refers to the fact that the algorithm is based on geometric operations using the features (i.e., faces, edges and vertices) of the rigid bodies being checked for collision. The *Gilbert-Johnson-Keerthi (GJK)* algorithm, on the other hand, is the best performing simplex-based algorithm known to date. The term simplex-based refers to the fact that the algorithm uses only the vertex information of the rigid bodies to construct a sequence of convex hulls. The operations are then carried out on subsets of points (i.e., simplices) that are part of such convex hulls.

4.7 The Voronoi Clip Algorithm

The basic idea of the Voronoi Clip algorithm is to subdivide the space around a given rigid body into a set of Voronoi regions, each associated with one rigid body's feature. A Voronoi region is defined as the region in space in which any point inside the region is closer to its associated feature than to any other feature of the rigid body. The proximity information encoded in the Voronoi region of each rigid body being tested for collision is used to determine the closest features between each pair of rigid bodies. The closest features are then used to estimate the collision frame (i.e., collision point, normal and tangent-plane vectors) if the bodies are not intersecting, or to determine the intersecting features otherwise.

In general, the Voronoi regions are bounded by at most two types of planes: *vertex-edge* and *face-edge* planes. Vertex-edge planes contain the vertex and are normal to an edge incident on the vertex, whereas face-edge planes contain the edge and are parallel to the normal vector of a face that contains the edge. More specifically, the Voronoi region associated with a vertex is bounded by a set of vertex-edge planes, each constructed from the vertex and an edge incident on it. The Voronoi region of an edge, on the other hand, is made of two face-edge planes (one for each face

---

[3] Here used to mean a face, an edge or a vertex of the rigid body.

sharing the edge) and two vertex-edge planes (one for each vertex defining the edge). Finally, the Voronoi region of a face is made of the face itself and face-edge planes, each constructed from the face and one of its edges. Figures 4.10, 4.11 and 4.12 show examples of Voronoi regions of a vertex, an edge and a face, respectively.

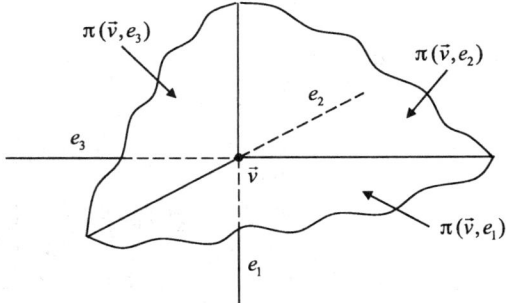

FIGURE 4.10. Voronoi region of a vertex.

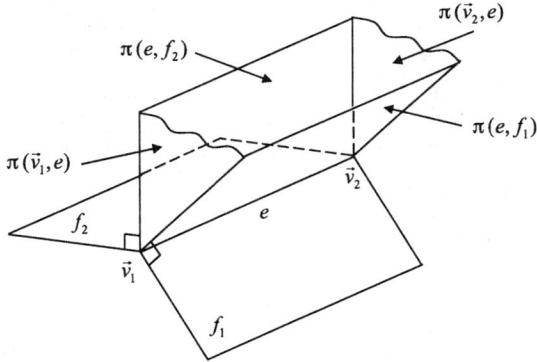

FIGURE 4.11. Voronoi region of an edge.

In Figure 4.10, vertex $\vec{v}$ has three incident edges, namely e_1, e_2 and e_3. Its associated Voronoi region is therefore made of three vertex-edge planes, that is, planes $\pi(\vec{v}, e_1)$, $\pi(\vec{v}, e_2)$ and $\pi(\vec{v}, e_3)$, where plane $\pi(\vec{v}, e_i)$ contains vertex $\vec{v}$ and is perpendicular to edge e_i. The normal vector $\vec{n}_i$ and plane constant d_i corresponding to each vertex-edge plane $\pi(\vec{v}, e_i)$ are computed as

$$\vec{n}_i = \frac{(\vec{v} - \vec{v}_i)}{|\vec{v} - \vec{v}_i|}$$

$$d_i = \vec{v} \cdot \vec{n}_i \, ,$$

(4.23)

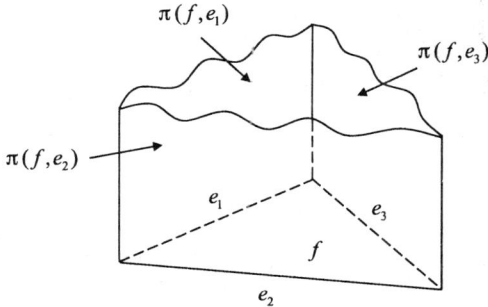

FIGURE 4.12. Voronoi region of a face.

with edge e_i being defined by vertices $\vec{v}$ and $\vec{v}_i$. Notice that the edge direction used in equation (4.23) is not necessarily the edge direction corresponding to an outward normal of the face containing the edge. Recall that rigid bodies are assumed to be described by vertices, edges and faces, and the edge direction is such that the normal of the face containing the edge is pointing from the inside to the outside of the body, using the right-hand coordinate system[4].

Figure 4.11 shows the Voronoi region of edge e. Since the edge is made up of two vertices and is shared by two faces, its corresponding Voronoi region is made of two face-edge planes ($\pi(f_1, e)$ and $\pi(f_2, e)$) and two vertex-edge planes ($\pi(\vec{v}_1, e)$ and $\pi(\vec{v}_2, e)$). The normal vector $\vec{n}_i$ and plane constant d_i corresponding to each face-edge plane $\pi(f, e)$ are computed as

$$\vec{n}_i = \vec{n}_f$$
$$d_i = \vec{v} \cdot \vec{n}_i ,$$

where $\vec{n}_f$ is the outward normal vector of the face and $\vec{v}$ is one of the two vertices defining edge e. An example of a Voronoi region associated with a face is shown in Figure 4.12. In this case, the face f is triangular and its associated Voronoi region is bounded by three face-edge planes ($\pi(f, e_1)$, $\pi(f, e_2)$ and $\pi(f, e_3)$) and the face itself.

Now, suppose we want to check for collisions between two convex rigid bodies B_1 and B_2. Let b_1 and b_2 be features of these rigid bodies such that $b_1 \in B_1$ and $b_2 \in B_2$. Also, let $\vec{p}_1$ and $\vec{p}_2$ be the closest points between b_1 and b_2, such that $\vec{p}_1 \in b_1$ and $\vec{p}_2 \in b_2$. It can be shown that, if point $\vec{p}_1$ is on the interior side of the Voronoi region of b_2 and, conversely, point $\vec{p}_2$ is on the interior side of the Voronoi region of b_1, then the points $\vec{p}_1$ and

---

[4]Since an edge is shared by two faces, the underlying implementation data structure representing the rigid body's face must have its own edge structure because the same edge has one direction for one of its faces, and the reverse of this direction for the adjacent face.

$\vec{p}_2$ are not only the closest points between b_1 and b_2, but also the closest points between the (convex) rigid bodies B_1 and B_2. We shall refer to these as the *closest feature* conditions.

Because the Voronoi regions are bounded by vertex-edge and face-edge planes that can be easily constructed from the current position and orientation of the rigid body, checking whether a point lies on the interior side of a given Voronoi region turns out to be equivalent to a simple sidedness check of the point against each plane bounding the Voronoi region. Let each plane π bounding a Voronoi region be defined by its normal vector $\vec{n}_\pi$ and a point $\vec{p}_\pi$. Any point $\vec{p} \in \pi$ satisfies the plane equation

$$\vec{p} \cdot \vec{n}_\pi = d_\pi ,$$

where d_π is the plane constant obtained from

$$d_\pi = \vec{p}_\pi \cdot \vec{n}_\pi .$$

Let $S_{\vec{p},\pi}$, defined as

$$S_{\vec{p},\pi} = \vec{p} \cdot \vec{n}_\pi - d_\pi , \qquad (4.24)$$

be the signed distance between point $\vec{p}$ and plane π. We have that $\vec{p}$ lies on the interior side of the Voronoi region bounded by plane π if

$$S_{\vec{p},\pi} > 0 \qquad (4.25)$$

and lies on the exterior side of the Voronoi region bounded by plane π if

$$S_{\vec{p},\pi} < 0 . \qquad (4.26)$$

So, if point $\vec{p}$ lies in the interior side of *all* planes bounding a Voronoi region, then it is said to be inside the Voronoi region. Otherwise, the point lies outside the Voronoi region (see Figure 4.13).

The Voronoi Clip algorithm works as follows. We start by arbitrarily selecting one feature of each rigid body being tested for collision. If the features satisfy the closest-feature conditions, then the algorithm terminates and the features are reported as being the ones that contain the closest points between the rigid bodies. However, if the closest-feature conditions are not satisfied, then the algorithm substitutes either one of the features failing the closest-feature conditions for one of its neighboring features. This process is repeated until the closest-feature conditions are satisfied.

In order to prevent cycles and guarantee the termination of the algorithm after a finite number of feature updates, it is necessary to ensure that the inter feature distance does not increase when a feature is substituted for one of its neighbors. It turns out that, if the new feature is of a higher dimension than the current feature being replaced, then the inter feature distance strictly decreases. Such substitutions include changing a vertex for an edge,

178   4. Rigid-Body Systems

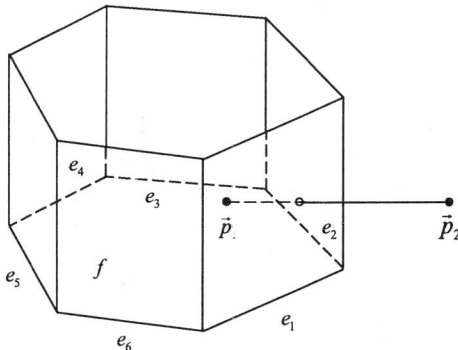

FIGURE 4.13. Testing whether points $\vec{p}_1$ and $\vec{p}_2$ lie inside the Voronoi region of face f consists of testing whether the points lie inside all face-edge planes bounding the region. In this particular case, point $\vec{p}_1$ lies on the inside of all planes, meaning it is inside the Voronoi region of f. Point $\vec{p}_2$, on the other hand, lies outside face-edge plane $\pi(f, e_1)$, meaning it is outside the Voronoi region associated with the face.

or an edge for a face. Moreover, if the new feature is of lower dimension than the current feature being replaced, then the inter feature distance remains unchanged. Examples of such substitutions include changing a face for an edge or vertex, and changing an edge for a vertex. Therefore, independent of the feature substitution made at intermediate steps of the algorithm, the inter feature distance will never increase[5] and the algorithm is guaranteed to terminate.

Basically, testing whether the closest-feature conditions are satisfied at each intermediate step of the algorithm consists of solving two instances of the following problem, one for each Voronoi region of the features being considered. Given features $b_1 \in B_1$ and $b_2 \in B_2$, we need to check whether the closest point $\vec{p}_2 \in b_2$ to b_1 lies inside the Voronoi region of b_1. If this is the case, then we repeat the test for the closest point $\vec{p}_1 \in b_1$ to b_2. If not, then we need to update b_1 such that the inter feature distance does not increase. This update consists of substituting b_1 for one of its neighbor features. The difficult part of the problem resides in determining which neighbor feature $(b_1)_{new}$ should replace b_1. We shall examine how this can be done on a case-by-case basis, for each possible combination of (b_1, b_2) features.

4.7.1 Feature b_2 is a Vertex

In this case, $b_2 = \vec{p}_2$ is a single point. We get the planes associated with the Voronoi region of b_1 and check for sidedness of vertex b_2 using equa-

---

[5]This owes to the fact that the bodies have convex shapes. Unfortunately, the same does not apply for the case in which the bodies have non-convex shapes.

tions (4.25) and (4.26). If b_2 lies on the inside of all planes of the Voronoi region of b_1, then we are done with this intermediate test. Otherwise, there exists at least one plane which b_2 lies on the outside of. We substitute b_1 for the feature associated with one of these violated planes[6].

Consider the particular example illustrated in Figure 4.14, with b_1 being a triangular face. Vertex b_2 is tested for sidedness against each of the three face-edge bounding planes, and the face itself. In Figure 4.14(a), vertex b_2 is on the interior side of all planes and we are done with this intermediate test. In Figure 4.14(b), vertex b_2 is on the exterior side of bounding planes $\pi(b_1, e_2)$ and $\pi(b_1, e_3)$. In this case, b_1 should be substituted for the feature associated with either one of these planes, namely edges e_2 or e_3.

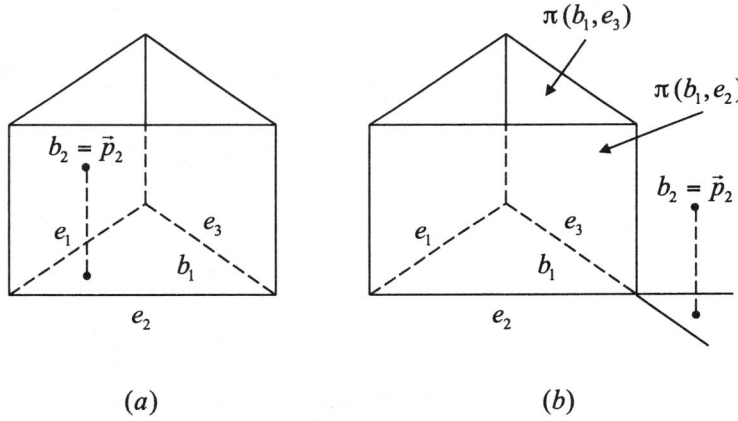

FIGURE 4.14. Example of an intermediate closest-feature test when b_2 is a vertex and b_1 is a face. (a) Vertex b_2 is inside all Voronoi planes associated with b_1; (b) Vertex b_2 lies behind Voronoi planes $\pi(f, e_2)$ and $\pi(f, e_3)$. In this case, b_1 should be substituted for either edge e_2 or e_3. Since the edge is of a lower dimension than the face, the inter feature distance remains the same after the substitution.

4.7.2 Feature b_2 is an Edge

Assume edge b_2 goes from vertex $\vec{v}_1$ to vertex $\vec{v}_2$. In this case, $\vec{p}_2$ can be any point satisfying b_2's edge equation

$$\vec{e}(\lambda) = (1 - \lambda)\, \vec{v}_1 + \lambda\, \vec{v}_2 \;, \tag{4.27}$$

with $0 \leq \lambda \leq 1$[7]. The idea is then to clip edge b_2 against all planes defining the Voronoi region of b_1 and check on which side of the clipped edge $\vec{p}_2$

---

[6]There is no particular preference for which violated plane should be used in the event that there is more than one.

[7]We shall use the parameter λ to index the points on edge b_2.

is located[8]. Let $[\lambda_{min}, \lambda_{max}]$ be the portion of edge b_2 that lies inside the Voronoi region of b_1. In other words, edge b_2 intersects two planes π_{min} and π_{max} bounding the Voronoi region of b_1 at points $\vec{e}(\lambda_{min})$ and $\vec{e}(\lambda_{max})$, respectively. Also, let b_{min} and b_{max} be the features associated with planes π_{min} and π_{max}. Figures 4.15, 4.16 and 4.17 illustrate possible situations for the case where b_1 is a triangular face.

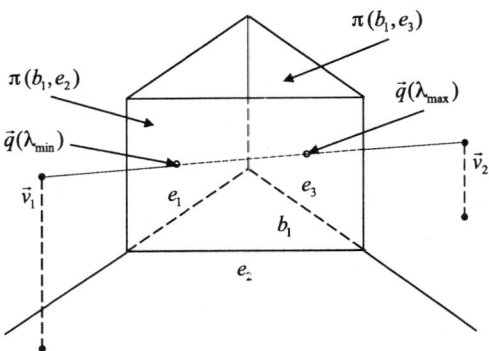

FIGURE 4.15. Edge b_2 intersects plane $\pi(b_1, e_2)$ at point $\vec{q}(\lambda_{min})$ and plane $\pi(b_1, e_3)$ at point $\vec{q}(\lambda_{max})$. Therefore, $b_{min} = e_2$ and $b_{max} = e_3$.

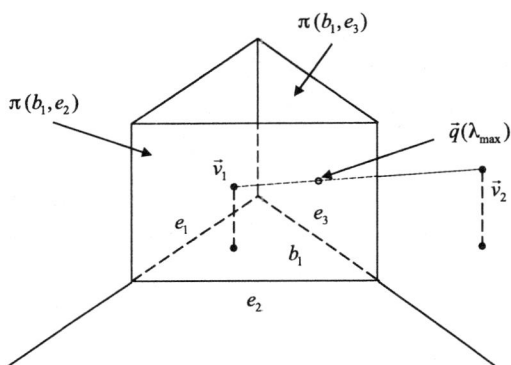

FIGURE 4.16. Vertex $\vec{v}_1$ lies inside the Voronoi region of b_1 and so, b_{min} is undefined.

Edge b_2 can be completely on the inside region of the bounding planes (i.e., $\lambda_{min} = 0$ and $\lambda_{max} = 1$), partially clipped by the bounding planes (i.e., $\lambda_{min} > 0$ or $\lambda_{max} < 1$), or totally clipped by the bounding planes (i.e., $\lambda_{min} = \lambda_{max} = 0$ or $\lambda_{min} > \lambda_{max}$, depending on how the edge is excluded, as will be explained later in this section.)

---

[8]Recall that $\vec{p}_2$ is the point on b_2 closest to b_1.

4.7 The Voronoi Clip Algorithm   181

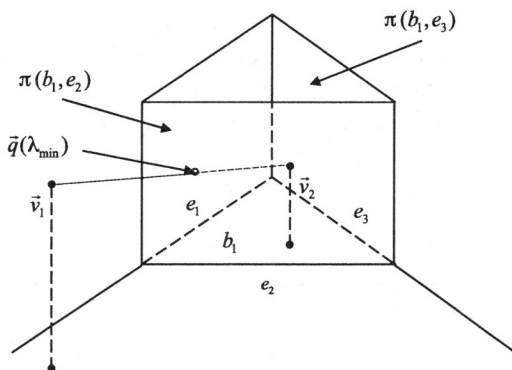

FIGURE 4.17. Vertex $\vec{v}_2$ lies inside the Voronoi region of b_1 and so, b_{max} is undefined. Notice that b_{min} and b_{max} are undefined whenever edge b_2 lies completely inside the Voronoi region of b_1.

If edge b_2 is completely on the inside region of the bounding planes, the algorithm reports edge b_2 and feature b_1 as the closest features between the bodies. However, if edge b_2 is clipped (either partly or completely), then we need to determine whether the closest point $\vec{p}_2$ on b_2 to b_1 lies on the part of the clipped edge b_2 that is on the inside region of the bounding planes, and if not, how to update it.

Case 1: Edge b_2 is partly clipped by the planes bounding the Voronoi region of b_1.

There are two approaches to check whether point $\vec{p}_2 \in b_2$ lies inside the Voronoi region of b_1. The first and more natural approach is to explicitly compute $\vec{p}_2$. This would entail solving one of the general geometric problems of computing the closest point of an edge to a vertex, an edge to another edge, or an edge to a face, depending on b_1 being a vertex, an edge or a face. Having computed $\vec{p}_2$, we would then proceed by checking whether its associated λ_{p_2} satisfies $\lambda_{min} \leq \lambda_{p_2} \leq \lambda_{max}$. If this is so, then $\vec{p}_2$ is inside the Voronoi region of b_1 and we are done. Otherwise, we have either $0 \leq \lambda_{p_2} < \lambda_{min}$ or $\lambda_{max} < \lambda_{p_2} \leq 1$, that is, $\vec{p}_2$ lies outside the Voronoi region of b_1.

One of the problems of explicitly computing the closest point $\vec{p}_2$ is the possible exposure to numerical round-off errors. Notice that we do not necessarily need to compute $\vec{p}_2$; we just need to determine in which interval $[0, \lambda_{min})$, $[\lambda_{min}, \lambda_{max}]$ or $(\lambda_{max}, 1]$ it lies. This is the fundamental motivation behind the second approach.

The second and more efficient approach considers the distance function from edge b_2 to feature b_1 defined by

$$D_{b_1,b_2}(\lambda) = \min_{x \in b_1} |x - \vec{e}(\lambda)|, \qquad (4.28)$$

which is a continuous, convex, and differentiable function of λ, provided that $\vec{e}(\lambda) \notin b_1$. Because $\vec{p}_2$ is the closest point of b_2 to b_1, its associated λ value is a minimum of the distance function. Therefore, checking in which interval $\vec{p}_2$ lies is equivalent to checking the sign of the derivative of the distance function at λ_{min} and λ_{max}. Let $\dot{D}_{b_1,b_2}(\lambda)$ be the derivative of the distance function defined in equation (4.28). We have:

1. If b_{min} is well defined (see Figure 4.16) and $\dot{D}_{b_1,b_2}(\lambda_{min}) > 0$, then the minimum (i.e., $\vec{p}_2$) lies in the interval $[0, \lambda_{min})$. In this case, we update b_1 to b_{min}.

2. If b_{max} is well defined (see Figure 4.17) and $\dot{D}_{b_1,b_2}(\lambda_{max}) < 0$, then the minimum lies in the interval $(\lambda_{max}, 1]$. In this case, we update b_1 to b_{max}.

3. Else, the minimum lies within the interval $[\lambda_{min}, \lambda_{max}]$ and we are done with this intermediate step.

The computation of the sign of the derivative of the distance function depends on b_1 being a vertex, an edge or a face. If b_1 is a vertex (i.e., $b_1 = \vec{v}$), then the sign of the derivative is immediately obtained from

$$\text{sign}\left(\dot{D}_{\vec{v},b_2}(\lambda)\right) = \text{sign}\left(\vec{u}_e \cdot (\vec{e}(\lambda) - \vec{v})\right), \tag{4.29}$$

where $\vec{u}_e$ is the unitary vector defining the direction of edge b_2 given by

$$\vec{u}_e = \frac{(\vec{v}_2 - \vec{v}_1)}{|\vec{v}_2 - \vec{v}_1|}.$$

In the case b_1 is a face (i.e., $b_1 = f$) with unit-normal vector $\vec{n}_f$ (pointing outwards) and plane-constant value d_f, then the sign of the derivative is given by

$$\text{sign}\left(\dot{D}_{f,b_2}(\lambda)\right) = \begin{cases} +\text{sign}\left(\vec{u}_e \cdot \vec{n}_f\right), & \text{if } S_{\vec{e}(\lambda),f} > 0 \\ -\text{sign}\left(\vec{u}_e \cdot \vec{n}_f\right), & \text{if } S_{\vec{e}(\lambda),f} < 0. \end{cases} \tag{4.30}$$

Finally, if b_1 is an edge, then we use its neighboring features b_{min} or b_{max} associated with λ_{min} and λ_{max} to determine the sign of the derivative on the intervals $[0, \lambda_{min})$ and $(\lambda_{max}, 1]$. Notice that, since b_1 is an edge, its neighboring features b_{min} and b_{max} must be a vertex or a face of the body and we can use equations (4.29) and (4.30) to compute the sign of the derivatives.

It is important to notice that the use of the neighboring features is only possible because the distance function is continuous. In other words, the computation of the sign of the derivative with respect to either b_{min} or b_{max} is equal to the sign of the derivative with respect to edge b_1 at the points where b_2 crosses (i.e., enters or leaves) the bounding planes of its Voronoi region, that is, at the points corresponding to λ_{min} and λ_{max}

(see Figure 4.15). The only cases in which the computation of the sign of the derivative fails is when the derivative function itself is undefined. This occurs whenever there exists a $\lambda \in [0, 1]$, such that

$$\vec{e}(\lambda) = b_1 \tag{4.31}$$

if b_1 is a vertex, or

$$S_{\vec{e}(\lambda), b_1} = 0 \tag{4.32}$$

if b_1 is a face. The geometric interpretation of being unable to compute the sign of the derivative is that edge b_2 intersects feature b_1. More specifically, equation (4.31) is satisfied whenever edge b_2 contains vertex b_1, and equation (4.32) is satisfied whenever edge b_2 intersects face b_1. In both cases, the Voronoi Clip algorithm terminates reporting the pair (b_1, b_2) as being interpenetrating.

One last remark on how to clip edge b_2 against the planes bounding the Voronoi region of b_1. If b_1 is a vertex, then all of its bounding planes are vertex-edge planes and we can clip against them in any order. However, if b_1 is an edge, then its Voronoi region is bounded by two vertex-edge and two face-edge planes. We should first clip edge b_2 against the two vertex-edge planes. If no clipping occurs, then we clip against the remaining face-edge planes. Finally, if b_1 is a face, its Voronoi region is bounded by several face-edge planes (one for each edge defining the face) and the face plane itself. In this case, we should first clip edge b_2 against the face-edge planes; if no clipping occurs, then we clip against the face plane.

Case 2: Edge b_2 is completely clipped by the planes bounding the Voronoi region of b_1.

There are two types of complete exclusion: simple exclusion and compound exclusion. Simple exclusion occurs whenever both vertices defining edge b_2 lie on the outside of a single plane bounding the Voronoi region of feature b_1. Compound exclusion occurs when both vertices defining edge b_2 lie on the outside of different planes bounding the Voronoi region of feature b_1.

Consider, for example, the simple-exclusion case illustrated in Figure 4.18, with b_1 being a triangular face. Edge b_2 is described by vertices $\vec{v}_1$ and $\vec{v}_2$ corresponding to $\lambda_{min} = 0$ and $\lambda_{max} = 1$, respectively. The inclusion test consists of checking whether edge b_2 is on the interior side of all planes bounding the Voronoi region of b_1. For the case illustrated in Figure 4.18, the result of the sidedness computation using equation (4.24) is

$$\begin{aligned} S_{\vec{v}_1, \pi(b_1, e_1)} &> 0 \\ S_{\vec{v}_2, \pi(b_1, e_1)} &> 0 \\ S_{\vec{v}_1, \pi(b_1, e_2)} &< 0 \end{aligned}$$

184   4. Rigid-Body Systems

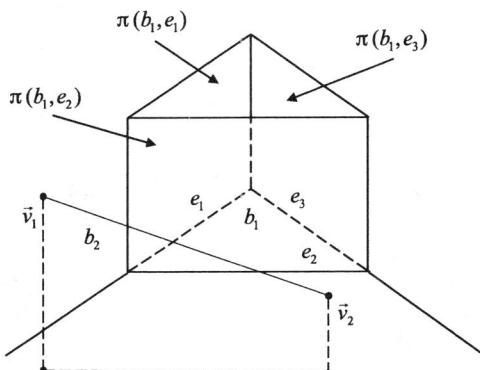

FIGURE 4.18. An example of a simple exclusion with b_1 being a triangular face.

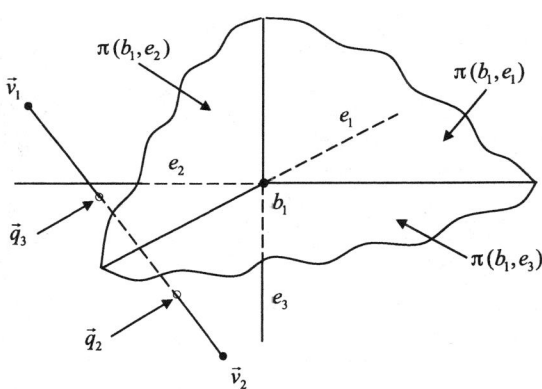

FIGURE 4.19. An example of a compound exclusion with b_1 being a vertex with three incident edges. Points $\vec{q}_2$ and $\vec{q}_3$ correspond to the intersections of edge b_2 (defined by points $\vec{v}_1$ and $\vec{v}_2$) with planes $\pi(b_1, e_2)$ and $\pi(b_1, e_3)$, respectively.

$$S_{\vec{v}_2,\pi(b_1,e_2)} < 0$$
$$S_{\vec{v}_1,\pi(b_1,e_3)} > 0$$
$$S_{\vec{v}_2,\pi(b_1,e_3)} > 0 \ .$$

Analyzing the results, we have that edge b_2 is on the interior side of plane $\pi(b_1, e_1)$ (the signed-distance computations resolve to a positive value), on the exterior side of plane $\pi(b_1, e_2)$, and on the interior side of plane $\pi(b_1, e_3)$. That is, edge b_2 is on the interior side of all bounding planes save $\pi(b_1, e_2)$, indicating the existence of simple exclusion. Feature b_1 is then substituted for the feature associated with plane $\pi(b_1, e_2)$, namely edge e_2. In summary, simple exclusion can be detected whenever the signed-distance computations resolve to a positive value for all bounding planes save one, which has a negative sign. Feature b_1 is then replaced by the feature associated with the violated plane.

Figure 4.19 illustrates an example of compound exclusion, with b_1 being a vertex with three incident edges. Again, edge b_2 is described by vertices $\vec{v}_1$ and $\vec{v}_2$ corresponding to $\lambda_{min} = 0$ and $\lambda_{max} = 1$, respectively. Evaluating the signed distance to plane $\pi(b_1, e_1)$ gives

$$S_{\vec{v}_1,\pi(b_1,e_1)} > 0$$
$$S_{\vec{v}_2,\pi(b_1,e_1)} > 0 \ ,$$

indicating that edge b_2 is completely on the interior side of plane $\pi(b_1, e_1)$. Evaluating the signed distance to plane $\pi(b_1, e_2)$ gives

$$S_{\vec{v}_1,\pi(b_1,e_2)} < 0$$
$$S_{\vec{v}_2,\pi(b_1,e_2)} > 0 \ ,$$

indicating that edge b_2 intersects plane $\pi(b_1, e_2)$. Because vertex $\vec{v}_1$ lies on the exterior side of $\pi(b_1, e_2)$, we update $\lambda_{min} = 0$ to $\lambda_{min} = \lambda_2 > 0$ corresponding to point $\vec{q}_2$. Effectuating the last signed-distance computation to plane $\pi(b_1, e_3)$ gives

$$S_{\vec{v}_1,\pi(b_1,e_3)} > 0$$
$$S_{\vec{v}_2,\pi(b_1,e_3)} < 0 \ ,$$

indicating that edge b_2 also intersects plane $\pi(b_1, e_3)$. Because vertex $\vec{v}_2$ lies on the outside region of $\pi(b_1, e_3)$, we update $\lambda_{max} = 1$ to $\lambda_{max} < 1$ corresponding to point $\vec{q}_3$. Notice, however, that point $\vec{q}_2$ corresponding to λ_{min} is closer to $\vec{v}_2$ than point $\vec{q}_3$ corresponding to λ_{max}. That is, $\lambda_{min} > \lambda_{max}$ which is clearly degenerate. In summary, a compound exclusion can be detected whenever $\lambda_{min} > \lambda_{max}$. Feature b_1 is then replaced by b_{max} if

186   4. Rigid-Body Systems

the last update that caused $\lambda_{min} > \lambda_{max}$ was triggered by a reduction of λ_{max}. On the other hand, feature b_1 is replaced by b_{min} if the last update that caused $\lambda_{min} > \lambda_{max}$ was triggered by an increase of λ_{min}.

The way feature b_1 is updated depends on b_2 lying on the exterior side of a vertex-edge or a face-edge plane bounding the Voronoi region of b_1. This in turn depends on b_1 being a vertex, an edge or a face. In the following paragraphs we shall examine all possible combinations and show how b_1 should be updated on each of them.

Case 2.1: Feature b_1 is a vertex and b_2 is a simple exclusion.

In this case, the Voronoi region of b_1 is bounded by vertex-edge planes only, and edge b_2 lies on the outside region of one of these planes, say plane $\pi(b_1, e_i)$. Vertex b_1 should then be substituted for the feature associated with $\pi(b_1, e_i)$, that is, edge e_i.

For example, consider the simple-exclusion case shown in Figure 4.20, with vertex b_1 having three incident edges.

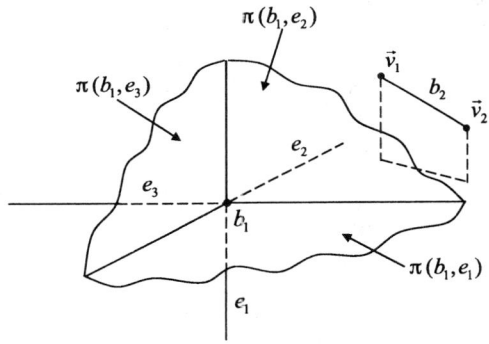

FIGURE 4.20. Simple exclusion with b_1 being a vertex with three incident edges, that is, its Voronoi region is bounded by three vertex-edge planes.

Evaluating the signed distance of the vertices defining edge b_2 to each of the bounding planes gives

$$\begin{aligned}
S_{\vec{v}_1, \pi(b_1, e_1)} &> 0 \\
S_{\vec{v}_2, \pi(b_1, e_1)} &> 0 \\
S_{\vec{v}_1, \pi(b_1, e_2)} &< 0 \\
S_{\vec{v}_2, \pi(b_1, e_2)} &< 0 \\
S_{\vec{v}_1, \pi(b_1, e_3)} &> 0 \\
S_{\vec{v}_2, \pi(b_1, e_3)} &> 0,
\end{aligned}$$

indicating that edge b_2 is completely on the exterior side of plane $\pi(b_1, e_2)$. So, we substitute b_1 for edge e_2, which is the feature associated with $\pi(b_1, e_2)$.

Case 2.2: Feature b_1 is a vertex and b_2 is a compound exclusion.

Again, the Voronoi region of b_1 is bounded by vertex-edge planes only, but edge b_2 spans the exterior side of two of these planes, say planes $\pi(b_1, e_i)$ and $\pi(b_1, e_j)$. We have that

$$0 < \lambda_{max} < \lambda_{min} < 1,$$

indicating that this is a compound exclusion. Assume $\lambda_i = \lambda_{min}$ and $\lambda_j = \lambda_{max}$. In other words, the intersection of edge b_2 with planes $\pi(b_1, e_i)$ and $\pi(b_1, e_j)$ corresponds to points $\vec{q}_i$ and $\vec{q}_j$ associated with λ_{min} and λ_{max}, respectively. Vertex b_1 is then updated as follows.

1. If the minimum lies in the interval $[0, \lambda_{max})$, that is, if

$$\dot{D}_{b_1,b_2}(\lambda_{max}) > 0$$
$$\dot{D}_{b_1,b_2}(\lambda_{min}) > 0,$$

   then we substitute b_1 for the feature associated with plane $\pi(b_1, e_i)$, namely edge e_i.

2. If the minimum lies in the interval $(\lambda_{min}, 1]$, that is, if

$$\dot{D}_{b_1,b_2}(\lambda_{max}) < 0$$
$$\dot{D}_{b_1,b_2}(\lambda_{min}) < 0,$$

   then we substitute b_1 for the feature associated with plane $\pi(b_1, e_j)$, namely edge e_j.

3. If those tests fail, then the minimum lies in the interval $[\lambda_{max}, \lambda_{min}]$ and we substitute b_1 for e_i if the last update was reducing λ_{max}, or e_j if the last update was increasing λ_{min}.

Figure 4.19 shows an example of a compound exclusion when b_1 is a vertex and b_2 is an edge.

Case 2.3: Feature b_1 is an edge and b_2 is a simple exclusion.

As mentioned before, the Voronoi region of an edge is bounded by two vertex-edge and two face-edge planes. In both cases, b_1 (i.e., the edge) is substituted for the feature of the plane excluding b_2. More specifically, if

188   4. Rigid-Body Systems

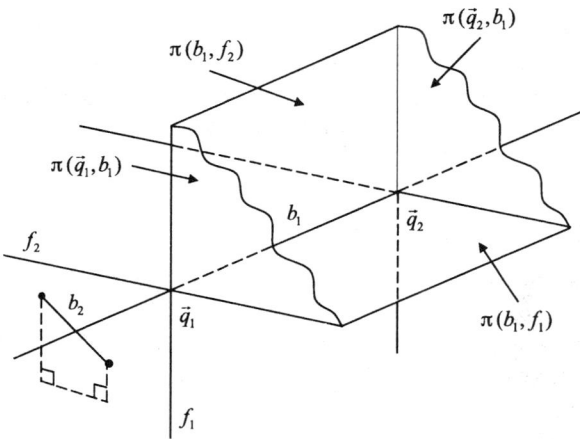

FIGURE 4.21. Edge b_2 is simply excluded from vertex-edge plane $\pi(\vec{q}_1, e_1)$ bounding the Voronoi region of edge b_1. In this case, b_1 is substituted for vertex $\vec{q}_1$.

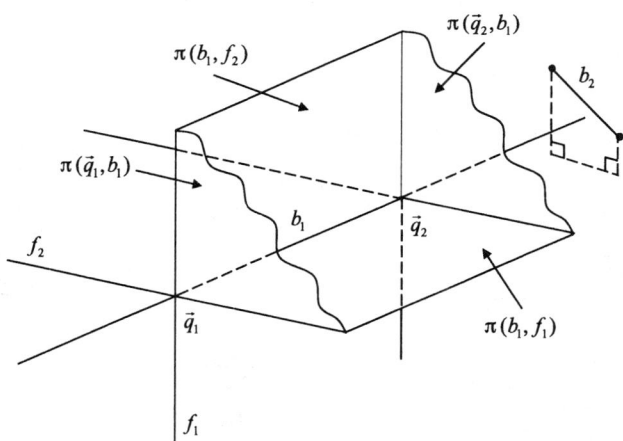

FIGURE 4.22. Edge b_2 is simply excluded from vertex-edge plane $\pi(\vec{q}_2, e_1)$ bounding the Voronoi region of edge b_1. In this case, b_1 is substituted for vertex $\vec{q}_2$.

b_2 is simply excluded from either vertex-edge planes, then b_1 is replaced by the vertex associated with the plane (see Figures 4.21 and 4.22).

Otherwise, b_2 is simply excluded from either face-edge planes, and b_1 is substituted for the face associated with the plane (see Figures 4.23 and 4.24).

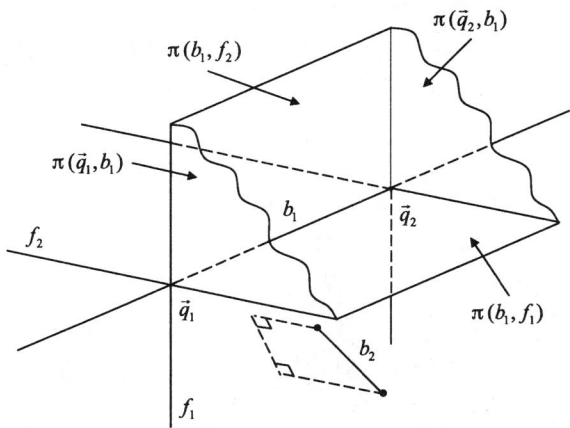

FIGURE 4.23. Edge b_2 is simply excluded from face-edge plane $\pi(b_1, f_1)$ bounding the Voronoi region of edge b_1. In this case, b_1 is substituted for face f_1.

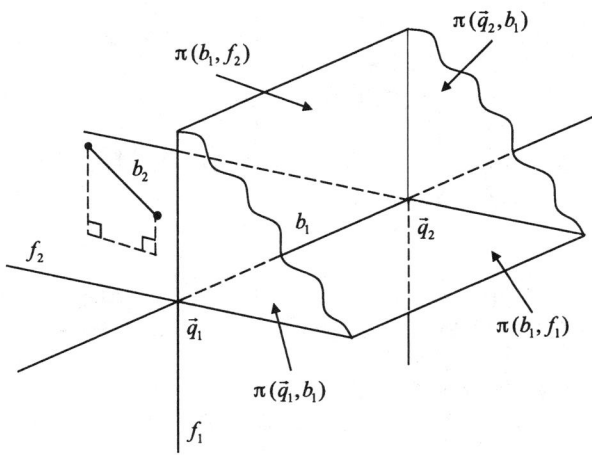

FIGURE 4.24. Edge b_2 is simply excluded from face-edge plane $\pi(b_1, f_2)$ bounding Voronoi region of edge b_1. In this case, b_1 is substituted for face f_2.

Case 2.4: Feature b_1 is an edge and b_2 is a compound exclusion.

Compound exclusions can occur with one vertex-edge and one face-edge plane, or the two face-edge planes. The compound exclusion cannot occur

190   4. Rigid-Body Systems

for the two vertex-edge planes because they are parallel. If the compound exclusion occurs with a vertex-edge and a face-edge plane, then edge b_1 is substituted for the vertex of the vertex-edge plane if the sign of the derivative at the vertex is positive, that is[9], if

$$\text{sign}\,(\dot{D}_{\vec{q}_2,b_2}(\lambda)) = \text{sign}\,(\vec{u}_e \cdot (\vec{e}(\lambda) - \vec{q}_2)) > 0 , \quad (4.33)$$

where $\vec{u}_e$ is the unitary vector defining the direction of edge b_1 (see Figure 4.25). Otherwise, b_1 is replaced by the face associated with the face-edge plane.

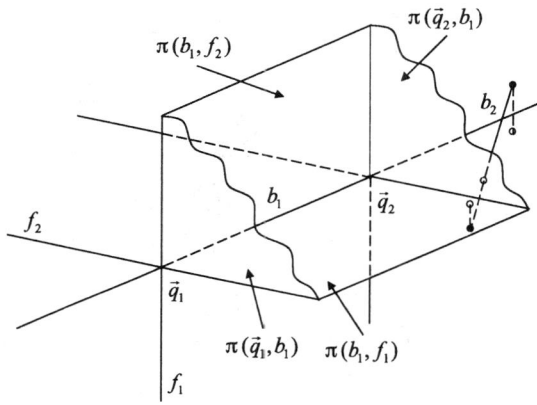

FIGURE 4.25. Example of a compound exclusion with a vertex-edge (i.e., π_2) and a face-edge (i.e., π_3) plane. Edge b_1 is substituted for vertex $\vec{q}_2$ if equation (4.28) is satisfied. Otherwise, it is substituted for face f_3.

However, if the compound exclusion occurs with two face-edge planes, say planes $\pi(b_1, f_i)$ and $\pi(b_1, f_j)$, then edge b_1 is replaced by the face of the face-edge plane as follows.

- If b_2 intersects plane $\pi(b_1, f_i)$ at point $\vec{q}_{min}$, then b_1 is substituted for face f_i associated with $\pi(b_1, f_i)$ if the sign of its derivative is negative.

- If b_2 intersects plane $\pi(b_1, f_i)$ at point $\vec{q}_{max}$, then b_1 is substituted for face f_i associated with $\pi(b_1, f_i)$ if the sign of its derivative is positive.

- Otherwise, b_1 is substituted for face f_j associated with plane $\pi(b_1, f_j)$.

The sign of the derivative at the faces is obtained by substituting f in equation (4.30) for the face being considered. Figure 4.26 shows an example of a compound exclusion with two face-edge planes.

---

[9] Equation (4.33) is the same as equation (4.29), and is repeated here for convenience.

4.7 The Voronoi Clip Algorithm   191

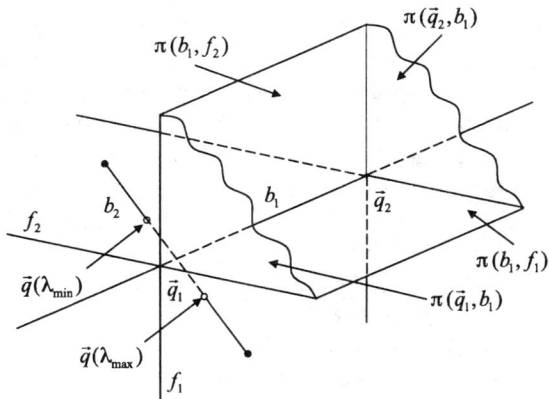

FIGURE 4.26. Example of a compound exclusion with two face-edge planes, namely planes π_3 and π_4. Edge b_1 is substituted for face f_3 if the sign of its derivative, evaluated using equation (4.30), is positive. Otherwise, b_1 is substituted for face f_4.

Case 2.5: Feature b_1 is a face and b_2 is either a simple or compound exclusion.

This is the most time-consuming update that can occur on the Voronoi Clip algorithm. The problem is that, even if the edge is reported to be a case of simple or compound exclusion, it can still span the exterior side of several face-edge planes bounding the Voronoi region of face b_1, as shown in Figure 4.27.

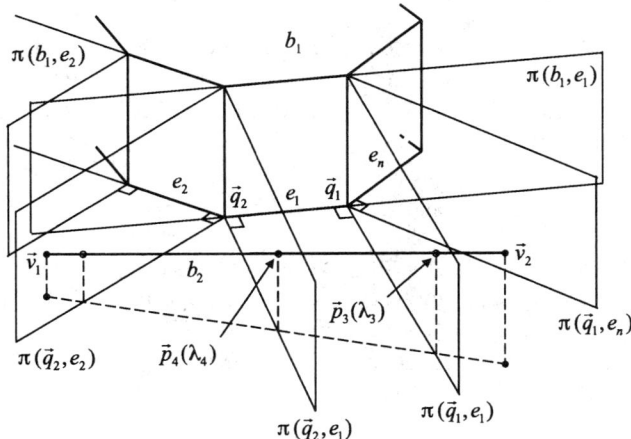

FIGURE 4.27. Edge b_2 is simply excluded from face-edge plane $\pi(e_1, b_1)$ associated with edge e_1. Nonetheless, it spans the vertex-edge planes $\pi(\vec{q}_1, e_1)$, $\pi(\vec{q}_2, e_1)$, $\pi(\vec{q}_2, e_2)$, as well as the face-edge plane $\pi(e_2, b_1)$ (not shown in the Figure).

In order to overcome this drawback, we need to clip edge b_2 against the vertex-edge planes bounding face b_1. For each vertex-edge plane considered, we check whether edge b_2 intersects the bounding plane, or lies on its interior or exterior sides. If b_2 intersects the plane, then we compute the sign of the derivative at the intersection point using equation (4.29), and use it to update the current vertex-edge plane by the neighboring vertex-edge plane associated with the interval in which the minimum point lies. On the other hand, if b_2 lies on the interior or exterior side of the plane, then we still need to check whether it lies, respectively, on the exterior or interior side of the corresponding neighboring vertex-edge plane. Depending on the result, face b_1 is updated to one of its vertices or edges.

Consider, for example, the situation shown in Figure 4.27. We start clipping edge b_2 against a randomly selected vertex-edge plane, say plane $\pi(\vec{q}_1, e_n)$ associated with vertex $\vec{q}_1$. Testing b_2 against $\pi(\vec{q}_1, e_n)$ gives

$$S_{\vec{v}_1, \pi(\vec{q}_1, e_n)} < 0$$
$$S_{\vec{v}_2, \pi(\vec{q}_1, e_n)} < 0,$$

indicating that edge b_2 lies completely on the interior side of $\pi(\vec{q}_1, e_n)$. Therefore, we consider clipping b_2 against vertex-edge plane $\pi(\vec{q}_1, e_1)$, which is the neighbor plane on the interior side of $\pi(\vec{q}_1, e_n)$. Clipping b_2 against $\pi(\vec{q}_1, e_1)$ make us find the intersection point $\vec{p}_3(\lambda_3)$ between plane $\pi(\vec{q}_1, e_1)$ and edge b_2. So, we need to compute the sign of the derivative to determine on which interval $[\vec{v}_1, \vec{p}_3]$ (i.e., $[0, \lambda_3]$) or $[\vec{p}_3, \vec{v}_2]$ (i.e., $[\lambda_3, 1]$) the minimum point lies. If the minimum point lies on the interval $[\vec{p}_3, \vec{v}_2]$, that is, if

$$\operatorname{sign} \dot{D}_{\vec{q}_1, b_2}(\lambda_3) = \operatorname{sign}(\vec{u}_{b_2} \cdot (\vec{p}_3 - \vec{q}_1)) < 0,$$

then the closest point on b_2 to b_1 lies somewhere between the bounding planes $\pi(\vec{q}_1, e_n)$ and $\pi(\vec{q}_1, e_1)$. In this case, we substitute face b_1 for vertex $\vec{q}_1$ (see Figure 4.27). Otherwise, the minimum point lies on the interval $[\vec{v}_1, \vec{p}_3]$, that is

$$\operatorname{sign} \dot{D}_{\vec{q}_1, b_2}(\lambda_3) > 0,$$

and we proceed clipping sub-edge $[\vec{v}_1, \vec{p}_3]$ against vertex-plane $\pi(\vec{q}_2, e_1)$. This, in turn, lets us determine the intersection point $\vec{p}_4(\lambda_4)$ between the sub-edge and plane $\pi(\vec{q}_2, e_1)$. Again, we need to check the sign of the derivative to determine on which interval $[\vec{v}_1, \vec{p}_4]$ or $[\vec{p}_4, \vec{p}_3]$ the minimum lies. If it lies on the interval $[\vec{p}_4, \vec{p}_3]$, that is

$$\operatorname{sign} \dot{D}_{\vec{q}_2, b_2}(\lambda_3) = \operatorname{sign}(\vec{u}_{b_2} \cdot (\vec{p}_4 - \vec{q}_2)) < 0,$$

then the closest point on b_2 to b_1 lies somewhere between the bounding planes $\pi(\vec{q}_2, e_1)$ and $\pi(\vec{q}_1, e_1)$. Therefore, we substitute face b_1 for edge e_1

(see Figure 4.27). However, if the minimum lies on the interval $[\vec{v}_1, \vec{p}_4]$, that is

$$\text{sign } \ddot{D}_{\vec{q}_2, b_2}(\lambda_3) = \text{sign } (\vec{u}_{b_2} \cdot (\vec{p}_4 - \vec{q}_2) > 0 \; ,$$

then we need to proceed by clipping sub-edge $[\vec{v}_1, \vec{p}_4]$ against vertex-edge plane $\pi(\vec{q}_2, e_2)$. This is analogous to the previously explained situation where we clipped edge b_2 against vertex-edge plane $\pi(\vec{q}_1, e_1)$. So, we continue clipping for as long as needed until we find the interval on b_2 that contains the point closest to b_1, and substitute b_1 for the closest feature associated with this interval.

4.7.3 Feature b_2 is a Face

The way the Voronoi Clip algorithm is set up, it can never compare a face with another face, with the exception of the initial choice of features done at the time the algorithm starts. So, in practice, we can not really select *any* pair b_1 and b_2 of features to start with, but a pair that contains at most one face of either rigid body. Consequently, if the initial selection for b_2 is a face, then b_1 should be either a vertex or an edge. The results of the previous sections are then directly applicable to this case (we just need to replace b_1 for b_2 and vice-versa in the derivations of the equations obtained so far).

4.7.4 Dealing with Interpenetration

Interpenetration of the bodies being tested for collision occurs only when b_1 is a face and b_2 is an edge. Recall that the Voronoi region in this case is bounded by a set of face-edge planes, one associated with each edge, and the face itself. So far, we have shown how to clip edge b_2 against the face-edge planes of b_1. Interpenetration can be detected only when we clip edge b_2 against the face plane itself. If the clip results on the determination of an intersection point between b_2 and the face (i.e., b_1), then the rigid bodies are intersecting and the Voronoi Clip algorithm reports features $b_1 \in B_1$ and $b_2 \in B_2$ as the violating features (see Figure 4.28).

4.7.5 Avoiding Local Minima

Up till now, we have shown how the Voronoi Clip algorithm updates feature b_1 according to the relative position of feature b_2 with respect to the bounding planes of b_1's Voronoi region. Unfortunately, there is a situation in which the algorithm can find itself trapped in a local minima. This occurs whenever:

- b_1 is a face, and

194   4. Rigid-Body Systems

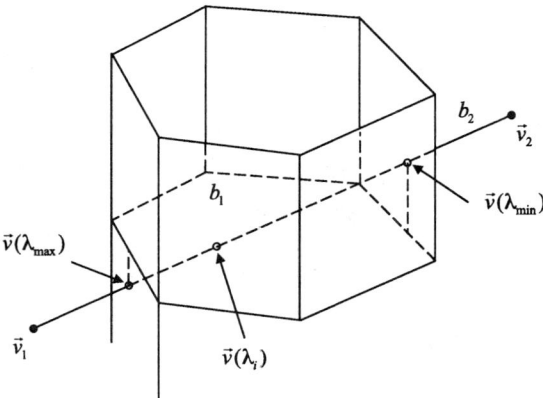

FIGURE 4.28. Edge b_2 was clipped against the face-edge planes bounding the Voronoi region of face b_1, resulting in points $\vec{v}(\lambda_{min})$ and $\vec{v}(\lambda_{max})$. Before the algorithm substitutes b_1 for b_{min} or b_{max}, it checks whether b_2 intersects b_1. In the case shown, an intersection point $\vec{v}(\lambda_i)$ was found and the bodies are reported as interpenetrating.

- b_2 is a vertex inside the Voronoi region of b_1, and
- b_2 lies "below" b_1, and
- all edges connected to b_2 are directed away from b_1.

Here, the algorithm, if implemented as is, reports vertex b_2 and face b_1 as the closest features between the rigid bodies. However, this is not always true, since the bodies can be interpenetrating, or there may exist another face $b_3 \in B_1$ closer to vertex b_2 than face b_1 (see Figure 4.29).

In order to avoid such local minima, we have to modify the Voronoi Clip algorithm with the following extra steps, applicable whenever b_1 is a face, b_2 is a vertex, and b_2 is in the Voronoi region of b_1:

1. Check whether vertex b_2 lies on the interior side of b_1, that is, if b_2 lies "below" b_1. If so, then move on to the next step. Otherwise, this is not a local minimum situation, and we are done.

2. For each edge e_i connected to vertex b_2, check whether it is directed away from b_1, that is, check whether

$$(\vec{v}_i - b_2) \cdot \vec{n}_{b_1} - d_{b_1} > 0 , \qquad (4.34)$$

where $\vec{v}_i$ is the other vertex of edge e_i, and $\vec{n}_{b_1}$ and d_{b_1} are the normal vector and plane constant of face b_1.

If both of the above conditions are true, then we are dealing with a potential local minimum. Therefore, we need to test whether vertex b_2 is

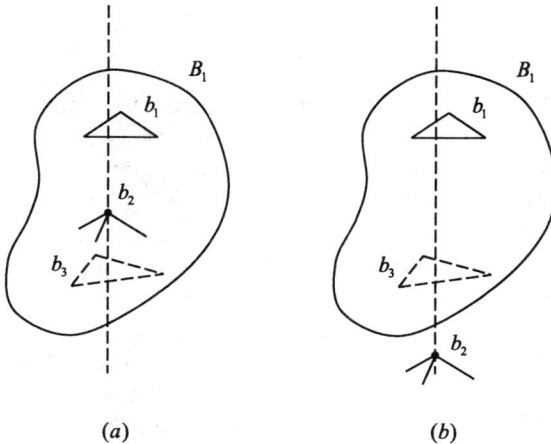

FIGURE 4.29. Examples of possible local minima. The algorithm reports vertex b_2 and face $b_1 \in B_1$ as closest features: (a) The bodies are actually interpenetrating; (b) There exists another face b_3 closer to vertex b_2 than face b_1.

inside rigid body B_1 by testing whether it is on the interior side of all B_1's faces. If so, then the bodies are interpenetrating and the algorithm terminates. Otherwise, the bodies are not intersecting, hence there exists a face $b_3 \neq b_1$ that is closer to vertex b_2 than b_1 is, and we substitute b_1 for b_3. Notice that b_3 can be obtained as a by-product of the inclusion test provided we also keep track of the face that has the smallest value obtained using equation (4.34).

Clearly, this is a computationally intense procedure, especially when we carry out the inclusion test of vertex b_2 with respect to rigid body B_1. Fortunately, the appearance of local minima is rare in practice, leading us to the conclusion that the average running time for the algorithm is rarely affected by this computation.

4.7.6 The GJK Algorithm

The Gilbert-Johnson-Keerthi (GJK) algorithm is used to compute the separation distance between two convex bodies, as well as a lower bound to their interpenetration distance if the bodies are intersecting. The general idea is to randomly select a point inside each convex body and use their distance as an initial value to the distance between the bodies. This distance is then iteratively refined as new pairs of points are found that are closer to each other than the ones already selected. The algorithm continues until there are no more pairs of points closer to each other than the current pair being considered, or a zero distance is found indicating that the bodies are intersecting. The rigid-body features that contain the closest pair of points are then reported back as the closest features between the bodies.

The distance between bodies B_1 and B_2 is defined as

$$d_{B_1,B_2} = \min |b_1 - b_2| \text{ with } b_1 \in B_1, b_2 \in B_2 \,. \tag{4.35}$$

Let $b_1 = b_1^*$ and $b_2 = b_2^*$ be the values associated with the minimum distance given by equation (4.35). In other words, b_1^* and b_2^* are the closest pairs of points between the bodies. The lowest dimensional features containing them are said to be the closest features between the bodies. For instance, if b_1^* lies on an edge of body B_1, then we report the edge as the closest feature associated with B_1, as opposed to reporting one of the faces containing the edge.

Since computing the distance for every possible combination of pairs of points is clearly impractical, the GJK algorithm computes successive approximations to points b_1^* and b_2^*. These approximated points are iteratively refined until their distance differs from a lower bound to the actual distance between the bodies (i.e., lower bound to $|b_1^* - b_2^*|$) by less than a tolerance value.

The mathematical foundation of the approximation algorithm consists of rewriting the distance between the bodies as their Minkowski difference Ψ, defined as

$$\Psi_{B_1,B_2} = \{(b_1 - b_2) \,|\, b_1 \in B_1 \text{ and } b_2 \in B_2\} \,.$$

If B_1 and B_2 are convex bodies, it can be shown that their Minkowski difference is also convex. The distance equation (4.35) can then be expressed as

$$d_{B_1,B_2} = \min |\Psi_{B_1,B_2}| \,. \tag{4.36}$$

From equation (4.36), we immediately have that the distance between the bodies is given by the point in the Minkowski difference closest to the origin. In other words, the GJK algorithm transforms the problem of computing the distance between two convex bodies into the problem of finding the point of the Minkowski difference nearest to the origin. Moreover, if the bodies are intersecting, then there exists a point b_I in both bodies such that $b_1 = b_I = b_2$, and the point of the Minkowski difference nearest to the origin is the origin itself.

The main idea behind the approximation algorithm is to construct a sequence of simplexes whose vertices are points of the Minkowski difference, such that at each iteration the current simplex is closer to the origin than any previously computed simplex.

At the first iteration, the initial simplex Q_i is set to empty, reflecting that no points of the Minkowski difference have been selected so far. We then pick an arbitrary point $\vec{p}_i \in \Psi_{B_1,B_2}$ and use it to compute an auxiliary point $\vec{q}_i \in \Psi_{B_1,B_2}$ such that:

$$\vec{q}_i = s_{\Psi_{B_1,B_2}}(-\vec{p}_i) \,, \tag{4.37}$$

4.7 The Voronoi Clip Algorithm

where $s_{\Psi_{B_1,B_2}}(-\vec{p}_i)$ is known as the support mapping of Ψ_{B_1,B_2} with respect to point $(-\vec{p}_i)$. It can be shown that the support mapping of the Minkowski difference can be computed as a function of the support mapping of each individual body as

$$s_{\Psi_{B_1,B_2}}(-\vec{p}_i) = s_{B_1}(-\vec{p}_i) - s_{B_2}(\vec{p}_i), \qquad (4.38)$$

where $s_B(\vec{p})$, with $B \in \{B_1, B_2\}$ and $\vec{p} = \pm \vec{p}_i$, is defined as

$$\begin{aligned} s_B(\vec{p}) &\in B, \\ \vec{p} \cdot s_B(\vec{p}) &= \max\{\vec{p} \cdot \vec{x} \mid \vec{x} \in B\}. \end{aligned} \qquad (4.39)$$

According to equations (4.39), the support mapping $s_B(\vec{p})$ with respect to point $\vec{p}$ is the point on B that has maximum projection along the direction defined by $\vec{p}$. In other words, the support mapping takes a point $\vec{p}$ in B and maps it to another point $s_B(\vec{p})$ also in B, such that its component along $\vec{p}$ is maximum for all points in B (see Figure 4.30).

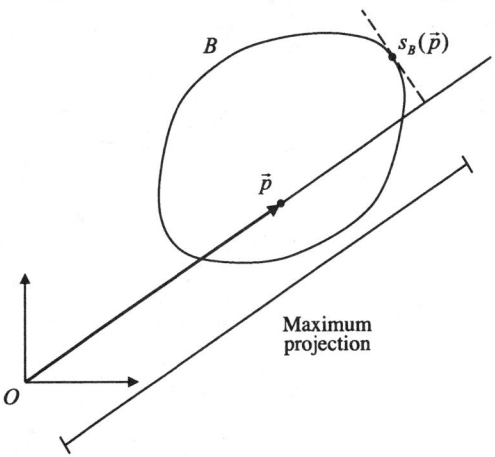

FIGURE 4.30. The support mapping $s_B(\vec{p})$ is the point on B with the maximum projection along $\vec{p}$.

There are three important issues that need to be mentioned before we can proceed with the explanation of the GJK algorithm. First, equation (4.38) says that there is no need to compute the Minkowski difference between the bodies. The support-mapping computations can be done independently for each body and then merged according to equation (4.38). Second, since the support mapping is the maximum projection along $\vec{p}$, it can only be a point on the boundary of the rigid body's shape. Therefore, we can limit our search for the support mapping to the vertices defining the rigid body's boundary, as opposed to looking for candidate points inside that body. Third, because the support mapping is on the rigid body's boundary, the initial point $\vec{p}_i$ in the Minkowski difference can be selected as the difference between any vertex of B_1 and any vertex of B_2. In other words, the determination of the initial point does not require the explicit computation of the Minkowski difference between the bodies.

Having determined $\vec{p}_i$ and Q_i, as well as the auxiliary point $\vec{q}_i$ at iteration i, we move on to compute $\vec{p}_{i+1}$ and Q_{i+1} associated with iteration $(i+1)$, such that

$$\vec{p}_{i+1} = \min\{convex(Q_i \cup \{\vec{q}_i\})\},$$

$$Q_{i+1} \subseteq (Q_i \cup \{\vec{q}_i\});$$

(4.40)

Q_{i+1} is chosen as the smallest nonempty set with $\vec{p}_{i+1} \in convex(Q_{i+1})$. In other words, the simplex Q_{i+1} corresponding to iteration $(i+1)$ is the convex hull of $(Q_i \cup \{\vec{q}_i\})$, with the added constraint that $\vec{p}_{i+1}$ must be one of its vertices.

The operation $convex(X)$ that appears in equation (4.40) is known as the *convex combination* of a finite set of points $\vec{x}_j$ of polyhedron X, and is defined as

$$convex(X) = \sum_{j=1}^{n} \lambda_j \vec{x}_j,$$

with

$$\sum_{j=1}^{n} \lambda_j = 1$$

$$\lambda_j \geq 0, \forall j \in \{1, 2, \ldots, n\}.$$

Another useful operation is the *affine combination* affine(X) of a finite set of points $\vec{x}_j$ of polyhedron X, expressed as

$$\text{affine}(X) = \sum_{j=1}^{n} \lambda_j \vec{x}_j,$$

with the restriction

$$\sum_{j=1}^{n} \lambda_j = 1.$$

By definition, the vertices of the simplex Q_{i+1} form a set of affinely independent points, that is, none of the points in Q_{i+1} can be written as an affine combination of the other vertices. Moreover, it can be shown that there exists only one Q_{i+1} satisfying equation (4.40). Figure 4.31 illustrates a few steps of the GJK algorithm, with polyhedron Ψ_{B_1,B_2} being the Minkowski difference of two hypothetical convex bodies (not shown in the figure).

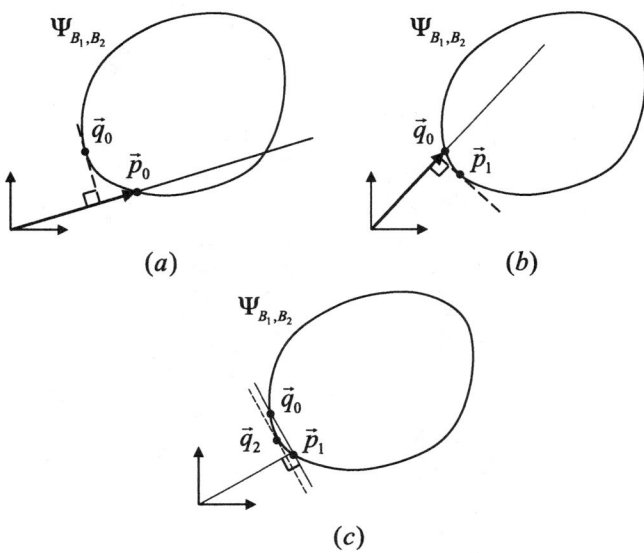

FIGURE 4.31. Visualizing a few steps of the GJK algorithm being executed.

On Figure 4.31(a), we start with an arbitrary point $\vec{p}_0$ that is the subtraction of a vertex in B_2 from a vertex in B_1, that is, a point on the boundary of the Minkowski difference, since both bodies are convex. The initial simplex Q_0 is set to empty. The next step is then to compute the auxiliar point $\vec{q}_0$ using the support mapping $s_{\Psi_{B_1,B_2}}(-\vec{p}_0)$ (see equation (4.37)). This is done by computing the support mapping for each individual body and merging the results using equation (4.38). Figure 4.31(a) shows the equivalent of this computation if we had to compute the convex polygon representing the Minkowski difference of the bodies. Point $\vec{q}_0$ is chosen such that it is the point on the Minkowski difference closest to the origin. It happens that computing $\vec{q}_0$ is equivalent to finding the point on the Minkowski difference that has minimum projection along $\vec{p}_0$. Having determined $\vec{p}_0$, $\vec{q}_0$ and Q_0, we use equation (4.40) to continue to the next step. In this case, $\vec{p}_1 = \vec{q}_0$ and $Q_1 = \{\vec{q}_0\}$. Figure 4.31(b) and (c) shows the procedure just explained applied to the next two iterations of the GJK algorithm.

The only remaining steps that need clarification are how to compute $\vec{p}_{i+1} = \min\{convex(Q_i \cup \{\vec{q}_i\})\}$, and how to construct the simplex $Q_{i+1} \subseteq (Q_i \cup \{\vec{q}_i\})$ such that $\vec{p}_{i+1} \in convex(Q_{i+1})$. Here, we shall focus on the results of how $\vec{p}_{i+1}$ and Q_{i+1} can be obtained. The interested reader is referred to Section 4.10 for references to the literature that presents a full derivation and proof of the results here described.

We shall first concentrate on how to determine the simplex Q_{i+1}. Later, we shall see that the point $\vec{p}_{i+1}$ can be directly obtained from the Q_{i+1} computations.

Let's assume the set $(Q_i \cup \{\vec{q}_i\})$ is represented by

$$(Q_i \cup \{\vec{q}_i\}) = \{\vec{x}_1, \vec{x}_2, \ldots, \vec{x}_n\} .$$

Clearly from equation (4.40), Q_{i+1} is restricted to being a subset of $(Q_i \cup \{\vec{q}_i\})$ at each iteration. We want to determine Q_{i+1} such that it is the smallest non-empty subset of $(Q_i \cup \{\vec{q}_i\})$ with $\vec{p}_{i+1} \in convex(Q_{i+1})$. Let $I_s \subseteq \{1, 2, \ldots, n\}$ be the set of indexes corresponding to the vertices of Q_{i+1}.

In general, any point $\vec{p}$ inside the simplex Q_{i+1} can be described as an affine combination of its vertices $\vec{x}_j$, that is

$$\vec{p} = \sum_{j \in I_s} \lambda_j \vec{x}_j ,$$

with

$$\sum_{j \in I_s} \lambda_j = 1 .$$

Since we are dealing with three-dimensional spaces, the maximum number of vertices a 3D simplex can have is four. That is, the maximum number of vertices in I_s is limited to four, corresponding to Q_{i+1} being a tetrahedron. So, if Q_i already has four vertices, then $(Q_i \cup \{\vec{q}_i\})$ will have five vertices, indicating that one of them can be written as an affine combination of the others. The simplex Q_{i+1} will then be characterized by the affinely independent vertices $\vec{x}_j$ of $(Q_i \cup \{\vec{q}_i\})$, that is, by the vertices $\vec{x}_j$ with $\lambda_j > 0$.

Putting it into equations, we have that Q_{i+1} is defined by the vertices $\vec{x}_j$ satisfying

$$\lambda_j > 0, \forall j \in I_s ,$$
$$\lambda_j \leq 0, \forall j \notin I_s .$$

In other words, the simplex Q_{i+1} is the convex hull of the affinely independent vertices of $(Q_i \cup \{\vec{q}_i\})$, and point $\vec{p}_{i+1}$ is the point in Q_{i+1} closest to the origin, given by

$$\vec{p}_{i+1} = \sum_{j \in I_s} \lambda_j \vec{x}_j , \qquad (4.41)$$

4.7 The Voronoi Clip Algorithm

with

$$\sum_{j \in I_s} \lambda_j = 1. \qquad (4.42)$$

Assume I_s has $r \leq n$ vertices, and let $\vec{x}_1, \vec{x}_2, \ldots, \vec{x}_r$ represent an arbitrary ordering of the vertices of I_s. In this context, equation (4.42) can be rewritten as

$$\lambda_1 = 1 - \sum_{j=2}^{r} \lambda_j.$$

Since we want $\vec{p}_{i+1}$ to be the point closest to the origin, the $\lambda_2, \ldots, \lambda_r$ are computed from the unconstrained minimization of

$$F(\lambda_2, \ldots, \lambda_r) = |\vec{x}_1 + \sum_{j=2}^{r} \lambda_j (\vec{x}_j - \vec{x}_1)|.$$

Since $F(\lambda_2, \ldots, \lambda_r)$ is a convex function, it is minimized whenever

$$\frac{\partial F(\lambda_2, \ldots, \lambda_r)}{\partial \lambda_j} = 0, \qquad (4.43)$$

for $j \in \{2, \ldots, r\}$. Equation (4.43) can then be written in matrix format

$$\mathbf{A_r}\,\vec{\lambda} = \vec{b},$$

with $\mathbf{A_r} \in \mathbb{R}^{r \times r}$ and $\vec{b} \in \mathbb{R}^r$, such that:

$$\mathbf{A_r} = \begin{pmatrix} 1 & \cdots & 1 \\ (\vec{x}_2 - \vec{x}_1)\cdot\vec{x}_1 & \cdots & (\vec{x}_2 - \vec{x}_1)\cdot\vec{x}_r \\ \vdots & \vdots & \vdots \\ (\vec{x}_r - \vec{x}_1)\cdot\vec{x}_1 & \cdots & (\vec{x}_r - \vec{x}_1)\cdot\vec{x}_r \end{pmatrix} \quad \vec{b} = \begin{pmatrix} 1 \\ 0 \\ \vdots \\ 0 \end{pmatrix}.$$

Using Cramer's rule, we can compute each λ_j as

$$\lambda_j = \frac{\triangle_j(Q_{i+1})}{\triangle(Q_{i+1})},$$

where

$$\begin{aligned}
\triangle_j(\{\vec{x}_j\}) &= 1 \\
\triangle_m(Q_{i+1} \cup \{\vec{x}_m\}) &= \sum_{j \in I_s} \triangle_j(Q_{i+1})(\vec{x}_j \cdot \vec{x}_k - \vec{x}_j \cdot \vec{x}_m) \quad (4.44) \\
\triangle(Q_{i+1}) &= \sum_{j \in I_s} \triangle_j(Q_{i+1}),
\end{aligned}$$

with $m \notin I_s$ and k being the minimum index in I_s. It happends that the smallest $Q_{i+1} \subset (Q_i \cup \{\vec{q_i}\})$ is such that

$$\begin{aligned} \Delta(Q_{i+1}) &> 0 \\ \Delta_j(Q_{i+1}) &> 0, \forall j \in I_s \\ \Delta_m(Q_{i+1} \cup \{\vec{x}_m\}) &\leq 0, \forall m \notin I_s . \end{aligned} \quad (4.45)$$

Notice that equations (4.44) are solved for each possible instance of I_s, that is, for each possible subset of $(Q_i \cup \{\vec{q_i}\})$. We then select as the solution the subset that satisfies the constraints described in equations (4.45). It can be shown that exactly one solution subset satisfies equations (4.45).

Since the maximum number of vertices n in $(Q_i \cup \{\vec{q_i}\})$ is small (i.e., $n \leq 4$), the determination of Q_{i+1} can be done by an exhaustive search among all possible nonempty subsets of $(Q_i \cup \{\vec{q_i}\})$. This translates into searching among

$$\sum_{m=1}^{n} \frac{n!}{m!(n-m)!}$$

candidate subsets that satisfying equations (4.45).

Having determined the subset Q_{i+1} that satisfies equations (4.45), point $\vec{p}_{i+1}$ follows directly from equation (4.41), that is

$$\vec{p}_{i+1} = \sum_{j \in I_s} \lambda_j \vec{x}_j .$$

We proceed to iteration $(i+2)$ to determine Q_{i+2} and $\vec{p}_{i+2}$, and so on, until we reach the termination condition explained in Section 4.7.7. Let t be the iteration at which the termination conditions are reached. We have

$$\vec{p}_t = \sum_{j \in I_s} \lambda_j \vec{x}_j . \quad (4.46)$$

Each point $\vec{x}_j$ in the Minkowski difference of B_1 and B_2 can then be expressed as

$$\vec{x}_j = (\vec{b}_1)^j - (\vec{b}_2)^j , \quad (4.47)$$

with $(\vec{b}_1)^j \in B_1$ and $(\vec{b}_2)^j \in B_2$. Substituting equation (4.47) into (4.46) gives:

$$\begin{aligned} \vec{p}_t &= \sum_{j \in I_s} \lambda_j ((\vec{b}_1)^j - (\vec{b}_2)^j) \\ &= \sum_{j \in I_s} \lambda_j (\vec{b}_1)^j - \sum_{j \in I_s} \lambda_j (\vec{b}_2)^j \\ &= \vec{b}_1^* - \vec{b}_2^* . \end{aligned}$$

Because B_1 and B_2 are convex, $\vec{b}_1^* \in B_1$ and $\vec{b}_2^* \in B_2$ are the closest points between the bodies.

4.7.7 Termination Condition

Even though it is guaranteed that the GJK algorithm will end in a finite number of iterations whenever bodies B_1 and B_2 are convex polyhedra, the presence of numerical round-off errors in a computer implementation makes it necessary to formulate a termination condition to be checked at the end of each iteration.

The termination condition consists of checking whether point $\vec{p}_i$ obtained at iteration i lies within a tolerance value from the origin. If so, then $\vec{p}_i$ is considered close enough to the origin and the algorithm terminates. The tolerance value used is a lower bound on the module of $\vec{p}_i$ computed as the signed distance of the supporting plane $\pi_{\vec{p}_i, \vec{q}_i}$ to the origin. The supporting plane is defined by its normal vector $\vec{n}_i$ and plane constant d_i as[10]

$$\vec{n}_i = -\vec{p}_i$$
$$d_i = \vec{p}_i \cdot \vec{q}_i .$$

The signed distance of the supporting plane to the origin is then

$$d = \frac{d_i}{|\vec{n}_i|} .$$

Notice that, for $d_i > 0$, the origin lies in the positive half space of $\pi_{\vec{p}_i, \vec{q}_i}$, that is, if we substitute $\vec{x} = \vec{0}$ in the plane equation, we get

$$\vec{n}_i \cdot \vec{x} + d_i = \vec{n}_i \cdot \vec{0} + d_i = d_i > 0 ,$$

whereas the Minkowski difference always lies in the negative half space of the plane. Therefore, at iteration i we use

$$L_b = \max\{0, d_0, d_1, \ldots, d_i\}$$

as the lower bound of $|\vec{p}_i|$, and we terminate the algorithm at iteration i whenever

$$|\vec{p}_i| - L_b \leq \mu ,$$

with μ being the error-tolerance value.

---

[10] The plane in this case is defined as $\pi_{\vec{p}_i, \vec{q}_i} = \{\vec{x} : (\vec{n}_i \cdot \vec{x} + d_i) = 0\}$, as opposed to $\{\vec{x} : (\vec{n}_i \cdot \vec{x} - d_i) = 0\}$. The latter is the definition used in all other sections of this book.

4.8 Rigid Body-Rigid Body Collision Response

Whenever a rigid body-rigid body collision is detected, the collision-response module is invoked to compute the appropriate collision impulses or contact forces that will prevent interpenetration between the colliding bodies. As explained in Section 4.4, the colliding rigid bodies are traced back in time to the moment before their collision. The collision point and normal are then determined from their geometric displacement.

The colliding bodies are arbitrarily assigned indexes 1 and 2, and the normal direction is selected such that the relative velocity $(\vec{v}_1 - \vec{v}_2)$ of the rigid bodies at the collision point along the collision normal is negative just before the collision, that is, we choose $\vec{n}$ such that

$$(\vec{v}_1 + \vec{\omega}_1 \times \vec{r}_1 - \vec{v}_2 - \vec{\omega}_2 \times \vec{r}_2) \cdot \vec{n} < 0 \qquad (4.48)$$

is satisfied just before the collision, indicating that the bodies are moving towards each other (see Figure 4.32). Notice that the velocities at the collision points $\vec{p}_1$ and $\vec{p}_2$ are computed using equation (4.6), described in Section 4.2.

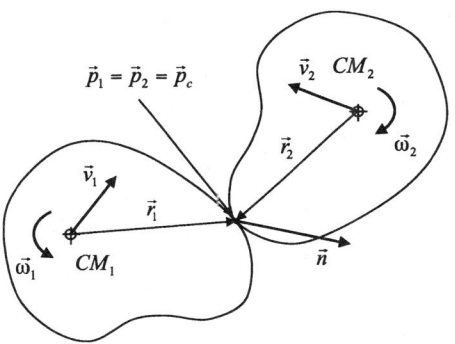

FIGURE 4.32. Rigid bodies B_1 and B_2 at the moment just before their collision. The relative velocity of the closest points $\vec{p}_1$ and $\vec{p}_2$ is used to determine the direction of the collision normal. Notice that at the exact moment of collision we have $\vec{p}_1 = \vec{p}_2 = \vec{p}_c$.

This index assignment is critical because, from Newton's principle of action and reaction, the collision impulses and contact forces between the rigid bodies have the same magnitude, but opposite directions. Following our convention, a positive impulse should be applied to the rigid body with index 1, whereas a negative impulse should be applied to the rigid body of index 2. Therefore, it is important to keep track of the index assigned to each rigid body, so as to later apply the collision impulses and contact forces on the correct direction (i.e., with the correct sign) to each body. Also, notice that, in the case of multiple rigid body-rigid body collisions, a rigid body might be assigned different indexes for each collision it is involved in.

The difference between a collision and a contact is determined from the module of the relative velocity along the collision normal, at the collision point. If the relative velocity of the rigid bodies at the collision point along the collision normal, at the moment before the collision, is less than a threshold value, then the rigid bodies are said to be in contact and a contact force is computed to prevent their interpenetration. Otherwise, the rigid bodies are said to be colliding and an impulsive force is applied to instantaneously change their direction of motion to avoid the imminent inter penetration.

Also, there may be situations in which several rigid bodies are involved in multiple collisions and contacts. If so, the collision-response module should first resolve all collisions by simultaneously computing all impulsive forces. Having determined all impulsive forces, the collision-response module proceeds by applying the impulses to the appropriate rigid bodies. By the time the impulses are applied, some of the contacts may or may not break, depending on whether the relative acceleration of the rigid bodies at their contact point, along the contact normal, is positive, zero or negative. A contact force is then simultaneously computed for all contacts that have a negative relative acceleration along their contact normal.

4.8.1 Computing Impulsive Forces for a Single Collision

Let's start by examining the case where we have one or more simultaneous collisions, each involving two different rigid bodies. Here, each collision can be dealt with separately and in parallel, since they do not have bodies in common.

Let collision C, involving rigid bodies B_1 and B_2, be defined by its collision normal $\vec{n}$ and tangent axes $\vec{t}$ and $\vec{k}$, as indicated in Figure 4.33. Let $\vec{v}_1 = ((v_1)_n, (v_1)_t, (v_1)_k)$ and $\vec{\omega}_1 = ((\omega_1)_n, (\omega_1)_t, (\omega_1)_k)$ be the linear and angular velocities of rigid body B_1 just before the collision. Analogously, let $\vec{v}_2 = ((v_2)_n, (v_2)_t, (v_2)_k)$ and $\vec{\omega}_2 = ((\omega_2)_n, (\omega_2)_t, (\omega_2)_k)$ be the linear and angular velocities of rigid body B_2 just before the collision. All these components are known quantities. We need to compute the linear and angular velocities of both bodies just after the collision, namely $\vec{V}_1 = ((V_1)_n, (V_1)_t, (V_1)_k)$, $\vec{\Omega}_1 = ((\Omega_1)_n, (\Omega_1)_t, (\Omega_1)_k)$, $\vec{V}_2 = ((V_2)_n, (V_2)_t, (V_2)_k)$ and $\vec{\Omega}_2 = ((\Omega_2)_n, (\Omega_2)_t, (\Omega_2)_k)$. These, together with the impulsive force $\vec{P} = (P_n, P_t, P_k)$, sums to a total of fifteen unknowns, thus requiring the solution of a system with fifteen equations.

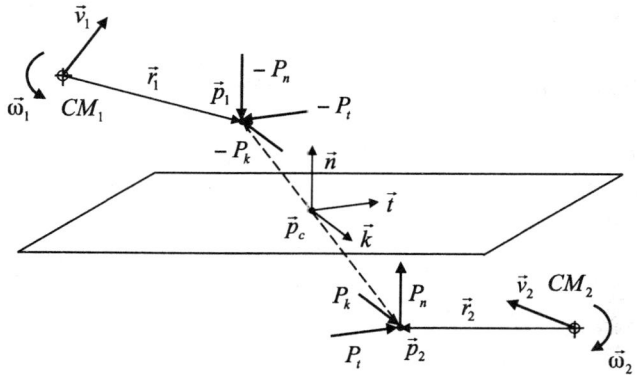

FIGURE 4.33. Single rigid-body collision, showing just the closest points $\vec{p}_1$ and $\vec{p}_2$, as well as the location of the center of mass of each colliding body, at the moment just before the collision. A negative and a positive impulse is applied to the body of indexes 2 and 1, respectively. Also, the distance between the closest points is exaggerated to facilitate drawing the tangent plane.

Applying the principle of impulse and linear momentum to each rigid body along the three axes defining the collision frame, we obtain six out of the fifteen equations needed[11]:

$$m_1(\vec{V}_1 - \vec{v}_1) = \vec{P} \qquad (4.49)$$
$$m_2(\vec{V}_2 - \vec{v}_2) = -\vec{P}. \qquad (4.50)$$

Applying the principle of impulse and angular momentum for each rigid body along the three axes defining the collision frame, we obtain another set of six equations:

$$\mathbf{I}_1(\vec{\Omega}_1 - \vec{\omega}_1) = \vec{r}_1 \times \vec{P} = \tilde{r}_1 \vec{P} \qquad (4.51)$$
$$\mathbf{I}_2(\vec{\Omega}_2 - \vec{\omega}_2) = -\vec{r}_2 \times \vec{P} = -\tilde{r}_2 \vec{P}, \qquad (4.52)$$

where $\tilde{r}_1 \vec{P}$ and $\tilde{r}_2 \vec{P}$ are the matrix-vector representation of their respective cross-products, as explained in detail in Section A.7 of Appendix A.

The next equation is obtained by the empirical relation involving the coefficient of restitution and the relative velocity of the rigid bodies at the collision point along the collision normal. Let e denote the coefficient of restitution along the normal direction. We have

---

[11] We will use the vector-based notation as much as possible to keep the equations concise. However, there are cases in which we do need to rewrite the equations using the individual components of each vector, such as when computing the critical-friction coefficient covered later in this section.

$$((\vec{V}_1 + \vec{\Omega}_1 \times \vec{r}_1) - (\vec{V}_2 + \vec{\Omega}_2 \times \vec{r}_2)) \cdot \vec{n} =$$
$$-e\left((\vec{v}_1 + \vec{\omega}_1 \times \vec{r}_1) - (\vec{v}_2 + \vec{\omega}_2 \times \vec{r}_2)\right) \cdot \vec{n} \quad ,$$

or equivalently

$$((V_1)_n + (r_1)_k (\Omega_1)_t - (r_1)_t (\Omega_1)_k) -$$
$$((V_2)_n + (r_2)_k (\Omega_2)_t - (r_2)_t (\Omega_2)_k) =$$
$$-e\,(((v_1)_n + (r_1)_k (\omega_1)_t - (r_1)_t (\omega_1)_k) -$$
$$((v_2)_n + (r_2)_k (\omega_2)_t - (r_2)_t (\omega_2)_k))\,. \qquad (4.53)$$

The remaining two equations are obtained from the Coulomb friction relations at the collision point. If the relative motion of the rigid bodies at the collision point along $\vec{t}$ and $\vec{k}$ is zero just before the collision, that is, if

$$((\vec{v}_1 + \vec{\omega}_1 \times \vec{r}_1) - (\vec{v}_2 + \vec{\omega}_2 \times \vec{r}_2)) \cdot \vec{t} = 0$$
$$((\vec{v}_1 + \vec{\omega}_1 \times \vec{r}_1) - (\vec{v}_2 + \vec{\omega}_2 \times \vec{r}_2)) \cdot \vec{k} = 0 \,,$$

then their relative motion will remain zero after the collision. More specifically, we use

$$(\vec{V}_1 + \vec{\Omega}_1 \times \vec{r}_1) \cdot \vec{t} = (\vec{V}_2 + \vec{\Omega}_2 \times \vec{r}_2) \cdot \vec{t} \qquad (4.54)$$
$$(\vec{V}_1 + \vec{\Omega}_1 \times \vec{r}_1) \cdot \vec{k} = (\vec{V}_2 + \vec{\Omega}_2 \times \vec{r}_2) \cdot \vec{k} \qquad (4.55)$$

as the two remaining equations to solve the system. However, if the relative motion is not zero, then the rigid bodies are sliding along $\vec{t}$ and $\vec{k}$ at the collision point. The collision impulse will then act on the opposite direction of motion, trying to prevent the sliding. If it succeeds, then equations (4.54) and (4.55) should be used. Otherwise, the rigid bodies continue sliding throughout the entire collision, and we use

$$P_t = (\mu_d)_t\, P_n \qquad (4.56)$$
$$P_k = (\mu_d)_k\, P_n \qquad (4.57)$$

as the two remaining equations to solve the system. Notice that $(\mu_d)_t$ and $(\mu_d)_k$ are the dynamic Coulomb friction coefficients along the $\vec{t}$ and $\vec{k}$ directions, respectively. Since P_t and P_k are always opposing the sliding motion, the coefficients of friction can be either positive or negative to reflect that condition. The actual signs of the coefficients depend on the relative velocity of the rigid bodies at the collision point along axes $\vec{t}$ and $\vec{k}$, just before the collision. The signs are directly obtained from

$$\text{sign}\left((\mu_d)_t\right) = \frac{((\vec{v}_1 + \vec{\omega}_1 \times \vec{r}_1) - (\vec{v}_2 + \vec{\omega}_2 \times \vec{r}_2)) \cdot \vec{t}}{((\vec{v}_1 + \vec{\omega}_1 \times \vec{r}_1) - (\vec{v}_2 + \vec{\omega}_2 \times \vec{r}_2)) \cdot \vec{n}} \quad (4.58)$$

$$\text{sign}\left((\mu_d)_k\right) = \frac{((\vec{v}_1 + \vec{\omega}_1 \times \vec{r}_1) - (\vec{v}_2 + \vec{\omega}_2 \times \vec{r}_2)) \cdot \vec{k}}{((\vec{v}_1 + \vec{\omega}_1 \times \vec{r}_1) - (\vec{v}_2 + \vec{\omega}_2 \times \vec{r}_2)) \cdot \vec{n}} . \quad (4.59)$$

As already explained in Chapter 3, this directional-friction model is a generalization of the widely used model of relating the tangential and normal impulses using just one omnidirectional coefficient of friction μ_d, as in

$$P_{tk} = \mu_d P_n , \quad (4.60)$$

where P_{tk} is the impulse on the tangent plane given by

$$P_{tk} = \sqrt{P_t^2 + P_k^2} .$$

For example, if friction is isotropic, that is, independent of direction, then we can write

$$(\mu_d)_t = \mu_d \cos\phi$$
$$(\mu_d)_k = \mu_d \sin\phi$$

for some angle ϕ, and so

$$\begin{aligned} P_{tk} &= \sqrt{P_t^2 + P_k^2} \\ &= \sqrt{\mu_d^2 P_n^2 \cos\phi^2 + \mu_d^2 P_n^2 \sin\phi^2} \\ &= \mu_d P_n , \end{aligned}$$

which is the same result obtained using the omnidirectional-friction model of equation (4.60). The main advantage of using this model is that the non-linear equation:

$$|P_{tk}| = \sqrt{P_t^2 + P_k^2} \le \mu_d P_n$$

that needs to be enforced when the rigid bodies are not sliding at the collision point can be substituted for two linear equations

$$|P_t| \le (\mu_d)_t P_n$$
$$|P_k| \le (\mu_d)_k P_n ,$$

4.8 Rigid Body-Rigid Body Collision Response

which are equivalent to the non-linear equation if friction is isotropic, and most important, are easier to handle in matrix form, as we shall see shortly.

As far as friction is concerned, we have to consider two possible cases. In the first, we assume the rigid bodies continue sliding along the tangent plane after collision, and we use equations (4.49) to (4.53) with equations (4.56) and (4.57) to compute the collision impulse and velocities after the collision. In the second, the rigid bodies are not sliding along the tangent plane after the collision, and we use equations (4.49) to (4.53) with equations (4.54) and (4.55) instead. For now, let's focus on the solution corresponding to the first case. Later, we shall consider the modifications needed to address the second case.

Instead of jumpstart-solving the fifteen-equation system, let's first consider its partitioned block-matrix representation and show how we can use linear-system methods to efficiently resolve the collision. Besides being extremely useful for the single-collision case, the block-matrix representation proves to be invaluable when dealing with multiple simultaneous collisions, as we shall see later in Section 4.8.2.

As mentioned, if the rigid bodies continue sliding along the tangent plane after collision, we need to use equations (4.49) to (4.53) with equations (4.56) and (4.57). Here, the fifteen-equation system can be put into the following matrix format:

$$\begin{pmatrix}
0 & 0 & 0 & 1 & 0 & 0 & 0 & (r_1)_k & -(r_1)_t \\
-(\mu_d)_t & 1 & 0 & 0 & 0 & 0 & 0 & 0 & 0 \\
-(\mu_d)_k & 1 & 0 & 0 & 0 & 0 & 0 & 0 & 0 \\
-1 & 0 & 0 & m_1 & 0 & 0 & 0 & 0 & 0 \\
0 & -1 & 0 & 0 & m_1 & 0 & 0 & 0 & 0 \\
0 & 0 & -1 & 0 & 0 & m_1 & 0 & 0 & 0 \\
0 & (r_1)_k & -(r_1)_t & 0 & 0 & 0 & (I_1)_{nn} & (I_1)_{nt} & (I_1)_{nk} \\
-(r_1)_k & 0 & (r_1)_n & 0 & 0 & 0 & (I_1)_{tn} & (I_1)_{tt} & (I_1)_{tk} \\
(r_1)_t & -(r_1)_n & 0 & 0 & 0 & 0 & (I_1)_{kn} & (I_1)_{kt} & (I_1)_{kk} \\
1 & 0 & 0 & 0 & 0 & 0 & 0 & 0 & 0 \\
0 & 1 & 0 & 0 & 0 & 0 & 0 & 0 & 0 \\
0 & 0 & 1 & 0 & 0 & 0 & 0 & 0 & 0 \\
0 & -(r_2)_k & (r_2)_t & 0 & 0 & 0 & 0 & 0 & 0 \\
(r_2)_k & 0 & -(r_2)_n & 0 & 0 & 0 & 0 & 0 & 0 \\
-(r_2)_t & (r_2)_n & 0 & 0 & 0 & 0 & 0 & 0 & 0
\end{pmatrix}$$

210   4. Rigid-Body Systems

$$\begin{pmatrix} -1 & 0 & 0 & 0 & -(r_2)_k & (r_2)_t \\ 0 & 0 & 0 & 0 & 0 & 0 \\ 0 & 0 & 0 & 0 & 0 & 0 \\ 0 & 0 & 0 & 0 & 0 & 0 \\ 0 & 0 & 0 & 0 & 0 & 0 \\ 0 & 0 & 0 & 0 & 0 & 0 \\ 0 & 0 & 0 & 0 & 0 & 0 \\ 0 & 0 & 0 & 0 & 0 & 0 \\ 0 & 0 & 0 & 0 & 0 & 0 \\ m_2 & 0 & 0 & 0 & 0 & 0 \\ 0 & m_2 & 0 & 0 & 0 & 0 \\ 0 & 0 & m_2 & 0 & 0 & 0 \\ 0 & 0 & 0 & (I_2)_{nn} & (I_2)_{nt} & (I_2)_{nk} \\ 0 & 0 & 0 & (I_2)_{tn} & (I_2)_{tt} & (I_2)_{tk} \\ 0 & 0 & 0 & (I_2)_{kn} & (I_2)_{kt} & (I_2)_{kk} \end{pmatrix} \begin{pmatrix} P_n \\ P_t \\ P_k \\ (V_1)_n \\ (V_1)_t \\ (V_1)_k \\ (\Omega_1)_n \\ (\Omega_1)_t \\ (\Omega_1)_k \\ (V_2)_n \\ (V_2)_t \\ (V_2)_k \\ (\Omega_2)_n \\ (\Omega_2)_t \\ (\Omega_2)_k \end{pmatrix} =$$

(4.61)

$$= \begin{pmatrix} -e\left((v_1)_n + (r_1)_k (\omega_1)_t - (r_1)_t (\omega_1)_k\right) - \\ \left((v_2)_n + (r_2)_k (\omega_2)_t - (r_2)_t (\omega_2)_k\right)) \\ 0 \\ 0 \\ m_1 (v_1)_n \\ m_1 (v_1)_t \\ m_1 (v_1)_k \\ (I_1)_{nn} (w_1)_n + (I_1)_{nt} (w_1)_t + (I_1)_{nk} (w_1)_k \\ (I_1)_{tn} (w_1)_n + (I_1)_{tt} (w_1)_t + (I_1)_{tk} (w_1)_k \\ (I_1)_{kn} (w_1)_n + (I_1)_{kt} (w_1)_t + (I_1)_{kk} (w_1)_k \\ m_2 (v_2)_n \\ m_2 (v_2)_t \\ m_2 (v_2)_k \\ (I_2)_{nn} (w_2)_n + (I_2)_{nt} (w_2)_t + (I_2)_{nk} (w_2)_k \\ (I_2)_{tn} (w_2)_n + (I_2)_{tt} (w_2)_t + (I_2)_{tk} (w_2)_k \\ (I_2)_{kn} (w_2)_n + (I_2)_{kt} (w_2)_t + (I_2)_{kk} (w_2)_k \end{pmatrix}$$

where the order in which the equations are laid out in each row is equation (4.53) first, then equations (4.56) and (4.57), followed by equations (4.49), (4.51), (4.50) and (4.52). This order is very important, since it simplifies the description of the above system using the partitioned block-matrix representation

$$\begin{pmatrix} \mathbf{A}_{1,2} & \mathbf{B}_{1,2} & \mathbf{C}_{1,2} & -\mathbf{B}_{1,2} & \mathbf{E}_{1,2} \\ -\mathbf{I} & m_1 \mathbf{I} & 0 & 0 & 0 \\ -\tilde{r}_1 & 0 & \mathbf{I}_1 & 0 & 0 \\ \mathbf{I} & 0 & 0 & m_2 \mathbf{I} & 0 \\ \tilde{r}_2 & 0 & 0 & 0 & \mathbf{I}_2 \end{pmatrix} \begin{pmatrix} \vec{P}_{1,2} \\ \vec{V}_1 \\ \vec{\Omega}_1 \\ \vec{V}_2 \\ \vec{\Omega}_2 \end{pmatrix} = \begin{pmatrix} \vec{d}_{1,2} \\ m_1 \vec{v}_1 \\ \mathbf{I}_1 \vec{\omega}_1 \\ m_2 \vec{v}_2 \\ \mathbf{I}_2 \vec{\omega}_2 \end{pmatrix},$$

(4.62)

where $\mathbf{0}$ is the 3×3 zero matrix, $\mathbf{I}$ is the 3×3 identity matrix, and $\vec{P}_{1,2} = \vec{P}$, with the index $(1, 2)$ indicating that the impulse corresponds to the collision between bodies B_1 and $B2$[12]. The matrices $\tilde{r}_1$ and $\tilde{r}_2$ are the matrix-vector representations of the cross-products $\vec{r}_1 \times \vec{P}$ and $\vec{r}_2 \times \vec{P}$, respectively. The other matrices are given by

$$\mathbf{A}_{1,2} = \begin{pmatrix} 0 & 0 & 0 \\ -(\mu_d)_t & 1 & 0 \\ -(\mu_d)_k & 1 & 0 \end{pmatrix} \qquad \mathbf{B}_{1,2} = \begin{pmatrix} 1 & 0 & 0 \\ 0 & 0 & 0 \\ 0 & 0 & 0 \end{pmatrix}$$

$$\mathbf{C}_{1,2} = \begin{pmatrix} 0 & (r_1)_k & -(r_1)_t \\ 0 & 0 & 0 \\ 0 & 0 & 0 \end{pmatrix} \qquad \mathbf{E}_{1,2} = \begin{pmatrix} 0 & -(r_2)_k & (r_2)_t \\ 0 & 0 & 0 \\ 0 & 0 & 0 \end{pmatrix}$$

and the vector $\vec{d}_{1,2}$ is computed as

$$\vec{d}_{1,2} = \begin{pmatrix} -e\left((\vec{v}_1 + \vec{\omega}_1 \times \vec{r}_1) - (\vec{v}_2 + \vec{\omega}_2 \times \vec{r}_2)\right) \cdot \vec{n} \\ 0 \\ 0 \end{pmatrix}. \qquad (4.63)$$

We can then solve the (often) sparse linear system described in equation (4.62) using Gaussian elimination, or a more sophisticated system solver suitable for sparse matrix only. The latter is more difficult to implement, but yields significant efficiency gains over the former method.

Notice that this partitioned-matrix representation is applicable only to the case where the colliding rigid bodies continue sliding throughout the collision. If the bodies are not sliding after collision, either because they were not sliding before the collision or the sliding motion stopped during the collision, then equations (4.54) and (4.55), namely

$$(\vec{V}_1 + \vec{\Omega}_1 \times \vec{r}_1) \cdot \vec{t} = (\vec{V}_2 + \vec{\Omega}_2 \times \vec{r}_2) \cdot \vec{t}$$
$$(\vec{V}_1 + \vec{\Omega}_1 \times \vec{r}_1) \cdot \vec{k} = (\vec{V}_2 + \vec{\Omega}_2 \times \vec{r}_2) \cdot \vec{k}$$

---

[12] Even though the use of the indexes is not particularly useful for the single-collision case, they will be extensively applied in the block-matrix representation of multiple collisions to distinguish between equations associated with collisions involving different rigid bodies.

should be used instead of equations (4.56) and (4.57). This in turn requires modifying the second and third rows of the partitioned-matrix representation, or equivalently, the second and third rows of matrices $\mathbf{A}_{1,2}$, $\mathbf{B}_{1,2}$, $\mathbf{C}_{1,2}$ and $\mathbf{E}_{1,2}$. The modification consists of updating the second row to

$$\begin{pmatrix} 0 & 0 & 0 & 0 & 1 & 0 & -(r_1)_k & 0 & (r_1)_n & (r_2)_k & 0 & -(r_2)_n \end{pmatrix}$$

if there is no sliding motion along the $\vec{t}$ direction, and updating the third row to

$$\begin{pmatrix} 0 & 0 & 0 & 0 & 0 & 1 & (r_1)_t & -(r_1)_n & 0 & -(r_2)_t & (r_2)_n & 0 \end{pmatrix}$$

if the bodies are not sliding along the $\vec{k}$ direction.

Notice that the sliding motion on the tangent plane is directly affected by the coefficients of restitution and friction, as well as by the relative velocities of the rigid bodies just before the collision. Intuitively, for a given coefficient of restitution and relative velocities, the sliding motion will continue if the coefficient of friction is small, or will stop if the coefficient of friction is sufficiently large. Therefore, there exists a critical-coefficient-of-friction value associated with a given coefficient of restitution and relative velocity. If the actual coefficient of friction is less than the critical coefficient of friction, then sliding continues throughout the collision and the system equations associated with the first case should be used. However, if the actual coefficient of friction is greater than or equal to the critical coefficient of friction, then the sliding stops somewhere during the collision and the system equations associated with the second case should be considered instead[13].

Let's derive an expression for computing the critical coefficients of friction $(\mu_d)_t^c$ and $(\mu_d)_k^c$ along the tangent-plane directions $\vec{t}$ and $\vec{k}$. This will be done by first expressing all velocity components in equation (4.53) as a function of the normal-impulse component P_n, and solving for P_n. The computed impulse component P_n will then be substituted back into the expressions for each velocity component. This in turn will give us all velocities as a function of the restitution and friction coefficients, as well as the velocities of the bodies just before their collision[14]. Lastly, we shall substitute these expressions into equation (4.54) to determine the critical-coefficient value $(\mu_d)_t^c$ along $\vec{t}$, and into equation (4.55) to determine the critical-coefficient value $(\mu_d)_k^c$ along $\vec{k}$.

So, from equations (4.49) and (4.50), we can write the linear velocity components as

---

[13] If the actual coefficient of friction is equal to the critical coefficient of friction, then the sliding motion will stop exactly at the instant corresponding to the end of the collision.

[14] Keep in mind that the velocities computed so far assume that the sliding motion continues throughout the collision.

4.8 Rigid Body-Rigid Body Collision Response

$$(V_1)_n = \frac{P_n}{m_1} + (v_1)_n \tag{4.64}$$

$$(V_1)_t = \frac{P_t}{m_1} + (v_1)_t \tag{4.65}$$

$$(V_1)_k = \frac{P_k}{m_1} + (v_1)_k \tag{4.66}$$

$$(V_2)_n = -\frac{P_n}{m_2} + (v_2)_n \tag{4.67}$$

$$(V_2)_t = -\frac{P_t}{m_2} + (v_2)_t \tag{4.68}$$

$$(V_2)_k = -\frac{P_k}{m_2} + (v_2)_k . \tag{4.69}$$

Expressing the angular velocities as a function of P_n is a bit more complicated. Using equations (4.51) and (4.52), each angular velocity can be written as

$$\vec{\Omega}_1 = \mathbf{I_1}^{-1} \tilde{r}_1 \vec{P} + \vec{w}_1$$
$$\vec{\Omega}_2 = -\mathbf{I_2}^{-1} \tilde{r}_2 \vec{P} + \vec{w}_2 .$$

Let $\mathbf{A_1} = \mathbf{I_1}^{-1} \tilde{r}_1$ and $\mathbf{A_2} = \mathbf{I_2}^{-1} \tilde{r}_2$ be the result of the multiplication of these four known matrices, where

$$\mathbf{A_i} = \left(\begin{array}{c|c|c} (\vec{A}_i)_n & (\vec{A}_i)_t & (\vec{A}_i)_k \end{array} \right) = \left(\begin{array}{c|c|c} (A_i)_{nn} & (A_i)_{nt} & (A_i)_{nk} \\ (A_i)_{tn} & (A_i)_{tt} & (A_i)_{tk} \\ (A_i)_{kn} & (A_i)_{kt} & (A_i)_{kk} \end{array} \right)$$

for $i \in \{1, 2\}$. Using the fact that $P_t = (\mu_d)_t P_n$ and $P_k = (\mu_d)_k P_n$, after considerable manipulation, we obtain the following expressions for the angular-velocity components.

$$(\Omega_1)_n = ((A_1)_{nn} + (\mu_d)_t (A_1)_{nt}$$
$$+ (\mu_d)_k (A_1)_{nk}) P_n + (\omega_1)_n \tag{4.70}$$

$$(\Omega_1)_t = ((A_1)_{tn} + (\mu_d)_t (A_1)_{tt}$$
$$+ (\mu_d)_k (A_1)_{tk}) P_n + (\omega_1)_t \tag{4.71}$$

$$(\Omega_1)_k = ((A_1)_{kn} + (\mu_d)_t (A_1)_{kt}$$
$$+ (\mu_d)_k (A_1)_{kk}) P_n + (\omega_1)_k \tag{4.72}$$

$$(\Omega_2)_n = -((A_2)_{nn} + (\mu_d)_t (A_2)_{nt}$$
$$+ (\mu_d)_k (A_2)_{nk}) P_n + (\omega_2)_n \tag{4.73}$$

$$(\Omega_2)_t = -((A_2)_{tn} + (\mu_d)_t (A_2)_{tt}$$

$$+ (\mu_d)_k (A_2)_{tk}) P_n + (\omega_2)_t \quad (4.74)$$

$$(\Omega_2)_k = -((A_2)_{kn} + (\mu_d)_t (A_2)_{kt}$$
$$+ (\mu_d)_k (A_2)_{kk}) P_n + (\omega_2)_k \quad (4.75)$$

Substituting equations (4.64), (4.67), (4.71), (4.72), (4.74) and (4.75) into equation (4.53), and grouping the terms that have P_n in common, we get

$$P_n \left(\bar{m} + ((r_1)_k (A_1)_{tn} - (r_1)_t (A_1)_{kn}) \right. +$$
$$((r_2)_k (A_2)_{tn} - (r_2)_t (A_2)_{kn}) + (\mu_d)_t (((r_1)_k (A_1)_{tt} - (r_1)_t (A_1)_{kt}) +$$
$$((r_2)_k (A_2)_{tt} - (r_2)_t (A_2)_{kt})) + (\mu_d)_k (((r_1)_k (A_1)_{tk} - (r_1)_t (A_1)_{kk}) +$$
$$\left. ((r_2)_k (A_2)_{tk} - (r_2)_t (A_2)_{kk}))) \right) =$$
$$-(1+e) ((\vec{v}_1 + \vec{\omega}_1 \times \vec{r}_1) - (\vec{v}_2 + \vec{\omega}_2 \times \vec{r}_2)) \cdot \vec{n} \quad (4.76)$$

where

$$\bar{m} = \left(\frac{1}{m_1} + \frac{1}{m_2} \right).$$

This equation can be further simplified if we observe that

$$\begin{aligned}
(r_i)_k (A_i)_{tn} - (r_i)_t (A_i)_{kn} &= ((\vec{A}_i)_n \times \vec{r}_i) \cdot \vec{n} \\
(r_i)_k (A_i)_{tt} - (r_i)_t (A_i)_{kt} &= ((\vec{A}_i)_t \times \vec{r}_i) \cdot \vec{n} \quad (4.77) \\
(r_i)_k (A_i)_{tk} - (r_i)_t (A_i)_{kk} &= ((\vec{A}_i)_k \times \vec{r}_i) \cdot \vec{n}
\end{aligned}$$

for $i \in \{1, 2\}$. Substituting these equations into (4.76), gives

$$P_n \left(\bar{m} + ((\vec{A}_1)_n \times \vec{r}_1 + (\vec{A}_2)_n \times \vec{r}_2) \cdot \vec{n} \right. +$$
$$(\mu_d)_t ((\vec{A}_1)_t \times \vec{r}_1 + (\vec{A}_2)_t \times \vec{r}_2) \cdot \vec{n} +$$
$$\left. (\mu_d)_k ((\vec{A}_1)_k \times \vec{r}_1 + (\vec{A}_2)_k \times \vec{r}_2) \cdot \vec{n} \right) =$$
$$-(1+e) ((\vec{v}_1 + \vec{\omega}_1 \times \vec{r}_1) - (\vec{v}_2 + \vec{\omega}_2 \times \vec{r}_2)) \cdot \vec{n} \quad . \quad (4.78)$$

Equation (4.78) can be reduced even further if we consider the following constants

$$\begin{aligned}
g_1^{ij} &= ((\vec{A}_1)_i \times \vec{r}_1 + (\vec{A}_2)_i \times \vec{r}_2) \cdot \vec{j} \\
& \quad (4.79) \\
g_2^j &= ((\vec{v}_1 + \vec{\omega}_1 \times \vec{r}_1) - (\vec{v}_2 + \vec{\omega}_2 \times \vec{r}_2)) \cdot \vec{j},
\end{aligned}$$

4.8 Rigid Body-Rigid Body Collision Response

with $i, j \in \{n, t, k\}$. Using these constants, equation (4.78) can then be rewritten as

$$P_n(\bar{m} + g_1^{nn} + (\mu_d)_t\, g_1^{tn} + (\mu_d)_k\, g_1^{kn}) = -(1+e)\, g_2^n ,$$

which can be solved for P_n, giving

$$P_n = \frac{-(1+e)\, g_2^n}{\bar{m} + g_1^{nn} + (\mu_d)_t\, g_1^{tn} + (\mu_d)_k\, g_1^{kn}} . \qquad (4.80)$$

Having computed P_n, we can substitute its value back into equations (4.64) to (4.75) to obtain the values of each velocity component after the collision, for the case where the sliding motion continues throughout the collision. We are now ready to derive the expressions for computing the critical coefficients of friction along the tangent directions $\vec{t}$ and $\vec{k}$.

Let's start by computing $(\mu_d)_t^c$. This coefficient can be obtained if we expand equation (4.54) to

$$(\vec{V_1} + \vec{\Omega}_1 \times \vec{r}_1) \cdot \vec{t} = (\vec{V_2} + \vec{\Omega}_2 \times \vec{r}_2) \cdot \vec{t}$$

that is

$$\begin{aligned}(V_1)_t + (\Omega_1)_k (r_1)_n - (\Omega_1)_n (r_1)_k &= \\ (V_2)_t + (\Omega_2)_k (r_2)_n - (\Omega_2)_n (r_2)_k & \end{aligned} \qquad (4.81)$$

and substitute the value of each velocity component computed as a function of P_n into this equation, giving

$$\begin{aligned}
\frac{P_t}{m_1} + (v_1)_t + (r_1)_n\, ((A_1)_{kn}\, P_n + \\
(A_1)_{kt}\, P_t + (A_1)_{kk}\, P_k) + (r_1)_n\, (w_1)_k - \\
(r_1)_k\, ((A_1)_{nn}\, P_n + (A_1)_{nt}\, P_t + (A_1)_{nk}\, P_k) - (r_1)_k\, (w_1)_n &= \\
-\frac{P_t}{m_2} + (v_2)_t - (r_2)_n\, ((A_2)_{kn}\, P_n + \\
(A_2)_{kt}\, P_t + (A_2)_{kk}\, P_k) + (r_2)_n\, (w_2)_k + \\
(r_2)_k\, ((A_2)_{nn}\, P_n + (A_2)_{nt}\, P_t + (A_2)_{nk}\, P_k) - (r_2)_k\, (w_2)_n &.
\end{aligned} \qquad (4.82)$$

From $P_t = (\mu_d)_t\, P_n$ and $P_k = (\mu_d)_k\, P_n$, we can group the terms of equation (4.82) so as to expose the constant values defined in equations (4.79). We have

$$(\mu_d)_t\, P_n \overbrace{\left(\frac{1}{m_1} + \frac{1}{m_2}\right)}^{\bar{m}} + (\vec{v}_1 + \vec{\omega}_1 \times \vec{r}_1 - \vec{v}_2 - \vec{\omega}_2 \times \vec{r}_2) \cdot \vec{t} \ +$$

216   4. Rigid-Body Systems

$$P_n \overbrace{((r_1)_n (A_1)_{kn} - (r_1)_k (A_1)_{nn} + (r_2)_n (A_2)_{kn} - (r_2)_k (A_2)_{nn})}^{-((\vec{A}_1)_n \times \vec{r}_1 + (\vec{A}_2)_n \times \vec{r}_2) \cdot \vec{t}} +$$

$$(\mu_d)_t P_n \overbrace{((r_1)_n (A_1)_{kt} - (r_1)_k (A_1)_{nt} + (r_2)_n (A_2)_{kt} - (r_2)_k (A_2)_{nt})}^{-((\vec{A}_1)_t \times \vec{r}_1 + (\vec{A}_2)_t \times \vec{r}_2) \cdot \vec{t}} +$$

$$(\mu_d)_k P_n \overbrace{((r_1)_n (A_1)_{kk} - (r_1)_k (A_1)_{nk} + (r_2)_n (A_2)_{kk} - (r_2)_k (A_2)_{nk})}^{-((\vec{A}_1)_k \times \vec{r}_1 + (\vec{A}_2)_k \times \vec{r}_2) \cdot \vec{t}} = 0,$$

that is

$$(\mu_d)_t P_n \, \bar{m} + \overbrace{(\vec{v}_1 + \vec{\omega}_1 \times \vec{r}_1 - \vec{v}_2 - \vec{\omega}_2 \times \vec{r}_2) \cdot \vec{t}}^{g_2^t} -$$

$$P_n \overbrace{((\vec{A}_1)_n \times \vec{r}_1 + (\vec{A}_2)_n \times \vec{r}_2) \cdot \vec{t}}^{g_1^{nt}} -$$

$$(\mu_d)_t P_n \overbrace{((\vec{A}_1)_t \times \vec{r}_1 + (\vec{A}_2)_t \times \vec{r}_2) \cdot \vec{t}}^{g_1^{tt}} -$$

$$(\mu_d)_k P_n \overbrace{((\vec{A}_1)_k \times \vec{r}_1 + (\vec{A}_2)_k \times \vec{r}_2) \cdot \vec{t}}^{g_1^{kt}} = 0,$$

which can be written in the condensed form

$$((\mu_d)_t \bar{m} - g_1^{nt} - (\mu_d)_t g_1^{tt} - (\mu_d)_k g_1^{kt}) P_n = -g_2^t. \quad (4.83)$$

Finally, substituting the value of P_n given in equation (4.80) into equation (4.83), and solving for $(\mu_d)_t$, we can compute the critical coefficient of friction along the tangent-plane direction $\vec{t}$ as

$$(\mu_d)_t = (\mu_d)_t^c = \frac{g_2^t (\bar{m} + g_1^{nn} + (\mu_d)_k g_1^{kn}) + (1+e) g_2^n (g_1^{nt} + (\mu_d)_k g_1^{kt})}{(\bar{m} - g_1^{tt})(1+e) g_2^n - g_2^t g_1^{tn}}, \quad (4.84)$$

where $(\mu_d)_t^c$ is the critical value of the coefficient of friction along $\vec{t}$ such that the sliding motion stops exactly at the end of the collision.

The derivation of the critical coefficient of friction $(\mu_d)_k^c$ along the tangent-plane direction $\vec{k}$ is very similar to that shown for $(\mu_d)_t^c$. We start by expanding equation (4.55) to

$$(V_1)_k + (\Omega_1)_n (r_1)_t - (\Omega_1)_t (r_1)_n =$$
$$(V_2)_k + (\Omega_2)_n (r_2)_t - (\Omega_2)_t (r_2)_n$$

and substitute the velocity-component values obtained from equations (4.66), (4.69) to (4.71), (4.73) and (4.74). We then group the terms in order to expose the constant-value expressions given in equations (4.79). After some manipulation we get

$$((\mu_d)_k \bar{m} + g_1^{nk} + (\mu_d)_t g_1^{tk} + (\mu_d)_k g_1^{kk}) P_n = -g_2^k .$$

Finally, substituting the value of P_n obtained from equation (4.80) into this expression, and solving for $(\mu_d)_k$, we can compute the critical coefficient of friction along the tangent-plane direction $\vec{k}$ as

$$(\mu_d)_k = (\mu_d)_k^c = \frac{g_2^k (\bar{m} + g_1^{nn} + (\mu_d)_t g_1^{tn}) - (1+e) g_2^n (g_1^{nk} + (\mu_d)_t g_1^{tk})}{(\bar{m} + g_1^{kk})(1+e) g_2^n - g_1^{kn} g_2^k} ,$$
(4.85)

where $(\mu_d)_k^c$ is the critical value of the coefficient of friction along $\vec{k}$ such that the sliding motion stops exactly at the end of the collision.

We do the following in practice. First, compute the critical coefficients of friction using equations (4.84) and (4.85). Next, compare the actual coefficient of friction $(\mu_d)_t$ and $(\mu_d)_k$ to their associated critical values. If $(\mu_d)_t < (\mu_d)_t^c$, then sliding continues along $\vec{t}$, and we use equation (4.56). Else, if $(\mu_d)_t \geq (\mu_d)_t^c$, then sliding along $\vec{t}$ stops during the collision, and we use equation (4.54) instead. The same analysis is used for comparing $(\mu_d)_k$ with $(\mu_d)_k^c$ and selecting the appropriate system equation.

As far as the block-matrix representation is concerned, the choice of the equation to use as a function of the critical-coefficient-of-friction values directly affects the rows of matrices $\mathbf{A}_{1,2}, \mathbf{B}_{1,2}, \mathbf{C}_{1,2}$ and $\mathbf{E}_{1,2}$ being used. We have, therefore, the following four possible cases to consider when building the system matrix:

1. If $(\mu_d)_t < (\mu_d)_t^c$ and $(\mu_d)_k < (\mu_d)_k^c$, then:

$$\mathbf{A}_{1,2} = \begin{pmatrix} 0 & 0 & 0 \\ -(\mu_d)_t & 1 & 0 \\ -(\mu_d)_k & 1 & 0 \end{pmatrix} \quad \mathbf{B}_{1,2} = \begin{pmatrix} 1 & 0 & 0 \\ 0 & 0 & 0 \\ 0 & 0 & 0 \end{pmatrix}$$

$$\mathbf{C}_{1,2} = \begin{pmatrix} 0 & (r_1)_k & -(r_1)_t \\ 0 & 0 & 0 \\ 0 & 0 & 0 \end{pmatrix} \quad \mathbf{E}_{1,2} = \begin{pmatrix} 0 & -(r_2)_k & (r_2)_t \\ 0 & 0 & 0 \\ 0 & 0 & 0 \end{pmatrix}$$

2. If $(\mu_d)_t \geq (\mu_d)_t^c$ and $(\mu_d)_k < (\mu_d)_k^c$, then:

$$\mathbf{A}_{1,2} = \begin{pmatrix} 0 & 0 & 0 \\ 0 & 0 & 0 \\ -(\mu_d)_k & 1 & 0 \end{pmatrix} \quad \mathbf{B}_{1,2} = \begin{pmatrix} 1 & 0 & 0 \\ 0 & 1 & 0 \\ 0 & 0 & 0 \end{pmatrix}$$

218   4. Rigid-Body Systems

$$\mathbf{C}_{1,2} = \begin{pmatrix} 0 & (r_1)_k & -(r_1)_t \\ -(r_1)_k & 0 & (r_1)_n \\ 0 & 0 & 0 \end{pmatrix}$$

$$\mathbf{E}_{1,2} = \begin{pmatrix} 0 & -(r_2)_k & (r_2)_t \\ (r_2)_k & 0 & -(r_2)_n \\ 0 & 0 & 0 \end{pmatrix}$$

3. If $(\mu_d)_t < (\mu_d)_t^c$ and $(\mu_d)_k \geq (\mu_d)_k^c$, then:

$$\mathbf{A}_{1,2} = \begin{pmatrix} 0 & 0 & 0 \\ -(\mu_d)_t & 1 & 0 \\ 0 & 0 & 0 \end{pmatrix} \qquad \mathbf{B}_{1,2} = \begin{pmatrix} 1 & 0 & 0 \\ 0 & 0 & 0 \\ 0 & 0 & 1 \end{pmatrix}$$

$$\mathbf{C}_{1,2} = \begin{pmatrix} 0 & (r_1)_k & -(r_1)_t \\ 0 & 0 & 0 \\ (r_1)_t & -(r_1)_n & 0 \end{pmatrix}$$

$$\mathbf{E}_{1,2} = \begin{pmatrix} 0 & -(r_2)_k & (r_2)_t \\ 0 & 0 & 0 \\ -(r_2)_t & (r_2)_n & 0 \end{pmatrix}$$

4. If $(\mu_d)_t \geq (\mu_d)_t^c$ and $(\mu_d)_k \geq (\mu_d)_k^c$, then:

$$\mathbf{A}_{1,2} = 0 \qquad \mathbf{B}_{1,2} = \mathbf{I}$$
$$\mathbf{C}_{1,2} = -\tilde{r}_1 \qquad \mathbf{E}_{1,2} = \tilde{r}_2$$

Having built the system matrix associated with the collision, we can use Gaussian-elimination or sparse-matrix techniques to solve the linear system of equations and determine the collision impulse, and linear and angular velocities, of the bodies at the moment after their collision.

4.8.2 Computing Impulsive Forces for Multiple Collisions

If three or more rigid bodies are simultaneously colliding with each other, the collision impulse of each individual collision will simultaneously affect the dynamics of the system. Therefore, instead of resolving one collision at a time ignoring the presence of the others, the simulation engine needs to group the rigid bodies into clusters that share at least one collision. The collisions within each cluster can then be simultaneously resolved independent of all other clusters (see Figure 4.34).

Consider the computation of the collision impulses associated with cluster G_1, as shown in Figure 4.34. Let collisions C_1 and C_2 be the collisions involving bodies $(B_1 - B_2)$ and $(B_2 - B_3)$, respectively. As far as rigid

4.8 Rigid Body-Rigid Body Collision Response

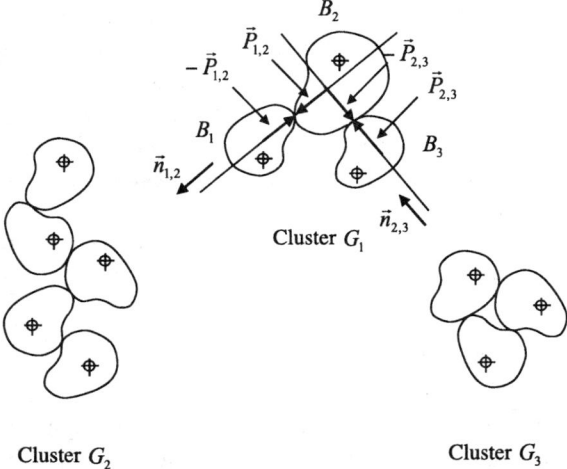

FIGURE 4.34. Multiple rigid-body collisions separated into three clusters. Body B_i is added to cluster G_j if it is colliding with at least one rigid body already in G_j. The collision-response module resolves each cluster in parallel, since they have no collisions in common and therefore can be viewed as independent groups of collisions.

body B_2 is concerned, the linear- and angular-momentum equations owing to both collisions are

$$m_2(\vec{V}_2 - \vec{v}_2) = -\vec{P}_{1,2} + \vec{P}_{2\mapsto 1, 3\mapsto 2}$$
$$I_2(\vec{\Omega}_2 - \vec{\omega}_2) = \tilde{r}_2(-\vec{P}_{1,2} + \vec{P}_{2\mapsto 1, 3\mapsto 2}),$$

where $P_{2\mapsto 1,3\mapsto 2}$ is the impulse $\vec{P}_{2,3}$ of collision C_2 expressed in the local-coordinate frame[15] associated with collision C_1.

Clearly, the impulse arising from collision C_2 will also affect the computation of the impulse arising from collision C_1, and vice-versa. The correct way to compute the collision impulses is, then, to take both collisions into account when solving the system equations. Recall from Section 4.8.1 that we adopted the convention that a positive impulse is applied to the rigid body with index 1, and a negative impulse is applied to the rigid body with index 2. The choice of indexes is related to the relative velocities of the rigid bodies at the collision point along the collision normal, such that equation (4.48) is satisfied at the moment before the collision.

Whenever a rigid body is involved in multiple collisions, it is possible to have it assigned to different indexes for each collision. For the particular situation of cluster G_1, rigid body B_2 has index 2 with respect to its collision with B_1, and index 1 with respect to its collision with B_3. This in turn,

---

[15] The local-coordinate frame is defined by the collision normal and tangent plane.

affects the choice of sign when combining the multiple-collision impulses in the system equations. For example, the minus sign on $\vec{P}_{1,2}$ indicates that rigid body B_2 has index 2 with respect to collision C_1, whereas the plus sign on $\vec{P}_{2\mapsto 1,3\mapsto 2}$ indicates body B_2 has index 1 with respect to collision C_2. Moreover, the collision normal and tangent plane are different for each collision. So, we also need to carry out a change of base between the collision impulses before combining them.

The best way to deal with multiple collisions is to represent the system equations associated with each cluster in its block-matrix form

$$\mathbf{A}\,\vec{x} = \vec{b}$$

where $\vec{x}$ is the state vector containing the variables that need to be determined. In the single collision case, the state vector is defined by the collision impulse, and the final linear and angular velocities of the colliding bodies. However, when dealing with multiple collisions, the state vector can be viewed as the concatenation of several single-collision state vectors, with the added complexity that no variables should be accounted for more than once. For instance, Figure 4.35(a) shows the result of a naive concatenation of state vectors for the multiple collisions associated with cluster G_1 of Figure 4.34.

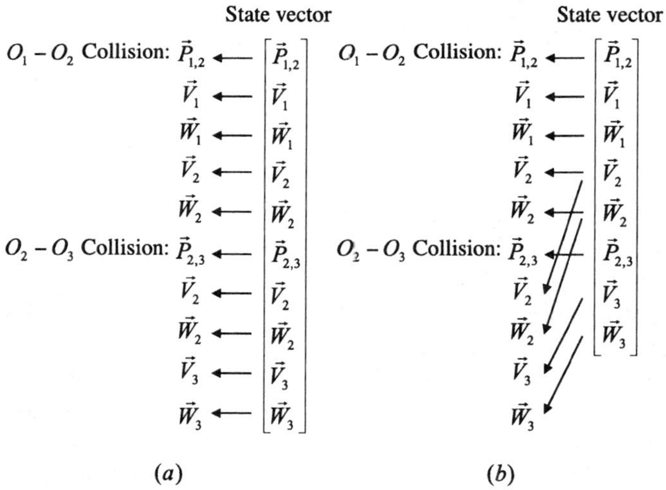

FIGURE 4.35. (a) A naive concatenation creates multiple entries for the final linear and angular velocities of all bodies involved in more than one collision; (b) The state-vector variables should have a link back to their collisions. More than one link is used for multiple collisions, as is the case of $\vec{V}_2$ and $\vec{\Omega}_2$.

Since rigid body B_2 is involved in both collisions, its final linear and angular velocities $\vec{V}_2$ and $\vec{\Omega}_2$ are accounted twice. The correct way to create the state vector is, then, to keep track of which variables were already

4.8 Rigid Body-Rigid Body Collision Response

added, and mark as "common" the ones added more than once. This is illustrated in Figure 4.35(b).

Having determined the state vector associated with a cluster, the next step is to fill in the rows of matrix $\mathbf{A}$ and vector $\vec{b}$. This can be done by considering the equation associated with the first link of each variable in the state vector. For example, for the G_1 cluster of Figure 4.34, the first variable of the state vector is $\vec{P}_{1,2}$. This variable is linked to the $(B_1 - B_2)$ collision. Its associated equations are those involving the coefficient of restitution and friction. Therefore, the first row of matrix $\mathbf{A}$ and vector $\vec{b}$ is:

$$\begin{pmatrix} \mathbf{A}_{1,2} & \mathbf{B}_{1,2} & \mathbf{C}_{1,2} & -\mathbf{B}_{1,2} & \mathbf{E}_{1,2} & 0 & 0 & 0 \\ x & x & x & x & x & x & x & x \\ x & x & x & x & x & x & x & x \\ x & x & x & x & x & x & x & x \\ x & x & x & x & x & x & x & x \\ x & x & x & x & x & x & x & x \\ x & x & x & x & x & x & x & x \\ x & x & x & x & x & x & x & x \end{pmatrix} \begin{pmatrix} \vec{P}_{1,2} \\ \vec{V}_1 \\ \vec{\Omega}_1 \\ \vec{V}_2 \\ \vec{\Omega}_2 \\ \vec{P}_{2,3} \\ \vec{V}_3 \\ \vec{\Omega}_3 \end{pmatrix} = \begin{pmatrix} \vec{d}_{1,2} \\ x \\ x \\ x \\ x \\ x \\ x \\ x \end{pmatrix}$$

The second variable of the state vector is $\vec{V}_1$. This variable is also linked to the $(B_1 - B_2)$ collision. Its associated equations are the conservation of linear momentum for body B_1. So, the second row of matrix $\mathbf{A}$ and vector $\vec{b}$ is:

$$\begin{pmatrix} \mathbf{A}_{1,2} & \mathbf{B}_{1,2} & \mathbf{C}_{1,2} & -\mathbf{B}_{1,2} & \mathbf{E}_{1,2} & 0 & 0 & 0 \\ -\mathbf{I} & m_1 \mathbf{I} & 0 & 0 & 0 & 0 & 0 & 0 \\ x & x & x & x & x & x & x & x \\ x & x & x & x & x & x & x & x \\ x & x & x & x & x & x & x & x \\ x & x & x & x & x & x & x & x \\ x & x & x & x & x & x & x & x \\ x & x & x & x & x & x & x & x \end{pmatrix}$$

222   4. Rigid-Body Systems

$$\begin{pmatrix} \vec{P}_{1,2} \\ \vec{V}_1 \\ \vec{\Omega}_1 \\ \vec{V}_2 \\ \vec{\Omega}_2 \\ \vec{P}_{2,3} \\ \vec{V}_3 \\ \vec{\Omega}_3 \end{pmatrix} = \begin{pmatrix} \vec{d}_{1,2} \\ m_1 \vec{v}_1 \\ x \\ x \\ x \\ x \\ x \\ x \end{pmatrix}$$

Doing the same for all other state variables, we obtain:

$$\begin{pmatrix} \mathbf{A}_{1,2} & \mathbf{B}_{1,2} & \mathbf{C}_{1,2} & -\mathbf{B}_{1,2} & \mathbf{E}_{1,2} & 0 & 0 & 0 \\ -\mathbf{I} & m_1 \mathbf{I} & 0 & 0 & 0 & 0 & 0 & 0 \\ -\tilde{r}_1 & 0 & \mathbf{I}_1 & 0 & 0 & 0 & 0 & 0 \\ \mathbf{I} & 0 & 0 & m_2 \mathbf{I} & 0 & 0 & 0 & 0 \\ \tilde{r}_2 & 0 & 0 & 0 & \mathbf{I}_2 & 0 & 0 & 0 \\ 0 & 0 & 0 & \mathbf{B}_{2,3} & \mathbf{C}_{2,3} & \mathbf{A}_{2,3} & -\mathbf{B}_{2,3} & \mathbf{E}_{2,3} \\ 0 & 0 & 0 & 0 & 0 & \mathbf{I} & m_3 \mathbf{I} & 0 \\ 0 & 0 & 0 & 0 & 0 & \tilde{r}_3 & 0 & \mathbf{T}_3 \end{pmatrix} \begin{pmatrix} \vec{P}_{1,2} \\ \vec{V}_1 \\ \vec{\Omega}_1 \\ \vec{V}_2 \\ \vec{\Omega}_2 \\ \vec{P}_{2,3} \\ \vec{V}_3 \\ \vec{\Omega}_3 \end{pmatrix} = \begin{pmatrix} \vec{d}_{1,2} \\ m_1 \vec{v}_1 \\ \mathbf{I}_1 \vec{\omega}_1 \\ m_2 \vec{v}_2 \\ \mathbf{I}_2 \vec{\omega}_2 \\ \vec{d}_{2,3} \\ m_3 \vec{v}_3 \\ \mathbf{I}_3 \vec{\omega}_3 \end{pmatrix} \quad (4.86)$$

Notice the difference in the order of the matrices displayed on rows 1 and 6 of the system matrix shown in equation (4.86). Since $\vec{V}_2$ and $\vec{\Omega}_2$ are common to both (B_1-B_2) and (B_2-B_3) collisions, the matrices $\mathbf{A}_{2,3}$, $\mathbf{B}_{2,3}$, $\mathbf{C}_{2,3}$ and $\mathbf{E}_{2,3}$ were rearranged to correctly multiply their associated state vector variables. The correct order is $\mathbf{B}_{2,3}$ multiplying the linear velocity of the body with index 1 (i.e., $\vec{V}_2$), $\mathbf{C}_{2,3}$ multiplying the angular velocity of the body with index 1 (i.e., $\vec{\Omega}_2$), $\mathbf{A}_{2,3}$ multiplying the impulse associated with collision $(B_2 - B_3)$ (i.e., $\vec{P}_{2,3}$), $(-\mathbf{B}_{2,3})$ multiplying the linear velocity of the body with index 2 (i.e., $\vec{V}_3$), and $\mathbf{E}_{2,3}$ multiplying the angular velocity of the body with index 2 (i.e., $\vec{\Omega}_3$.)

Also, notice that equation (4.86) was built following *only* the first link of each state-vector variable. We still need to update equation (4.86) with the multiple-collision terms. This can be done by considering the state variables that have more than one associated link. The first link was used to define

4.8 Rigid Body-Rigid Body Collision Response

the row. The following links are used to update some elements of this row with the multiple-collision terms.

In general, if body B_i is involved in more than one collision, then the rows associated with $\vec{V}_i$ and $\vec{\Omega}_i$, that is, the rows associated with its final linear and angular velocities, need to be updated. Say, for example, that body B_i has a second link to body B_j. Let $\vec{P}_{i,j}$ designate the state-vector variable corresponding to the impulse associated with this collision. So, the indexes of $\vec{V}_i$ and $\vec{\Omega}_i$ in the state vector define the rows of the system matrix to be updated, and the index of $\vec{P}_{i,j}$ in the state vector defines the column of the system matrix that needs to be updated. Therefore, we need to update the elements

$$[\text{index of } \vec{V}_i][\text{index of } \vec{P}_{i,j}]$$

$$[\text{index of } \vec{\Omega}_i][\text{index of } \vec{P}_{i,j}]$$

of the system matrix given in equation (4.86).

The actual update consists of accounting for $\vec{P}_{i,j}$ in the linear- and angular-momentum equations associated with body B_i. This can be done by expressing $\vec{P}_{i,j}$ with respect to the local-coordinate frame of the collision corresponding to the first link of the state variables $\vec{V}_i$ and $\vec{\Omega}_i$.

Say, for example, that the first link of the state variables $\vec{V}_i$ and $\vec{\Omega}_i$ is associated with collision C_m involving bodies B_m and B_i. Let the local-coordinate frame $\mathcal{F}_{m,i}$ of collision $(B_m - B_i)$ be defined by vectors $\vec{n}_{m,i}$, $\vec{t}_{m,i}$ and $\vec{k}_{m,i}$.

Let the second link of the state variables $\vec{V}_i$ and $\vec{\Omega}_i$ be associated with collision C_j involving bodies B_i and B_j. Let the local-coordinate frame $\mathcal{F}_{i,j}$ of collision $(B_i - B_j)$ be defined by vectors $\vec{n}_{i,j}$, $\vec{t}_{i,j}$ and $\vec{k}_{i,j}$. The collision impulse $\vec{P}_{i,j}$ defined in the local frame $\mathcal{F}_{i,j}$ is expressed in the local frame $\mathcal{F}_{m,i}$ as

$$\vec{P}_{i \mapsto m, j \mapsto i} = \mathbf{M}_{i \mapsto m, j \mapsto i} \vec{P}_{i,j}$$

with

$$\mathbf{M}_{i \mapsto m, j \mapsto i} = \lambda \begin{pmatrix} \vec{n}_{i,j} \cdot \vec{n}_{m,i} & \vec{n}_{i,j} \cdot \vec{t}_{m,i} & \vec{n}_{i,j} \cdot \vec{k}_{m,i} \\ \vec{t}_{i,j} \cdot \vec{n}_{m,i} & \vec{t}_{i,j} \cdot \vec{t}_{m,i} & \vec{t}_{i,j} \cdot \vec{k}_{m,i} \\ \vec{k}_{i,j} \cdot \vec{n}_{m,i} & \vec{k}_{i,j} \cdot \vec{t}_{m,i} & \vec{k}_{i,j} \cdot \vec{k}_{m,i} \end{pmatrix}.$$

The variable λ can be either 1 or -1, depending on whether body B_i is assigned to index 2 or 1 in collision C_j. The necessary updates are then

$$[\text{index of } \vec{V}_i][\text{index of } \vec{P}_{i,j}] = \vec{P}_{i \mapsto m, j \mapsto i},$$

224   4. Rigid-Body Systems

$$[\text{index of } \vec{\Omega}_i][\text{index of } \vec{P}_{i,j}] = \lambda \tilde{r}_i .$$

As an example, let's apply this multiple-collision-terms update to the G_1 cluster example of Figure 4.34. In this example, the second link of $\vec{V}_2$ and $\vec{\Omega}_2$ points to the collision between bodies B_2 and B_3. Therefore, we need to update the elements at

$$[\text{index of } \vec{V}_2][\text{index of } \vec{P}_{2,3}] = [4, 6]$$

$$[\text{index of } \vec{\Omega}_2][\text{index of } \vec{P}_{2,3}] = [5, 6]$$

in the system matrix of equation 4.86. The actual update will be to substitute the current $\mathbf{0}$ matrix at position $[4, 6]$ for

$$\mathbf{M}_{2\mapsto 1, 3\mapsto 2} = \lambda \begin{pmatrix} \vec{n}_{2,3} \cdot \vec{n}_{1,2} & \vec{n}_{2,3} \cdot \vec{t}_{1,2} & \vec{n}_{2,3} \cdot \vec{k}_{1,2} \\ \vec{t}_{2,3} \cdot \vec{n}_{1,2} & \vec{t}_{2,3} \cdot \vec{t}_{1,2} & \vec{t}_{2,3} \cdot \vec{k}_{1,2} \\ \vec{k}_{2,3} \cdot \vec{n}_{1,2} & \vec{k}_{2,3} \cdot \vec{t}_{1,2} & \vec{k}_{2,3} \cdot \vec{k}_{1,2} \end{pmatrix} , \quad (4.87)$$

where frame $\mathcal{F}_{1,2}$ is defined by vectors $\vec{n}_{1,2}$, $\vec{t}_{1,2}$ and $\vec{k}_{1,2}$, and frame $\mathcal{F}_{2,3}$ is defined by vectors $\vec{n}_{2,3}$, $\vec{t}_{2,3}$ and $\vec{k}_{2,3}$. We also need to substitute the element at position $[5, 6]$ for

$$\lambda \tilde{r}_2 . \quad (4.88)$$

Also, since body B_2 is assigned to index 1 in its collision with body B_3 (see Figure 4.34), we should use $\lambda = +1$ in equations (4.87) and (4.88). The final system matrix for this particular example is then:

$$\begin{pmatrix} \mathbf{A}_{1,2} & \mathbf{B}_{1,2} & \mathbf{C}_{1,2} & -\mathbf{B}_{1,2} & \mathbf{E}_{1,2} & 0 & 0 & 0 \\ -\mathbf{I} & m_1 \mathbf{I} & 0 & 0 & 0 & 0 & 0 & 0 \\ -\tilde{r}_1 & 0 & \mathbf{I}_1 & 0 & 0 & 0 & 0 & 0 \\ \mathbf{I} & 0 & 0 & m_2 \mathbf{I} & 0 & \mathbf{M}_{2\mapsto 1, 3\mapsto 2} & 0 & 0 \\ \tilde{r}_2 & 0 & 0 & 0 & \mathbf{I}_2 & \tilde{r}_2 & 0 & 0 \\ 0 & 0 & 0 & \mathbf{B}_{2,3} & \mathbf{C}_{2,3} & \mathbf{A}_{2,3} & -\mathbf{B}_{2,3} & \mathbf{E}_{2,3} \\ 0 & 0 & 0 & 0 & 0 & \mathbf{I} & m_3 \mathbf{I} & 0 \\ 0 & 0 & 0 & 0 & 0 & \tilde{r}_3 & 0 & \mathbf{T}_3 \end{pmatrix} \begin{pmatrix} \vec{P}_{1,2} \\ \vec{V}_1 \\ \vec{\Omega}_1 \\ \vec{V}_2 \\ \vec{\Omega}_2 \\ \vec{P}_{2,3} \\ \vec{V}_3 \\ \vec{\Omega}_3 \end{pmatrix} = \begin{pmatrix} \vec{d}_{1,2} \\ m_1 \vec{v}_1 \\ \mathbf{I}_1 \vec{\omega}_1 \\ m_2 \vec{v}_2 \\ \mathbf{I}_2 \vec{\omega}_2 \\ \vec{d}_{2,3} \\ m_3 \vec{v}_3 \\ \mathbf{I}_3 \vec{\omega}_3 \end{pmatrix}$$

In summary, for each state-vector variable with more than one link, we need to update the elements of the system matrix corresponding to each of these collisions. When all elements are updated, we solve the resulting linear system using, for example, Gaussian elimination techniques. Another option would be to use specialized methods to solve sparse linear systems, since the system matrix is often sparse. The solution would then give the correct values of the state-vector variables to be used by the collision-response module to prevent the objects from interpenetrating after colliding.

4.8.3 Computing Contact Forces for a Single Contact

Two rigid bodies are said to be in contact whenever their relative velocities along the collision normal is either zero, or less than a threshold value. In such situations, a contact force should be applied, instead of the impulsive force described in Section 4.8.1.

In the case of computing impulsive forces for rigid body-rigid body collisions, the system is described by equations of conservation of linear and angular momentum, and the coefficients of friction and restitution. Unfortunately, these equations are no longer valid for the contact-force computation. Therefore, we need to derive other conditions to compute the contact forces, based on the contact geometry[16] and dynamic state of each rigid body. These conditions are exactly the same as those described in Chapter 3 for the particle-particle contact. They are rephrased here for convenience.

The first condition states that the relative acceleration of the rigid bodies at the contact point, along the contact normal, should be greater than or equal to zero, assuming that a negative value indicates that the bodies are accelerating towards each other. In this case, if the computed contact force is such that the relative acceleration at the contact point along the contact normal is zero, then the bodies remain in contact. However, if their relative acceleration is greater than zero, then contact is about to break.

The second condition implies that the contact-force component along the contact normal should be greater than or equal to zero, indicating that the rigid bodies are being pushed away from each other. The contact force is not allowed to have a negative value, that is, is not allowed to keep the bodies connected to each other, preventing their separation.

The third and last condition states that the contact force should be set to zero if the contact between the rigid bodies is about to break. In other words, if the relative acceleration at the contact point, along the contact normal, is greater than zero, then contact is about to break and the contact force should be set to zero.

---

[16]Whenever a collision becomes a contact, the collision normal will be referred to as the contact normal.

Let's translate these three conditions into equations that can be used to compute the contact force. Figure 4.36 illustrates a typical situation in which rigid bodies B_1 and B_2 are shown at the moment before contact, in contact, and interpenetrating if a contact force is not applied.

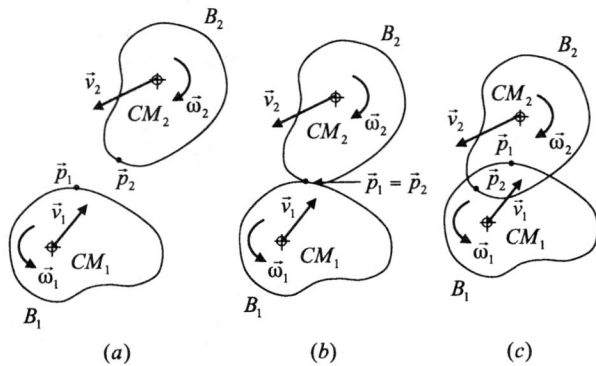

FIGURE 4.36. (a) Rigid bodies B_1 and B_2 are about to touch each other at points $\vec{p}_1$ and $\vec{p}_2$; (b) Contact is established whenever $\vec{p}_1 = \vec{p}_2$; (c) Interpenetration occurs if $(\vec{p}_1 - \vec{p}_2) \cdot \vec{n} < 0$, where $\vec{n}$ is the contact normal.

Let $\vec{p}_1(t)$ and $\vec{p}_2(t)$ be the points on bodies B_1 and B_2, respectively, that are about to be in contact. Consider the vector $\vec{q}(t)$ defined as

$$\vec{q}(t) = \begin{pmatrix} q_n(t) \\ q_t(t) \\ q_k(t) \end{pmatrix} = \begin{pmatrix} (\vec{p}_1(t) - \vec{p}_2(t)) \cdot \vec{n}(t) \\ (\vec{p}_1(t) - \vec{p}_2(t)) \cdot \vec{t}(t) \\ (\vec{p}_1(t) - \vec{p}_2(t)) \cdot \vec{k}(t) \end{pmatrix}, \quad (4.89)$$

where $\vec{n}(t)$ is the contact normal, pointing from body B_2 to body B_1, and $\vec{t}(t)$ and $\vec{k}(t)$ are vectors defining the tangent plane at the contact. Clearly, $q_n(t)$ defines a distance measure between points $\vec{p}_1(t)$ and $\vec{p}_2(t)$, along the contact normal, as a function of time. We have $q_n(t) > 0$ if the bodies are separated, $q_n(t) = 0$ if the bodies are in contact, and $q_n(t) < 0$ if the bodies are interpenetrating (see Figure 4.36). Let t_c be the instant at which contact is established, that is

$$\vec{q}(t_c) = \vec{0}.$$

The first condition states that the relative acceleration at the contact point, along the contact normal, should be greater than or equal to zero. This is equivalent to assuring

$$\left. \frac{d^2 q_n(t)}{dt^2} \right|_{t=t_c} \geq 0. \quad (4.90)$$

If we let $\vec{a}(t) = (a_n(t), a_t(t), a_k(t))$ be the relative acceleration at the contact point, we can rewrite equation (4.90) as

$$a_n(t_c) \geq 0 \,. \tag{4.91}$$

The components $a_t(t)$ and $a_k(t)$ define the relative acceleration at the contact point on the tangent plane of the contact. They are used only if static or dynamic friction are considered at the contact point, as will be explained later in this section.

The second condition states that the contact-force component along the contact normal should be non-negative, that is

$$F_n \geq 0 \,, \tag{4.92}$$

where $\vec{F} = (F_n, F_t, F_k)$ is the contact force to be determined. If friction is taken into account, the tangent components F_t and F_k of the contact force are computed according to the Coulomb friction model. To elaborate, if the relative velocity of the points $\vec{p}_1$ and $\vec{p}_2$ along $\vec{t}$ is zero, or less than a threshold value, then there is no sliding at the contact point. In this case, the component F_t will assume values in the range

$$-(\mu_s)_t \, F_n \leq F_t \leq (\mu_s)_t \, F_n \,,$$

depending on the relative acceleration component $a_t(t)$ being positive or negative. In other words, F_t will do its best to prevent the bodies from sliding at the contact point by always opposing the relative acceleration $a_t(t)$[17]. On the other hand, if the relative velocity along $\vec{t}$ is greater than the threshold value, then the bodies are sliding at the contact point and

$$F_t = +(\mu_d)_t \, F_n \quad \text{or} \quad F_t = -(\mu_d)_t \, F_n \,,$$

depending on the relative acceleration $a_t(t)$ being negative or positive. Here, $(\mu_d)_t$ is the dynamic coefficient of friction along direction $\vec{t}$. A similar analysis holds for $\vec{k}$.

The third and last condition states that the contact force is zero if the contact is breaking away, that is, if the relative acceleration along the contact normal is positive. Equivalently, we have

$$F_n \, a_n(t_c) = 0 \,, \tag{4.93}$$

meaning that, if F_n is greater than zero, then the bodies are in contact and the relative acceleration is zero. Otherwise, if a_n is greater than zero, then the contact is about to break and the contact force should be zero. Putting it all together, we have that the computation of the contact force involves solving the following system of equations:

---

[17] Notice that F_t is zero if $a_t(t)$ is zero.

$$a_n(t_c) \geq 0$$
$$F_n \geq 0 \qquad (4.94)$$
$$F_n \, a_n(t_c) = 0 \, .$$

Here, we adopt the convention that a positive contact force $+\vec{F}$ is applied to the rigid body B_1 (i.e., the body with index 1) and a negative contact force $-\vec{F}$ is applied to the rigid body B_2 (i.e., the body with index 2).

According to equation (4.90), the relative acceleration $a_n(t)$, along the contact normal, can be obtained by differentiating equation (4.89) twice with respect to time. The first time derivative of equation (4.89) gives

$$\frac{d\,q_n(t)}{dt} = \left(\frac{d\,\vec{p}_1(t)}{dt} - \frac{d\,\vec{p}_2(t)}{dt}\right) \cdot \vec{n}(t) +$$
$$(\vec{p}_1 - \vec{p}_2) \cdot \frac{d\,\vec{n}(t)}{dt} \, , \qquad (4.95)$$

or equivalently

$$v_n(t) = (\vec{v}_{p_1}(t) - \vec{v}_{p_2}(t)) \cdot \vec{n}(t) +$$
$$(\vec{p}_1 - \vec{p}_2) \cdot \frac{d\,\vec{n}(t)}{dt} \, , \qquad (4.96)$$

where $\vec{v}_{p_1}(t)$ and $\vec{v}_{p_2}(t)$ are the velocity vectors of points $\vec{p}_1(t)$ and $\vec{p}_2(t)$. This gives us an expression for the relative velocity $v_n(t) = d\,q(t)/dt$ of points $\vec{p}_1(t)$ and $\vec{p}_2(t)$ along the contact normal, as a function of their velocities and collision normal. The time derivative of the collision normal indicates its rate of change in direction as a function of time.

Differentiating equation (4.95) once more with respect to time, we obtain

$$\frac{d^2\,q(t)}{dt^2} = \left(\frac{d^2\,\vec{p}_1(t)}{dt^2} - \frac{d^2\,\vec{p}_2(t)}{dt^2}\right) \cdot \vec{n}(t) +$$
$$2\left(\frac{d\,\vec{p}_1(t)}{dt} - \frac{d\,\vec{p}_2(t)}{dt}\right) \cdot \frac{d\,\vec{n}(t)}{dt} +$$
$$(\vec{p}_1 - \vec{p}_2) \cdot \frac{d^2\,\vec{n}(t)}{dt^2} \, , \qquad (4.97)$$

or equivalently

$$a_n(t) = (\vec{a}_{p_1}(t) - \vec{a}_{p_2}(t)) \cdot \vec{n}(t) + 2\,(\vec{v}_{p_1}(t) - \vec{v}_{p_2}(t)) \cdot \frac{d\,\vec{n}(t)}{dt} +$$
$$(\vec{p}_1 - \vec{p}_2) \cdot \frac{d^2\,\vec{n}(t)}{dt^2} \, , \qquad (4.98)$$

where $\vec{a}_{p_1}(t)$ and $\vec{a}_{p_2}(t)$ are the acceleration vectors of points $\vec{p}_1(t)$ and $\vec{p}_2(t)$. This gives us an expression for the relative acceleration $a_n(t) = d^2 q(t)/dt^2$ of points $\vec{p}_1(t)$ and $\vec{p}_2(t)$ along the contact normal, as a function of their accelerations, velocities, contact normal and rate of change in direction of the contact normal.

At the instant of contact $t = t_c$, points $\vec{p}_1(t)$ and $\vec{p}_2(t)$ are coincident, that is

$$\vec{p}_1(t_c) = \vec{p}_2(t_c) \ . \tag{4.99}$$

Substituting equation (4.99) into (4.98), we obtain an expression for the relative acceleration along the contact normal at the instant of contact:

$$a_n(t_c) = (\vec{a}_{p_1}(t_c) - \vec{a}_{p_2}(t_c)) \cdot \vec{n}(t_c) + 2(\vec{v}_{p_1}(t_c) - \vec{v}_{p_2}(t_c)) \cdot \frac{d\vec{n}(t_c)}{dt} \ . \tag{4.100}$$

According to equation (4.100), the relative acceleration at the instant of contact has two terms. The first depends on the accelerations of the contact points, which in turn are related to the contact force using Newton's law. The second depends on the velocities of the contact points and the rate of change in direction of the collision normal.

For now, let's assume the contact is frictionless, that is

$$\vec{F} = F_n \, \vec{n} \ .$$

Later in this section, we shall relax this assumption and show how the system of equations used in the frictionless case can be extended to cope with friction.

If we isolate the terms that depend on the contact force from the terms that do not, we can rewrite equation (4.100) as

$$a_n(t_c) = (a_{11})_n F_n + b_1 \ . \tag{4.101}$$

Substituting equation (4.101) into (4.94) we obtain

$$\begin{aligned} ((a_{11})_n F_n + b_1) &\geq 0 \\ F_n &\geq 0 \\ F_n \, ((a_{11})_n F_n + b_1) &= 0 \ . \end{aligned} \tag{4.102}$$

Thus, the computation of the contact force involves solving the system of equations defined in (4.102), which is quadratic on F_n. One way of doing so is to use quadratic-programming techniques. However, such techniques are difficult to implement, often requiring the use of sophisticated numerical software packages.

230   4. Rigid-Body Systems

Fortunately, the system of equations defined in (4.102) is also formally similar to a numerical programming technique called *linear complementarity*. The implementation using linear-complementarity techniques is significantly easier than the implementation of a quadratic program, and is discussed in detail in Appendix G. There, we start presenting solution methods for the frictionless case, and show how to modify them to cope with static and dynamic friction at the contacts. These modifications on the solution method require that equation (4.102) be expanded to also consider the relation between the relative-acceleration and contact-force components on the tangent plane of the contact.

In the general case where friction is taken into account, the system of equations becomes

$$\begin{pmatrix} a_n(t_c) \\ a_t(t_c) \\ a_k(t_c) \end{pmatrix} = \begin{pmatrix} (a_{11})_n & (a_{12})_t & (a_{13})_k \\ (a_{21})_n & (a_{22})_t & (a_{23})_k \\ (a_{31})_n & (a_{32})_t & (a_{33})_k \end{pmatrix} \begin{pmatrix} F_n \\ F_t \\ F_k \end{pmatrix} + \begin{pmatrix} (b_1)_n \\ (b_1)_t \\ (b_1)_k \end{pmatrix}$$
$$= \mathbf{A}\vec{F} + \vec{b},$$

where

$$a_t(t_c) = (\vec{a}_{p_1}(t_c) - \vec{a}_{p_2}(t_c)) \cdot \vec{t}(t_c)$$
$$+ 2(\vec{v}_{p_1}(t_c) - \vec{v}_{p_2}(t_c)) \cdot \frac{d\vec{t}(t_c)}{dt} \qquad (4.103)$$

$$a_k(t_c) = (\vec{a}_{p_1}(t_c) - \vec{a}_{p_2}(t_c)) \cdot \vec{k}(t_c)$$
$$+ 2(\vec{v}_{p_1}(t_c) - \vec{v}_{p_2}(t_c)) \cdot \frac{d\vec{k}(t_c)}{dt}. \qquad (4.104)$$

The solution method presented in Appendix G assumes both matrix **A** and vector $\vec{b}$ are known constants computed from the geometric displacement and dynamic state of the bodies at the instant of contact. Therefore, we need to determine the coefficients of matrix **A** and vector $\vec{b}$, before we can apply the linear-complementarity techniques of Appendix G.

The first row of matrix **A** and vector $\vec{b}$ is obtained by expressing the normal relative acceleration $a_n(t_c)$ at the instant of contact as a function of the contact-force components F_n, F_t and F_k. This can be done using equations (4.100), (4.103) and (4.104). Let's start by examining the second term of these equations, namely

$$2(\vec{v}_{p_1}(t_c) - \vec{v}_{p_2}(t_c)) \cdot \frac{d\vec{n}(t_c)}{dt}$$

$$2(\vec{v}_{p_1}(t_c) - \vec{v}_{p_2}(t_c)) \cdot \frac{d\vec{t}(t_c)}{dt}$$

$$2\left(\vec{v}_{p_1}(t_c) - \vec{v}_{p_2}(t_c)\right) \cdot \frac{d\vec{k}(t_c)}{dt}.$$

The velocities of points $\vec{p}_1$ and $\vec{p}_2$ are known quantities independent of the contact force. We still need to compute the rate of change in direction of the contact normal as a function of time. Section E.3.2 of Appendix E presents a detailed description of how the time derivative of the contact normal for the rigid body-rigid body case can be computed. There are two possible ways of computing the time derivative of the normal vector, depending on the type of contact being a vertex-face or edge-edge contact. In either case, the result of the time derivative of the contact normal is independent of the contact force. So, the contribution of the second term of equations (4.100), (4.103) and (4.104) to matrix $\mathbf{A}$ is none, and to vector $\vec{b}$ is

$$(b_1)_n = 2\left(\vec{v}_1 + \vec{\omega}_1 \times (\vec{p}_1 - \vec{r}_1) - \vec{v}_2 - \vec{\omega}_2 \times (\vec{p}_2 - \vec{r}_2)\right) \cdot \frac{d\vec{n}}{dt}$$

$$(b_1)_t = 2\left(\vec{v}_1 + \vec{\omega}_1 \times (\vec{p}_1 - \vec{r}_1) - \vec{v}_2 - \vec{\omega}_2 \times (\vec{p}_2 - \vec{r}_2)\right) \cdot \frac{d\vec{t}}{dt}$$

$$(b_1)_k = 2\left(\vec{v}_1 + \vec{\omega}_1 \times (\vec{p}_1 - \vec{r}_1) - \vec{v}_2 - \vec{\omega}_2 \times (\vec{p}_2 - \vec{r}_2)\right) \cdot \frac{d\vec{k}}{dt},$$

(4.105)

where $d\vec{t}/dt$ and $d\vec{k}/dt$ are the time derivatives of the tangent-plane directions computed following the techniques presented in Section E.4 of Appendix E.

Now, let's focus on the first term of equations (4.100), (4.103) and (4.104), namely the terms

$$\begin{aligned}(\vec{a}_{p_1}(t_c) - \vec{a}_{p_2}(t_c)) &\cdot \vec{n}(t_c) \\ (\vec{a}_{p_1}(t_c) - \vec{a}_{p_2}(t_c)) &\cdot \vec{t}(t_c) \\ (\vec{a}_{p_1}(t_c) - \vec{a}_{p_2}(t_c)) &\cdot \vec{k}(t_c).\end{aligned} \quad (4.106)$$

The acceleration $\vec{a}_{p_1}$ of point $\vec{p}_1$ is obtained directly from equation (4.8) as

$$\vec{a}_{p_1} = \vec{\alpha}_1 \times (\vec{p}_1 - \vec{r}_1) + \vec{\omega}_1 \times (\vec{\omega}_1 \times (\vec{p}_1 - \vec{r}_1)) + \vec{a}_1,$$

where $\vec{\alpha}_1$, $\vec{\omega}_1$ and $\vec{a}_1$ are the angular acceleration, angular velocity and linear acceleration of body B_1 (see Figure 4.36). Using equation (4.11), the linear acceleration $\vec{a}_1$ can be obtained from the net force $(\vec{F}_1)_{net}$ acting at contact point $\vec{p}_1$ as

$$\vec{a}_1 = \frac{(\vec{F}_1)_{net}}{m_1} = \left(\frac{\vec{F} + (\vec{F}_1)_{ext}}{m_1}\right),$$

where $(\vec{F}_1)_{ext}$ is the net external force (such as gravity, spring forces, spatially dependent forces, etc.) acting on body B_1 at $t = t_c$, and $\vec{F}$ is the contact force to be determined. Also, using equation (4.17), the angular acceleration $\vec{\alpha}_1$ can be computed from the net torque $(\vec{\tau}_1)_{net}$ acting at contact point $\vec{p}_1$ as

$$\vec{\alpha}_1 = \mathbf{I}_1^{-1}(\vec{\tau}_1)_{net} + \mathbf{I}_1^{-1}\vec{H}_1 \times \vec{\omega}_1 , \qquad (4.107)$$

where $\mathbf{I}_1$ and $\vec{H}_1$ are the inertia tensor and angular momentum of body B_1, respectively. The net torque acting on body B_1 is computed by summing the torque induced by all external forces, that is

$$(\vec{\tau}_1)_{net} = (\vec{\tau}_1)_{ext} + \overbrace{(\vec{p}_1 - \vec{r}_1) \times \vec{F}}^{\text{torque resulting from contact force}} , \qquad (4.108)$$

where

$$(\vec{\tau}_1)_{ext} = \sum_i (\vec{p}_i - \vec{r}_1) \times (\vec{F}_i)_{ext}$$

with $\vec{p}_i$ being the point on body B_1 at which the external force $(\vec{F}_i)_{ext}$ is being applied. Substituting equation (4.108) into (4.107), we obtain

$$\vec{\alpha}_1 = \mathbf{I}_1^{-1}(\vec{p}_1 - \vec{r}_1) \times \vec{F} + \mathbf{I}_1^{-1}(\vec{\tau}_1)_{ext} + \vec{H}_1 \times \vec{\omega}_1 . \qquad (4.109)$$

The acceleration $\vec{a}_{p_1}$ of point $\vec{p}_1$ is then

$$\begin{aligned}
\vec{a}_{p_1} &= (\mathbf{I}_1^{-1}(\vec{p}_1 - \vec{r}_1) \times \vec{F}) \times (\vec{p}_1 - \vec{r}_1) \\
&+ (\mathbf{I}_1^{-1}(\vec{\tau}_1)_{ext} + \vec{H}_1 \times \vec{\omega}_1) \times (\vec{p}_1 - \vec{r}_1) \\
&+ \vec{\omega}_1 \times (\vec{\omega}_1 \times (\vec{p}_1 - \vec{r}_1)) + \left(\frac{\vec{F} + (\vec{F}_1)_{ext}}{m_1}\right) .
\end{aligned} \qquad (4.110)$$

Using the general cross-product relations

$$\begin{aligned}
\vec{a} \times \vec{b} &= -\vec{b} \times \vec{a} \\
\vec{a} \times \vec{b} &= \tilde{a}\vec{b}
\end{aligned}$$

and letting

$$\vec{x}_1 = \vec{p}_1 - \vec{r}_1 ,$$

we can further simplify the first term of equation (4.110) as follows:

$$\begin{aligned}
(\mathbf{I_1}^{-1}(\vec{p}_1-\vec{r}_1)\times\vec{F})\times(\vec{p}_1-\vec{r}_1) &= \\
(\mathbf{I_1}^{-1}\vec{x}_1\times\vec{F})\times\vec{x}_1 &= \\
-\vec{x}_1\times(\mathbf{I_1}^{-1}\vec{x}_1\times\vec{F}) &= \\
-\tilde{x}_1(\mathbf{I_1}^{-1}\vec{x}_1\times\vec{F}) &= \\
-(\tilde{x}_1\mathbf{I_1}^{-1})\vec{x}_1\times\vec{F} &= \\
-(\tilde{x}_1\mathbf{I_1}^{-1})\tilde{x}_1\vec{F}\,.
\end{aligned} \qquad (4.111)$$

Substituting equation (4.111) into (4.110), we have

$$\begin{aligned}
\vec{a}_{p_1} &= \left(\frac{1}{m_1}\mathbf{I}-\tilde{x}_1\mathbf{I_1}^{-1}\tilde{x}_1\right)\vec{F} \\
&+ \frac{1}{m_1}(\vec{F}_1)_{ext}+(\mathbf{I_1}^{-1}(\vec{\tau}_1)_{ext}+\vec{H}_1\times\vec{\omega}_1)\times\vec{x}_1 \\
&+ \vec{\omega}_1\times(\vec{\omega}_1\times\vec{x}_1)\,,
\end{aligned} \qquad (4.112)$$

which can be written as

$$\vec{a}_{p_1} = \mathbf{A_1}\vec{F}+\vec{b}_1 \qquad (4.113)$$

with

$$\begin{aligned}
\mathbf{A_1} &= \left(\frac{1}{m_1}\mathbf{I}-\tilde{x}_1\mathbf{I_1}^{-1}\tilde{x}_1\right) \\
\vec{b}_1 &= \frac{1}{m_1}(\vec{F}_1)_{ext}+(\mathbf{I_1}^{-1}(\vec{\tau}_1)_{ext}+\vec{H}_1\times\vec{\omega}_1)\times\vec{x}_1 \\
&+ \vec{\omega}_1\times(\vec{\omega}_1\times\vec{x}_1)\,.
\end{aligned}$$

Analogously, the acceleration $\vec{a}_2$ of point $\vec{p}_2$ is given by

$$\begin{aligned}
\vec{a}_{p_2} &= (\mathbf{I_2}^{-1}(\vec{p}_2-\vec{r}_2)\times(-\vec{F}))\times(\vec{p}_2-\vec{r}_2) \\
&+ (\mathbf{I_2}^{-1}(\vec{\tau}_2)_{ext}+\vec{H}_2\times\vec{\omega}_2)\times(\vec{p}_2-\vec{r}_2) \\
&+ \vec{\omega}_2\times(\vec{\omega}_2\times(\vec{p}_2-\vec{r}_2))+\left(\frac{(-\vec{F})+(\vec{F}_2)_{ext}}{m_2}\right),
\end{aligned} \qquad (4.114)$$

which can be further simplified to

$$\vec{a}_{p_2} = -\mathbf{A_2}\vec{F}+\vec{b}_2 \qquad (4.115)$$

with

$$\mathbf{A_2} = \left(\frac{1}{m_2}\mathbf{I} - \tilde{x}_2\,\mathbf{I_2}^{-1}\,\tilde{x}_2\right)$$

$$\vec{b}_2 = \frac{1}{m_2}(\vec{F_2})_{ext} + (\mathbf{I_2}^{-1}\,(\vec{\tau}_2)_{ext} + \vec{H}_2 \times \vec{\omega}_2) \times \vec{x}_2$$
$$+ \ \vec{\omega}_2 \times (\vec{\omega}_2 \times \vec{x}_2)\,.$$

The relative acceleration at the contact point is therefore

$$(\vec{a}_{p_1} - \vec{a}_{p_2}) = (\mathbf{A_1} + \mathbf{A_2})\,\vec{F} + (\vec{b}_1 - \vec{b}_2)\,. \tag{4.116}$$

The final contribution to the elements of vector $\vec{b}$ is obtained by summing the individual contributions of equations (4.105) and (4.116). The matrix $\mathbf{A}$ is obtained by adding up $\mathbf{A_1}$ and $\mathbf{A_2}$, as indicated in equation (4.116).

Applying the linear-complementarity techniques of Appendix G, we can determine the components of the contact-force vector $\vec{F}$. Having $\vec{F}$, we update the dynamic state of each rigid body by applying $+\vec{F}$ on body B_1 and $-\vec{F}$ on body B_2.

4.8.4 Computing Contact Forces for Multiple Contacts

The principle behind the computation of multiple rigid body-rigid body contact forces is the same as that behind the computation of multiple rigid body-rigid body collision impulses. Again, the simulation engine needs to group the rigid bodies into clusters that share at least one contact. The contacts within each cluster can then be simultaneously resolved independent of all other clusters (see Figure 4.37).

Whenever a rigid body is involved in multiple contacts, it is possible to have it assigned to different indexes for each contact. For the particular situation of cluster G_2 in Figure 4.37, body B_2 has index 2 with respect to its contact with body B_1, and index 1 with respect to its contact with body B_3. This in turn affects the choice of sign when combining the multiple contact forces in the system equations. Moreover, the contact normal and tangent plane are different for each contact. So, we also need to carry out a change of base between the contact forces before combining them.

In the single rigid body-rigid body contact, the contact-force computation taking friction into account was done using linear-complementarity techniques to solve a system of equations of the form

$$\begin{aligned} a_n(t_c) &\geq 0 \\ F_n &\geq 0 \\ \vec{F}^t(\mathbf{A}\vec{F} + \vec{b}) &= 0\,, \end{aligned}$$

where

4.8 Rigid Body-Rigid Body Collision Response

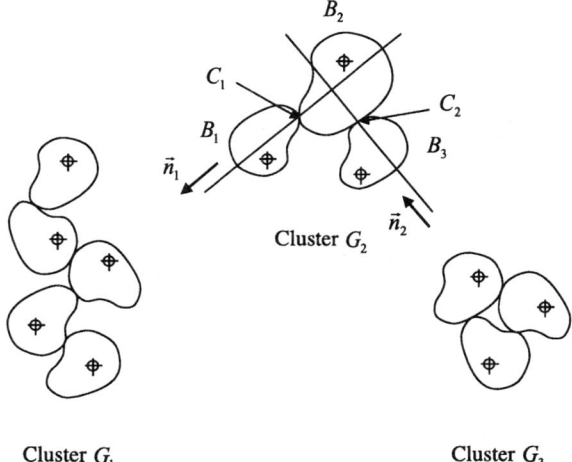

FIGURE 4.37. A multiple rigid body-rigid body contact-force computation. In the situation shown, the rigid bodies are grouped into three clusters that can be resolved in parallel.

$$\mathbf{A} = \begin{pmatrix} (a_{11})_n & (a_{12})_t & (a_{13})_k \\ (a_{21})_n & (a_{22})_t & (a_{23})_k \\ (a_{31})_n & (a_{31})_t & (a_{33})_k \end{pmatrix}$$
$$\vec{F} = (F_n, F_t, F_k)^t$$
$$\vec{b} = ((b_1)_n, (b_1)_t, (b_1)_k)^t .$$

This solution method can be extended to the case of multiple-contact-force computations. The main difference between multiple- and single-contact-force computation involving a given body is that the contact force at contact C_i can affect the computation of the contact force at contact C_j. So, instead of solving one contact at a time, we need to simultaneously solve all contacts having a body in common. This in turn has the same effect as merging the several individual systems of equations for each contact into one larger system, and then applying the linear-complementarity techniques to the merged system.

For example, suppose we have a cluster with m simultaneous contacts. Each contact C_i is defined by its contact-normal $(\vec{n})_i$ and tangent-plane vectors $(\vec{t})_i$ and $(\vec{k})_i$. The contact force at contact C_i is then expressed as

$$\vec{F}_i = ((F_i)_{n_i}, (F_i)_{t_i}, (F_i)_{k_i})^t .$$

The contact-force vector for the multiple-collision system is obtained by concatenating the contact-force vectors for each of the m contacts, that is

$$\vec{F} = ((F_1)_{n_1}, (F_1)_{t_1}, (F_1)_{k_1}, \ldots, (F_m)_{n_m}, (F_m)_{t_m}, (F_m)_{k_m})^t .$$

The vector $\vec{b}$ becomes

$$\vec{b} = ((b_1)_n, (b_1)_t, (b_1)_k, \ldots, (b_m)_n, (b_m)_t, (b_m)_k)^t$$

and the matrix $\mathbf{A}$ is enlarged to accommodate all contact forces. Its partitioned representation is given by

$$\mathbf{A} = \begin{pmatrix} \mathbf{A_{11}} & \mathbf{A_{12}} & \cdots & \mathbf{A_{1m}} \\ \mathbf{A_{21}} & \mathbf{A_{22}} & \cdots & \mathbf{A_{2m}} \\ & \cdots & & \cdots \\ \mathbf{A_{m1}} & \mathbf{A_{m2}} & \cdots & \mathbf{A_{mm}} \end{pmatrix},$$

where each sub-matrix is given by

$$\mathbf{A_{ij}} = \begin{pmatrix} (a_{ij})_{n_i} & (a_{i(j+1)})_{t_i} & (a_{i(j+2)})_{k_i} \\ (a_{(i+1)j})_{n_i} & (a_{(i+1)(j+1)})_{t_i} & (a_{(i+1)(j+2)})_{k_i} \\ (a_{(i+2)j})_{n_i} & (a_{(i+2)(j+1)})_{t_i} & (a_{(i+2)(j+2)})_{k_i} \end{pmatrix}.$$

If contacts C_i and C_j have no bodies in common, the sub-matrix $\mathbf{A_{ij}}$ is set to $\mathbf{0}$, indicating that their contact forces do not affect each other. However, if contacts C_i and C_j do have a body in common, then the coefficients a_{ij} are the contribution of the contact force at contact C_j to the relative acceleration at contact C_i. More specifically, the coefficient $(a_{ij})_{n_i}$ is the contribution of the contact-force component $(F_j)_{n_j}$ to the relative acceleration at the contact C_i. Analogously, the coefficients $(a_{ij})_{t_i}$ and $(a_{ij})_{k_i}$ are, respectively, the contribution of the contact-force components $(F_j)_{t_j}$ and $(F_j)_{k_j}$ to the relative acceleration at the contact C_i.

Also, notice that the contact force $\vec{F}_j$ is given with respect to the contact frame of C_j, whereas the relative acceleration $\vec{a}_i$ is given with respect to the contact frame of C_i. Therefore, a change of basis is required when computing the coefficients of matrix $\mathbf{A_{ij}}$ and vector $\vec{b}_i$.

Suppose contact C_i involves bodies B_1 and B_2, and contact C_j involves bodies B_2 and B_3, that is, they have body B_2 in common. We want to determine the contribution of the contact force $\vec{F}_j$ of contact C_j to the relative acceleration of contact C_i. This in turn involves determining the coefficients of the sub-matrix $\mathbf{A_{ij}}$ and the components $(b_i)_{n_i}$, $(b_i)_{t_i}$ and $(b_i)_{k_i}$ of vector $\vec{b}$. The relative acceleration at contact C_i between bodies B_1 and B_2 is given by

$$\begin{aligned} (a_i)_{n_i} &= (\vec{a}_1 - \vec{a}_2) \cdot \vec{n}_i + 2(\vec{v}_1 - \vec{v}_2) \cdot \frac{d\vec{n}_i}{dt} \\ (a_i)_{t_i} &= (\vec{a}_1 - \vec{a}_2) \cdot \vec{t}_i + 2(\vec{v}_1 - \vec{v}_2) \cdot \frac{d\vec{t}_i}{dt} \qquad (4.117) \\ (a_i)_{k_i} &= (\vec{a}_1 - \vec{a}_2) \cdot \vec{k}_i + 2(\vec{v}_1 - \vec{v}_2) \cdot \frac{d\vec{k}_i}{dt}. \end{aligned}$$

As explained in the single-contact case, only the first term of equations (4.117) depends on the forces exerted at contact C_i. The second term depends on the linear and angular velocities, and is added to $(b_i)_{n_i}$, $(b_i)_{t_i}$ and $(b_i)_{k_i}$, as appropriate. Thus, the contribution of the contact force $\vec{F}_j$ of contact C_j does not affect the components of vector $\vec{b}$. In other words, the expressions used to compute vector $\vec{b}$ for the single-contact case are still valid for the multiple-contact case, that is, the components $(b_i)_{n_i}$, $(b_i)_{t_i}$ and $(b_i)_{k_i}$ of vector $\vec{b}$ are given by summing equations (4.105) and (4.116).

Using equation (4.61), the contribution of the contact force $\vec{F}_j$ of contact C_j to the acceleration $\vec{a}_1$ of body B_1 involved in collision C_i is

$$(\mathbf{I_1}^{-1}(\vec{p}_1 - \vec{r}_1) \times \vec{F}_j) \times (\vec{p}_1 - \vec{r}_1) + \frac{\vec{F}}{m_1}.$$

Conversely, the contribution of $\vec{F}_j$ to $\vec{a}_2$ is

$$(\mathbf{I_2}^{-1}(\vec{p}_2 - \vec{r}_2) \times (-\vec{F}_j)) \times (\vec{p}_2 - \vec{r}_2) - \frac{\vec{F}}{m_2}.$$

The net contribution of $\vec{F}_j$ to the relative acceleration $(\vec{a}_1 - \vec{a}_2)$ at contact C_i is then

$$\begin{aligned} g_j^i &= (\mathbf{I_1}^{-1}(\vec{p}_1 - \vec{r}_1) \times \vec{F}_j) \times (\vec{p}_1 - \vec{r}_1) \\ &+ (\mathbf{I_2}^{-1}(\vec{p}_2 - \vec{r}_2) \times \vec{F}_j) \times (\vec{p}_2 - \vec{r}_2) \\ &+ \left(\frac{1}{m_1} + \frac{1}{m_2}\right) \vec{F}. \end{aligned}$$

Substituting this into the first terms of equations (4.117), we obtain the contributions of $\vec{F}_j$ to each relative-acceleration component at contact C_i as

$$\begin{aligned} \text{contribution to } (a_i)_{n_i} &= g_j^i \cdot \vec{n}_i \\ \text{contribution to } (a_i)_{t_i} &= g_j^i \cdot \vec{t}_i \\ \text{contribution to } (a_i)_{k_i} &= g_j^i \cdot \vec{k}_i. \end{aligned}$$

Using the fact that the contact force $\vec{F}_j$ is expressed with respect to the contact frame C_j as

$$\vec{F}_j = (F_j)_{n_j} \vec{n}_j + (F_j)_{t_j} \vec{t}_j + (F_j)_{k_j} \vec{k}_j,$$

it can be written with respect to the contact frame C_i as

$$\vec{F}_{j\mapsto i} = \mathbf{M}_{j\mapsto i}\,\vec{F}_j$$

with

$$\mathbf{M}_{j\mapsto i} = \begin{pmatrix} \vec{n}_j\cdot\vec{n}_i & \vec{t}_j\cdot\vec{n}_i & \vec{k}_j\cdot\vec{n}_i \\ \vec{n}_j\cdot\vec{t}_i & \vec{t}_j\cdot\vec{t}_i & \vec{k}_j\cdot\vec{t}_i \\ \vec{n}_j\cdot\vec{k}_i & \vec{t}_j\cdot\vec{k}_i & \vec{k}_j\cdot\vec{k}_i \end{pmatrix}.$$

Therefore, the coefficients of the sub-matrix $\mathbf{A_{ij}}$ can be immediately obtained after carrying out the matrix multiplication

$$\mathbf{A_{ij}} = (\mathbf{A_1} + \mathbf{A_2})\,\mathbf{M}_{j\mapsto i}. \tag{4.118}$$

Notice that, if $i = j$, then the matrix $\mathbf{M}_{j\mapsto i}$ becomes the identity matrix, and the matrix $\mathbf{A_{ij}}$ in (4.118) is the same as that obtained in equation (4.116) for the single-contact case. Also, if friction is not taken into account, the sub-matrix $\mathbf{A_{ij}}$ is reduced to

$$\mathbf{A_{ij}} = (a_{ij})_{n_i},$$

since the contact-force components $(F_j)_{t_j}$ and $(F_j)_{k_j}$ are zero in the frictionless case. This result is also compatible with that obtained for the frictionless single-contact-force computation explained in Section 4.8.3.

Having computed the contact force $\vec{F}_i$ for each contact C_i, $1 \leq i \leq m$, we update the dynamic state of each rigid body involved in contact C_i by applying $+\vec{F}_i$ to body B_1 (i.e., the body with index 1) and $-\vec{F}_i$ to body B_2 (i.e., the body with index 2.)

When a rigid body is involved in multiple contacts, it is possible to have it assigned to different indexes for each contact. For the particular situation of cluster G_2 in Figure 4.37, body B_2 has index 2 with respect to its contact C_1 with body B_1, and index 1 with respect to its contact C_2 with body B_3. So, the net contact force actually applied to body B_2 after all contact forces have been computed is

$$(\vec{F}_2 - \vec{F}_1),$$

with $\vec{F}_1$ and $\vec{F}_2$ being the contact forces associated with contacts C_1 and C_2, respectively.

4.9 Particle-Rigid Body Contact Revisited

As mentioned in Section 3.7 of Chapter 3, the contact between a particle and a rigid body is modeled as a particle-particle contact between the

4.9 Particle-Rigid Body Contact Revisited

particle itself and another particle on the rigid body's surface. Modeling the contact in this way has the advantage of letting us use techniques similar to those applied to the particle-particle case. The main differences are:

1. The velocity and acceleration of the particle associated with the rigid body are computed using the rigid-body dynamic equations derived in Section 4.2, as opposed to using the particle dynamic equations of Section 3.2.

2. The normal and tangent-plane directions are determined from the rigid-body geometry. If the particle on the rigid body lies on a face, edge or vertex, then the contact normal is assigned to the face, edge or vertex normal, respectively. The actual computation of these normals was already covered in Section 4.4.

Consider, for example, the particle-rigid body contact illustrated in Figure 4.38. Assume particle O_1 is in contact with particle O_2 of rigid body B_2.

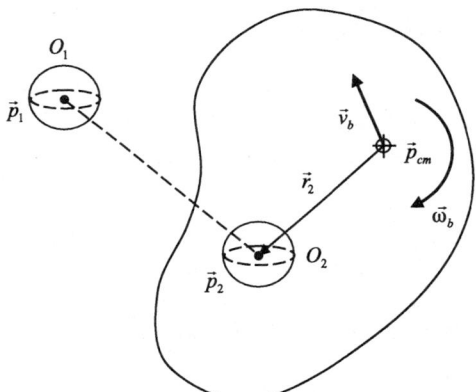

FIGURE 4.38. Particle O_1 is colliding with rigid body B_1 at O_2. The velocity $\vec{v}_{p_2}$ and acceleration $\vec{a}_{p_2}$ of point $\vec{p}_2$ are computed using the rigid body's equations of motion.

Let $\vec{p}_1(t)$ and $\vec{p}_2(t)$ be the points representing the particles in contact. Analogously to both particle-particle and rigid body-rigid body single contact cases, we consider the vector $\vec{q}(t)$ defined as

$$\vec{q}(t) = \begin{pmatrix} q_n(t) \\ q_t(t) \\ q_k(t) \end{pmatrix} = \begin{pmatrix} (\vec{p}_1(t) - \vec{p}_2(t)) \cdot \vec{n}(t) \\ (\vec{p}_1(t) - \vec{p}_2(t)) \cdot \vec{t}(t) \\ (\vec{p}_1(t) - \vec{p}_2(t)) \cdot \vec{k}(t) \end{pmatrix}, \quad (4.119)$$

where $\vec{n}(t)$ is the contact normal, pointing from particle O_2 to particle O_1, and $\vec{t}(t)$ and $\vec{k}(t)$ are vectors defining the tangent plane at the contact.

Clearly, $q_n(t)$ defines a distance measure between points $\vec{p}_1(t)$ and $\vec{p}_2(t)$, along the contact normal, as a function of time. We have $q_n(t) > 0$ if the particles are separated, $q_n(t) = 0$ if the particles are in contact, and $q_n(t) < 0$ if the particles are interpenetrating (see Figure 4.39).

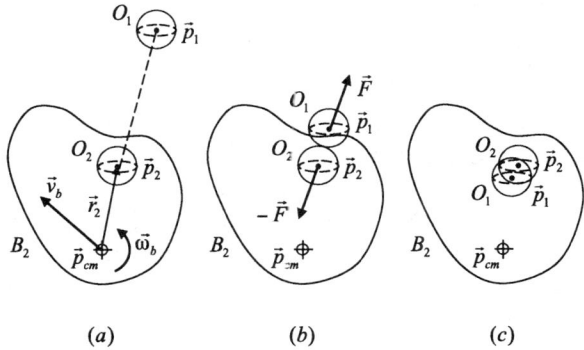

(a) (b) (c)

FIGURE 4.39. (a) Particles O_1 and $O_2 \in B_2$ are about to touch each other at points $\vec{p}_1$ and $\vec{p}_2$; (b) Contact is established whenever $\vec{p}_1 = \vec{p}_2$. In this case, a positive contact force $\vec{F}$ is applied to particle O_1 and a negative contact force $-\vec{F}$ is applied to particle O_2; (c) Interpenetration occurs if $(\vec{p}_1 - \vec{p}_2) \cdot \vec{n} < 0$, where $\vec{n}$ is the contact normal.

The relative normal acceleration at the contact point is therefore that obtained for the particle-particle single contact, given by

$$a_n(t) = (\vec{a}_{p_1}(t) - \vec{a}_{p_2}(t)) \cdot \vec{n}(t) + 2(\vec{v}_{p_1}(t) - \vec{v}_{p_2}(t)) \cdot \frac{d\vec{n}(t)}{dt}. \qquad (4.120)$$

The contact conditions derived in Section 3.6.3 still hold here:

$$a_n = ((a_{11})_n F_n + (b_1)_n) \geq 0$$
$$F_n \geq 0$$
$$F_n ((a_{11})_n F_n + (b_1)_n) = 0$$

and the contact force can be computed using the linear-complementarity techniques presented in Appendix G. The relative acceleration can be written as a linear function of the contact force, that is

$$\vec{a} = \mathbf{A}\vec{F} + \vec{b} \qquad (4.121)$$

and we need to determine the coefficients of matrix $\mathbf{A}$ and vector $\vec{b}$ corresponding to the particle-rigid body contact case. Equation (4.121) can be expanded to

$$\begin{pmatrix} a_n(t_c) \\ a_t(t_c) \\ a_k(t_c) \end{pmatrix} = \begin{pmatrix} (a_{11})_n & (a_{12})_t & (a_{13})_k \\ (a_{21})_n & (a_{22})_t & (a_{23})_k \\ (a_{31})_n & (a_{32})_t & (a_{33})_k \end{pmatrix} \begin{pmatrix} F_n \\ F_t \\ F_k \end{pmatrix} + \begin{pmatrix} (b_1)_n \\ (b_1)_t \\ (b_1)_k \end{pmatrix}$$
$$= \mathbf{A}\vec{F} + \vec{b}, \qquad (4.122)$$

where

$$a_t(t_c) = (\vec{a}_{p_1}(t_c) - \vec{a}_{p_2}(t_c)) \cdot \vec{t}(t_c) +$$
$$2(\vec{v}_{p_1}(t_c) - \vec{v}_{p_2}(t_c)) \cdot \frac{d\vec{t}(t_c)}{dt} \qquad (4.123)$$

$$a_k(t_c) = (\vec{a}_{p_1}(t_c) - \vec{a}_{p_2}(t_c)) \cdot \vec{k}(t_c) +$$
$$2(\vec{v}_{p_1}(t_c) - \vec{v}_{p_2}(t_c)) \cdot \frac{d\vec{k}(t_c)}{dt} \qquad (4.124)$$

and

$$\vec{F} = (F_n, F_t, F_k)^t$$

is the associated contact force. From the results already obtained in Sections 3.6.3 and 4.8.3, we know that the contributions of the contact force to the relative acceleration come only from the first term of equation (4.120). More specifically, the acceleration of particle O_1 can be expressed as

$$\vec{a}_{p_1} = \frac{(\vec{F}_1)_{net}}{m_1} = \frac{\vec{F}}{m_1} + \frac{(\vec{F}_1)_{ext}}{m_1}, \qquad (4.125)$$

where $(\vec{F}_1)_{ext}$ is the sum of all external forces acting on particle O_1. The acceleration of particle O_2 can be determined from the rigid-body motion as

$$\vec{a}_{p_2} = -\mathbf{A_2}\vec{F} + \vec{b}_2,$$

where matrix $\mathbf{A_2}$ and vector $\vec{b}_2$ are obtained from equation (4.115). The relative acceleration $(\vec{a}_{p_1} - \vec{a}_{p_2})$ is then

$$\vec{a}_{p_1} - \vec{a}_{p_2} = \frac{1}{m_1}(\vec{F} + (\vec{F}_1)_{ext} + \mathbf{A_2}\vec{F} - \vec{b}_2)$$
$$= \left(\frac{\mathbf{I}}{m_1} + \mathbf{A_2}\right)\vec{F} + \left(\frac{(\vec{F}_1)_{ext}}{m_1} - \vec{b}_2\right), \qquad (4.126)$$

which is already in the desired matrix format of equation (4.122).

242   4. Rigid-Body Systems

Now, let's examine the second term of equations (4.120), (4.123) and (4.124), namely:

$$2(\vec{v}_{p_1} - \vec{v}_{p_2}) \cdot \frac{d\vec{n}}{dt}$$
$$2(\vec{v}_{p_1} - \vec{v}_{p_2}) \cdot \frac{d\vec{t}}{dt} \qquad (4.127)$$
$$2(\vec{v}_{p_1} - \vec{v}_{p_2}) \cdot \frac{d\vec{k}}{dt} .$$

From the results of Sections 3.6.3 and 4.8.3, we already know that these terms are independent of the contact force, meaning they only affect the coefficients of vector $\vec{b}$. The computation of the time derivatives of the contact frame, that is, the derivatives of the normal and tangent vectors $\vec{n}$, $\vec{t}$ and $\vec{k}$ is covered in Sections E.3.2 and E.4 of Appendix E. Therefore, in the following derivations we shall assume these to be known quantities.

The velocity of particle O_2 is computed from the rigid-body motion as

$$\vec{v}_{p_2} = \vec{v}_2 + \vec{\omega}_2 \times (\vec{p}_2 - \vec{r}_2) . \qquad (4.128)$$

Substituting equation (4.128) into (4.127), we have that the contribution of the second term of equations (4.120), (4.123) and (4.124) to the coefficients of vector $\vec{b}$ is:

$$(b_1)_n = 2(\vec{v}_1 - \vec{v}_2 - \vec{\omega}_2 \times (\vec{p}_2 - \vec{r}_2)) \cdot \frac{d\vec{n}}{dt}$$
$$(b_1)_t = 2(\vec{v}_1 - \vec{v}_2 - \vec{\omega}_2 \times (\vec{p}_2 - \vec{r}_2)) \cdot \frac{d\vec{t}}{dt} \qquad (4.129)$$
$$(b_1)_k = 2(\vec{v}_1 - \vec{v}_2 - \vec{\omega}_2 \times (\vec{p}_2 - \vec{r}_2)) \cdot \frac{d\vec{k}}{dt} .$$

The final coefficients of vector $\vec{b}$ are obtained by summing equations (4.129) with the components of the $\vec{b}$ vector of equation (4.126). Having computed the contact force, we apply $+\vec{F}$ to particle O_1 and $-\vec{F}$ to rigid body B_2 at point $\vec{p}_2$.

4.10   Notes and Comments

Nowadays, there is a substantial number of books and journal articles in the literature that address several aspects of the dynamics of rigid bodies. In this chapter, we used the classic Goldstein [Gol50], and the more recent Beer et al. [BJ77b] and Shabana [Sha94], books as the main references to rigid-body dynamics. Another excellent reference is Baraff et al. [BW98b]

SIGGRAPH course notes. For example, the derivations of the dynamic state of a rigid body using the position, rotation matrix, linear and angular momenta follows the same line of thought as that presented by Baraff *et al.* in their course notes.

The collision-detection and response algorithms presented in this chapter assume rigid bodies to be described by their boundary representation, that is, a list of vertices, edges and faces that make up the rigid body's contour. Campagna *et al.* [CKS98] present a data-structure representation especially tailored for triangle meshes using this boundary representation. It trades memory usage for access time by adding redundant linking information. Using this representation, we can access in constant time every edge and vertex of a given face, every edge incident on a given vertex and every face that contains this vertex, as well as every vertex of a given edge and the faces that share an edge. As far as the collision normal computation is concerned, Thürmer *et al.* [TW98] present an alternate method for computing vertex normals that reduces the dependence of the normal on the underlying mesh representation. This is achieved by adding weights to the contribution of each face using the angle under which the face is incident to the vertex in question (i.e., it uses the average normal computation weighted by incidence angle).

We also discussed in detail two collision-detection algorithms especially tailored for convex bodies, namely the Voronoi Clip and the GJK algorithms. The Voronoi Clip algorithm was developed by Mirtich [Mir97], whereas the GJK algorithm was developed by Gilbert, Johnson and Keerthi (see [GJK88]). The original references have pointers to implementations of these algorithms provided by their authors. In the case of the GJK algorithm, Bergen [vdB99], Cameron [Cam97] and Ong *et al.* [OG97] describe more robust and efficient implementations than that provided by the original authors. Even though the Minkowski difference is never explicitly computed in the GJK algorithm, the interested reader is referred to Berg *et al.* [dBvKOS97], Rourke [O'R98] or Skiena [Ski97] for an in-depth description of how the Minkowski sum and difference are computed.

The collision-response module described in this chapter can be subdivided into two sub-modules: one to compute collision impulses, and another to compute contact forces. Hahn [Hah88] and Mirtich [Mir96b] used the concept of micro-collisions to simulate bodies in contact. That is, contact is simulated as a series of several consecutive collisions. Mirtich went one step further and modeled the relative slipping and sticking of the collision point through the (very short) time interval the bodies are colliding. Another interesting approach to model the relative slipping and sticking of the collision point was developed by Keller [Kel86].

The approach developed in this chapter to deal with friction in collisions is based on the critical-friction-coefficient formulation of Brach [Bra91]. We have extended his work in several ways, however. First, we have explicitly derived the computation of the critical-friction coefficient for single collisions. Second, we present an innovative matrix representation with columns and rows arranged in such way that it can be easily extended to multiple collisions. Last, but not least, our method reduces the multiple-collision problem to the solution of a large sparse linear system (see Duff *et al.* [DER86] for a comprehensive treatment of sparse matrix methods).

The formulation of the contact-force computation as a quadratic-programming problem was originally introduced by Lötstedt [Löt84]. Baraff (see [Bar92, Bar89, Bar90]) initially modeled the contact-force computation as a quadratic program, but used a heuristic formulation to solve the problem by linear-programming techniques. He then extended his heuristic approach to cope with friction (see Baraff [Bar91]), and later presented other algorithms to solve the quadratic program using the linear-complementarity (see Baraff [Bar94]) formulation.

In this book, we focused our approach to computing contact forces using Baraff's linear-complementarity formulation. We have modified Baraff's formulation to cope with directional friction at the contact points. This in turn required some modification of the linear-complementarity algorithm used to compute the contact forces, described in detail in Appendix G.

Preliminary results on an alternate contact-force computation technique using singular value decomposition are presented by Mirtich [Mir98]. Another interesting work in applying the linear-complementarity formulation to compute multiple collision impulses with friction was presented by Kawachi *et al.* [KSK97].

Lastly, the integration of particle and rigid-body systems proposed in this book is at the simulation-engine level, that is, embedding the necessary functionality into a single simulation engine. However, there may be situations in which we want to merge different simulation engines, perhaps developed by different teams. In this case, a simulation-engine level integration may be very difficult, and a higher-level integration technique is needed. Such a technique can be found in Baraff *et al.* [BW97].

5
Articulated Rigid-Body Systems

5.1 Introduction

The dynamic simulation of rigid-body systems covered in the previous chapter can be further extended to the case of articulated rigid-body systems, where bodies are attached to each other using joints. There are several types of joints that can be used to connect bodies, and they differ from each other by the degree of freedom of the relative motion allowed. Several methods have been proposed to address the dynamics of articulated systems, and most of them fall into one of the following two categories. In the first category, the dynamic equations describing the system's motion are formulated using a reduced set of variables. This is the so called *reduced coordinate* formulation. The reduced set of variables, also known as generalized coordinates, is obtained by removing all degrees of freedom constrained by the joints. The result is a set of parameterized coordinates that fully describes the motion of the entire articulated system while assuring the joint constraints. In the second category, additional constraint forces are introduced in the system to assure the joint constraints throughout motion. This method is known as the *Lagrangian* formulation. The idea is to formulate equations relating the constraint forces (also referred to as the Lagrangian multipliers) with the dynamic state of the articulated system. In the case of articulated rigid-body systems, the formulation consists of building and solving a linear system (often sparse) for the joint forces. Sparsity can then be used advantageously to derive $\mathcal{O}(n)$ algorithms, where n is the total number of articulated bodies being considered.

246   5. Articulated Rigid-Body Systems

In this book we shall direct our analysis to techniques based solely on the Lagrangian formulation. Even though the reduced-coordinate formulation is in some cases more effective than the Lagrangian approach, there are still several reasons for using the Lagrangian formulation as opposed to the reduced-coordinate formulation in a software implementation. The most important reason in our view is modularity, in the sense that, once the joints are specified, their constraints can be formulated in terms of acceleration conditions at the joint points, and a linear system relating the joint forces with the dynamic state of the system is readily obtained from such acceleration conditions. In other words, as soon as the acceleration conditions are determined from the type of joints being used, the mathematical framework for computing the joint-constraint forces is exactly the same for all types of joints. For simplicity, we shall focus on the underlying mathematical framework for the case in which the articulated system is connected by spherical joints only. Section 5.5 contains pointers to the literature where the derivation of the acceleration conditions for other types of joints can be found. Such conditions can be easily applied to the mathematical framework presented in this chapter.

By definition, spherical joints do not allow relative translations, only relative rotations between the interconnected bodies. Therefore, they have three degrees of freedom to rotate the bodies with respect to the coordinate axis defined by the joint axis and two orthogonal axes on the plane perpendicular to the joint axis. Figure 5.1 shows the schematic representation of a spherical joint.

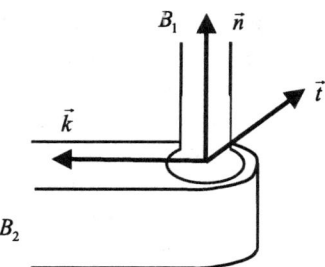

FIGURE 5.1. Bodies B_1 and B_2 connected by a spherical joint. The joint constrains the motion to relative rotations only (relative translations are disallowed).

Most of the notions discussed in Chapter 4 for rigid-body systems can be directly applied to articulated rigid-body systems. The main difference in the mathematical formulation between rigid bodies and articulated rigid bodies consists of enforcing the joint constraints throughout the motion, especially when the articulated body is colliding, or in contact with itself or other articulated bodies.

5.2 Articulated Rigid-Body Dynamics

The dynamic state of an articulated rigid body can always be viewed as the concatenation of the dynamic state of each of its links. Consider the articulated rigid body shown in Figure 5.2, containing eight links (i.e., eight rigid bodies) and eight spherical joints.

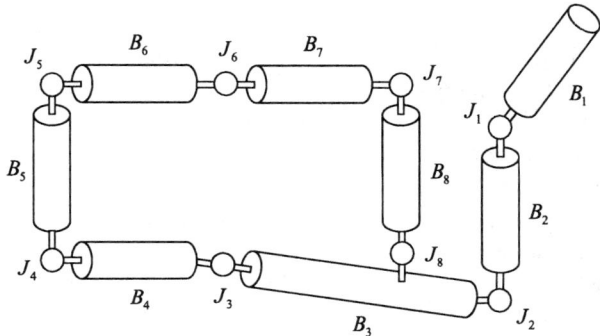

FIGURE 5.2. An example of an articulated body with eight bodies connected by eight spherical joints.

The dynamic state of the articulated rigid body can be expressed as

$$\vec{y}(t) = (\vec{y}_1(t), \vec{y}_2(t), \ldots, \vec{y}_8(t))^t , \qquad (5.1)$$

where each $\vec{y}_i(t)$ for $i \in \{1, \ldots, 8\}$, represents the dynamic state of body i given by

$$\vec{y}_i(t) = \begin{pmatrix} \vec{r}_i(t) \\ \mathbf{R}_i(t) \\ \vec{L}_i(t) \\ \vec{H}_i(t) \end{pmatrix},$$

with $\vec{r}_i(t)$, $\mathbf{R}_i(t)$, $\vec{L}_i(t)$ and $\vec{H}_i(t)$ being the position and orientation of body B_i's center of mass, and the body's linear and angular momenta, respectively. To numerically integrate the equations of motion of the articulated body, we first need to compute the time derivative of its dynamic state. Deriving equation (5.1) with respect to time, we get

$$\frac{d\vec{y}(t)}{dt} = \left(\frac{d\vec{y}_1(t)}{dt}, \frac{d\vec{y}_2(t)}{dt}, \ldots, \frac{d\vec{y}_8(t)}{dt} \right)^t , \qquad (5.2)$$

where each $d\vec{y}_i(t)/dt$ is given by

$$\frac{d\vec{y}_i(t)}{dt} = \begin{pmatrix} \vec{v}_i(t) \\ \vec{\omega}_i(t)\,\mathbf{R}_i(t) \\ \vec{F}_i(t) \\ \vec{\tau}_i(t) \end{pmatrix}. \tag{5.3}$$

The variables $\vec{v}_i(t)$, $\vec{\omega}_i(t)$, $\vec{F}_i(t)$ and $\vec{\tau}_i(t)$ in equation (5.3) are body B_i's linear velocity, angular velocity, net force and net torque acting on its center of mass.

Clearly from equations (5.2) and (5.3), the numerical integration of the equations of motion can only be carried out if all external forces and torques acting on each body (i.e., each link) are known. In the case of articulated bodies, the motion of the bodies is constrained by the joints attached to them. This constraint is represented by a joint force $\vec{F}_i$ associated with joint J_i, which acts as an external force applied to the connected bodies. So, in order to numerically integrate the equations of motion of an articulated body, we need to first determine the constraint forces associated with each joint in the system. These constraint forces are then summed with all other external forces acting on each interconnected body, thus completely defining the net external force and torque on each link. Having determined the net external force and torque acting on each body, we can proceed with the numerical integration of equation (5.2) and determine the position and orientation of each link at the end of the current time step being considered.

Since the motion of each body influences the motion of all other bodies it is connected to, all constraint forces need to be simultaneously computed to ensure that the connected bodies will remain connected after all external forces and torques are applied to all bodies in the articulation. The goal is then to derive an expression that relates how the dynamics of each interconnected rigid body are affected by the application of the joint force. This expression will then be used to simultaneously compute all joint forces.

As far as notation is concerned, the notation in this section significantly differs from that used in upcoming sections 5.4.1 and 5.4.2, for computing impulsive and contact forces between connected links. In those sections, the index of the linked bodies is used to generate the correct index of the impulsive or contact forces associated with each joint. For example, the joint connecting bodies B_i and B_j is referred to as joint J_{ij}, and its associated contact and impulsive forces are $\vec{F}_{ij}$ and $\vec{P}_{ij}$. In this section, the index of each joint is used to generate the correct index of the linked bodies. For instance, the bodies connected by joint J_i are referred to as bodies $(B_1)_i$ and $(B_2)_i$, and the $\vec{F}_i$ represents the joint force constraining their relative motion[1]. These different notations for computing the dynamics of the articulated system and responding to collisions and contacts requires a

---

[1] By convention, a positive joint force $+\vec{F}_i$ is applied to body $(B_1)_i$, whereas a negative joint force $-\vec{F}_i$ is applied to body $(B_2)_i$.

redundant underlying representation of the system. More specifically, the software data structures should be such that given a link (i.e., a body) we can efficiently obtain the joints it is connected to, and given a joint we can quickly return the two links attached to it.

For the sake of clarity, we shall first focus on the derivation of the expression relating the joint force and the dynamics of the interconnected bodies for the two-body articulated system shown in Figure 5.3. Later in this section, we shall generalize the derivation for the case of a n-body articulated system.

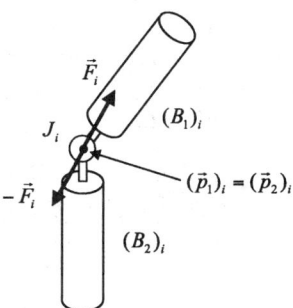

FIGURE 5.3. Bodies $(B_1)_i$ and $(B_2)_i$ are connected by the spherical joint J_i at points $(\vec{p}_1)_i \in (B_1)_i$ and $(\vec{p}_2)_i \in (B_2)_i$.

Let $(\vec{p}_1)_i$ and $(\vec{p}_2)_i$ be the points on bodies $(B_1)_i$ and $(B_2)_i$ to which joint J_i is attached. Since the relative motion of the interconnected bodies in the case of spherical joints is constrained to rotations only, the relative position of the joint points on each body should remain the same throughout the motion, that is

$$(\vec{p}_1)_i(t) = (\vec{p}_2)_i(t), \ \forall t \ . \tag{5.4}$$

Using techniques similar to those in Section 4.8.3 for computing the contact force between two rigid bodies, we can relate the joint force to the dynamic state of the connected bodies by differentiating equation (5.4) twice in time. Differentiating once we have

$$(\vec{v}_{p_1})_i(t) = (\vec{v}_{p_2})_i(t) \ , \tag{5.5}$$

that is, the velocity at the joint points should remain the same throughout the movement. Differentiating one more time, we obtain

$$(\vec{a}_{p_1})_i(t) = (\vec{a}_{p_2})_i(t) \ , \tag{5.6}$$

indicating that the accelerations at the joint points should be the same as well. According to equation (4.110) on page 232, the acceleration $(\vec{a}_{p_1})_i$ of the joint point $(\vec{p}_1)_i$ is given by

$$(\vec{a}_{p_1})_i = (\mathbf{I_1}^{-1}((\vec{p}_1)_i - \vec{r}_1) \times \vec{F}_i) \times ((\vec{p}_1)_i - \vec{r}_1)$$
$$+ (\mathbf{I_1}^{-1}(\vec{\tau}_1)_{ext} + \vec{H}_1 \times \vec{\omega}_1) \times ((\vec{p}_1)_i - \vec{r}_1)$$
$$+ \vec{\omega}_1 \times (\vec{\omega}_1 \times ((\vec{p}_1)_i - \vec{r}_1)) + \left(\frac{\vec{F}_i + (\vec{F}_1)_{ext}}{m_1}\right), \quad (5.7)$$

where $\vec{F}_i$, $(\vec{F}_1)_{ext}$ and $(\vec{\tau}_1)_{ext}$ are, respectively, the joint force, the net external force and the net external torque acting on body $(B_1)_i$. The variables $\mathbf{I_1}$, $\vec{r}_1$, $\vec{H}_1$ and $\vec{\omega}_1$ are the inertia tensor, center of mass position, angular momentum and angular velocity of body $(B_1)_i$, all computed with respect to the world-coordinate frame.

Equation (5.7) can be further simplified to (see derivation of equation (4.113) on page 233):

$$(\vec{a}_{p_1})_i = (\mathbf{A_1})_i \vec{F}_i + (\vec{b}_1)_i, \quad (5.8)$$

with

$$(\mathbf{A_1})_i = \left(\frac{1}{m_1}\mathbf{I} - \tilde{x}_1 \mathbf{I_1}^{-1} \tilde{x}_1\right)$$
$$(\vec{b}_1)_i = \frac{1}{m_1}(\vec{F}_1)_{ext} + (\mathbf{I_1}^{-1}(\vec{\tau}_1)_{ext} + \vec{H}_1 \times \vec{\omega}_1) \times \vec{x}_1$$
$$+ \vec{\omega}_1 \times (\vec{\omega}_1 \times \vec{x}_1)$$
$$\vec{x}_1 = (\vec{p}_1)_i - \vec{r}_1 .$$

Analogously, the acceleration at joint point $(\vec{p}_2)_i$ can be expressed as

$$(\vec{a}_{p_2})_i = -(\mathbf{A_2})_i \vec{F}_i + (\vec{b}_2)_i, \quad (5.9)$$

with

$$(\mathbf{A_2})_i = \left(\frac{1}{m_2}\mathbf{I} - \tilde{x}_2 \mathbf{I_2}^{-1} \tilde{x}_2\right)$$
$$(\vec{b}_2)_i = \frac{1}{m_2}(\vec{F}_2)_{ext} + (\mathbf{I_2}^{-1}(\vec{\tau}_2)_{ext} + \vec{H}_2 \times \vec{\omega}_2) \times \vec{x}_2$$
$$+ \vec{\omega}_2 \times (\vec{\omega}_2 \times \vec{x}_2)$$
$$\vec{x}_2 = (\vec{p}_2)_i - \vec{r}_2 .$$

Substituting equations (5.8) and (5.9) into equation (5.6), we get

$$(\mathbf{A_1})_i \vec{F}_i + (\vec{b}_1)_i = -(\mathbf{A_2})_i \vec{F}_i + (\vec{b}_2)_i ,$$

that is

5.2 Articulated Rigid-Body Dynamics

$$((\mathbf{A_1})_i + (\mathbf{A_2})_i)\, \vec{F}_i = -(\vec{b}_1)_i + (\vec{b}_2)_i \,. \tag{5.10}$$

Clearly, the joint force $\vec{F}_i$ can be immediately obtained by solving the linear system given in equation (5.10). Also, it can be shown that matrix $((\mathbf{A_1})_i + (\mathbf{A_2})_i)$ is always invertible, and that the linear system in equation (5.10) is always solvable.

Now, consider the situation in which the articulated system is composed of n bodies connected by q spherical joints. For each joint J_i in the system, we still have to assure

$$(\vec{a}_{p_1})_i = (\vec{a}_{p_2})_i \,, \tag{5.11}$$

where the indexes 1 and 2 refer to bodies $(B_1)_i$ and $(B_2)_i$ connected by joint J_i. However, the accelerations at the joint points (i.e., points $(\vec{p}_1)_i$ and $(\vec{p}_2)_i$) may involve one or more joint forces, as opposed to just one joint force, as in equations (5.8) and (5.9) for the two-body case. More specifically, the accelerations at the joint points will involve as many joint forces as there are joints attached to the body. Consider the articulated body of Figure 5.2, repeated in Figure 5.4 with the body index assignment for each joint.

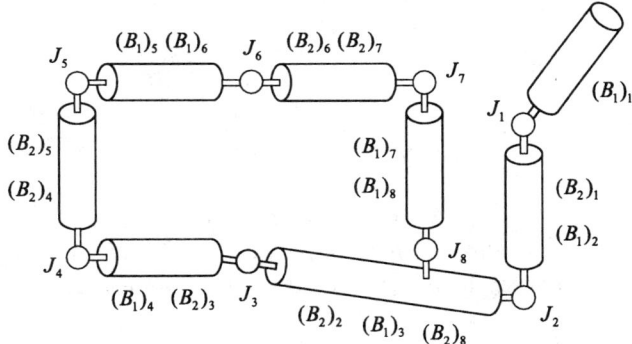

FIGURE 5.4. For each joint in the system, we assign indexes to the bodies attached to it. Bodies attached to more than one joint are assigned multiple indexes. We then use this assignment to build the linear system containing all joint equations.

With this arbitrary index assignment, bodies attached to multiple joints may have different indexes associated with each joint. For instance, body B_3 has index 1 with respect to joint J_3 and index 2 with respect to joints J_2 and J_8. Recall that the index relates the sign of the joint force applied to the body, namely index 1 indicates a positive sign and index 2 indicates a negative sign. So, the constraint equation (5.11) applied to joint J_1 gives

$$(\mathbf{A_1})_1^1 \vec{F}_1 + (\vec{b}_1)_1^1 = -(\mathbf{A_2})_1^1 \vec{F}_1 + (\mathbf{A_1})_2^1 \vec{F}_2 + (\vec{b}_2)_1^1 + (\vec{b}_1)_2^1 \,, \tag{5.12}$$

252   5. Articulated Rigid-Body Systems

that is, the acceleration $(\vec{a}_{p_1})_1$ depends only on the joint force $\vec{F}_1$, whereas the acceleration $(\vec{a}_{p_2})_1$ depends on the joint forces $\vec{F}_1$ and $\vec{F}_2$, since body B_2 is attached to joints J_1 and J_2 (see Figure 5.4). Since body B_2 has index 2 with respect to joint J_1, and index 1 with respect to joint J_2, we use a negative sign preceeding $\vec{F}_1$ and a positive sign preceeding $\vec{F}_2$ on the right term of equation (5.12).

Notice the extra superscript used on all matrices and vectors in equation (5.12). This superscript indicates the joint point the computation is referring to. For example, matrix $(\mathbf{A_2})_1^1$ is computed as the coefficient matrix of the body with index 2 of joint J_1, applied to the joint point on J_1. Matrix $(\mathbf{A_1})_2^1$, on the other hand, is the coefficient matrix of the body with index 1 of joint J_2, applied to the joint point on J_1.

In general, the terms $(\mathbf{A_i})_j^q \vec{F}_j$ and $(\vec{b}_i)_j^q$ refer to the coefficients of the body with index $i \in \{1,2\}$ of joint J_j, applied to the joint point on J_q, and are computed as

$$(\mathbf{A_i})_j^q = \left(\frac{1}{m_i} \mathbf{I} - \tilde{x}_i^q \, \mathbf{I_i}^{-1} \, \tilde{x}_i^q \right)$$

$$(\vec{b}_i)_j^q = \frac{1}{m_i} (\vec{F}_i)_{ext} + (\mathbf{I_i}^{-1} (\vec{\tau}_i)_{ext} + \vec{H}_i \times \vec{\omega}_i) \times \tilde{x}_i^q$$
$$+ \ \vec{\omega}_i \times (\vec{\omega}_i \times \tilde{x}_i^q) \, ,$$

with

$$\tilde{x}_i^q = (\vec{p}_i)_q - \vec{r}_i \, .$$

Notice that, for the two-body case, we have just one joint, that is, $q = i$ and the superscript index q can be ignored.

Moving all unknown forces to the left side of equation (5.12), we can rewrite it as

$$((\mathbf{A_1})_1^1 + (\mathbf{A_2})_1^1) \, \vec{F}_1 - (\mathbf{A_1})_2^1 \, \vec{F}_2 = (\vec{b}_1)_1^1 - (\vec{b}_2)_1^1 - (\vec{b}_1)_2^1 \, .$$

Now, applying the constraint equation (5.11) to joint J_2 gives

$$-(\mathbf{A_2})_1^2 \, \vec{F}_1 + (\vec{b}_2)_1^2 + (\mathbf{A_1})_2^2 \, \vec{F}_2 + (\vec{b}_1)_2^2 =$$
$$-(\mathbf{A_2})_2^2 \, \vec{F}_2 + (\vec{b}_2)_2^2 + (\mathbf{A_1})_3^2 \, \vec{F}_3 + (\vec{b}_1)_3^2 - (\mathbf{A_2})_8^2 \, \vec{F}_8 + (\vec{b}_2)_8^2 \, ,$$

which can be rearranged as

$$-(\mathbf{A_2})_1^2 \, \vec{F}_1 + ((\mathbf{A_1})_2^2 + (\mathbf{A_2})_2^2) \, \vec{F}_2 - (\mathbf{A_1})_3^2 \, \vec{F}_3 + (\mathbf{A_2})_8^2 \, \vec{F}_8 =$$
$$(\vec{b}_2)_2^2 + (\vec{b}_1)_3^2 + (\vec{b}_2)_8^2 - (\vec{b}_1)_2^2 - (\vec{b}_2)_1^2 \, .$$

Doing the same for all other joints, we can group the eight joint equations into the following linear system:

$$\begin{pmatrix}
(\mathbf{A}_1)_1^1 + (\mathbf{A}_2)_1^1 & -(\mathbf{A}_1)_2^1 & 0 & 0 \\
-(\mathbf{A}_2)_1^2 & (\mathbf{A}_1)_2^2 + (\mathbf{A}_2)_2^2 & -(\mathbf{A}_1)_3^2 & 0 \\
0 & -(\mathbf{A}_2)_2^3 & (\mathbf{A}_1)_3^3 + (\mathbf{A}_2)_3^3 & -(\mathbf{A}_1)_4^3 \\
0 & 0 & -(\mathbf{A}_2)_3^4 & (\mathbf{A}_1)_4^4 + (\mathbf{A}_2)_4^4 \\
0 & 0 & 0 & (\mathbf{A}_2)_4^5 \\
0 & 0 & 0 & 0 \\
0 & 0 & 0 & 0 \\
0 & (\mathbf{A}_2)_2^8 & -(\mathbf{A}_1)_3^8 & 0 \\
\end{pmatrix}$$

$$\begin{pmatrix}
0 & 0 & 0 & 0 \\
0 & 0 & 0 & (\mathbf{A}_2)_8^2 \\
0 & 0 & 0 & -(\mathbf{A}_2)_8^3 \\
(\mathbf{A}_2)_5^4 & 0 & 0 & 0 \\
(\mathbf{A}_1)_5^5 + (\mathbf{A}_2)_5^5 & (\mathbf{A}_1)_6^5 & 0 & 0 \\
(\mathbf{A}_1)_5^6 & (\mathbf{A}_1)_6^6 + (\mathbf{A}_2)_6^6 & (\mathbf{A}_2)_7^6 & 0 \\
0 & (\mathbf{A}_2)_6^7 & (\mathbf{A}_1)_7^7 + (\mathbf{A}_2)_7^7 & (\mathbf{A}_1)_8^7 \\
0 & 0 & (\mathbf{A}_1)_7^8 & (\mathbf{A}_1)_8^8 + (\mathbf{A}_2)_8^8 \\
\end{pmatrix}$$

$$\begin{pmatrix} \vec{F}_1 \\ \vec{F}_2 \\ \vec{F}_3 \\ \vec{F}_4 \\ \vec{F}_5 \\ \vec{F}_6 \\ \vec{F}_7 \\ \vec{F}_8 \end{pmatrix} = \begin{pmatrix}
-(\vec{b}_1)_1^1 + (\vec{b}_2)_2^1 + (\vec{b}_1)_2^1 \\
-(\vec{b}_1)_2^2 - (\vec{b}_2)_1^2 + (\vec{b}_2)_2^2 + (\vec{b}_1)_3^2 + (\vec{b}_2)_8^2 \\
-(\vec{b}_2)_2^3 - (\vec{b}_1)_3^3 - (\vec{b}_2)_3^3 + (\vec{b}_1)_4^3 + (\vec{b}_2)_3^3 \\
-(\vec{b}_1)_4^4 - (\vec{b}_2)_3^4 + (\vec{b}_2)_4^4 + (\vec{b}_2)_5^4 \\
-(\vec{b}_1)_5^5 - (\vec{b}_1)_6^5 + (\vec{b}_2)_5^5 + (\vec{b}_2)_4^5 \\
-(\vec{b}_1)_5^6 - (\vec{b}_1)_6^6 + (\vec{b}_2)_6^6 + (\vec{b}_2)_7^6 \\
-(\vec{b}_1)_7^7 - (\vec{b}_1)_8^7 + (\vec{b}_2)_7^7 + (\vec{b}_2)_6^7 \\
-(\vec{b}_1)_7^8 - (\vec{b}_1)_8^8 + (\vec{b}_2)_2^8 + (\vec{b}_1)_3^8 + (\vec{b}_2)_8^8 \\
\end{pmatrix}$$

Solving this linear system for the joint forces, we can obtain the contact forces at each joint and compute the net force and net torque acting on each body. This will let us numerically integrate the articulated system's equation of motion.

5.3 Collision Detection

The collision detection between articulated bodies and other rigid bodies, articulated rigid bodies and particles, can be viewed as an extension of the collision-detection techniques already explained in Chapter 4. Detecting collisions between articulated bodies and other bodies is the roughly same as detecting collisions between each individual link of the articulation and other bodies. Efficiency is achieved by taking advantage of the hierarchical

254   5. Articulated Rigid-Body Systems

tree representation of each link in the articulation to build a hierarchical tree representation of the articulated system as a whole.

The idea is to use a two-level hierarchical tree representation of the articulated system to efficiently carry out the collision checks. At the first level, each individual link is represented by its own hierarchical tree. The leaves of such trees are the triangular faces defining the geometric boundary of the link. At the second level, the individual hierarchical tree representations of each link are substituted for the root node of their tree, and another hierarchical tree is built having the root nodes of each link as its leaves. In other words, the second-level hierarchical tree is a tree where each of its leaves is a hierarchical tree itself. Figures 5.5, 5.6 and 5.7 show this two-level hierarchical tree representation for a simple articulated body.

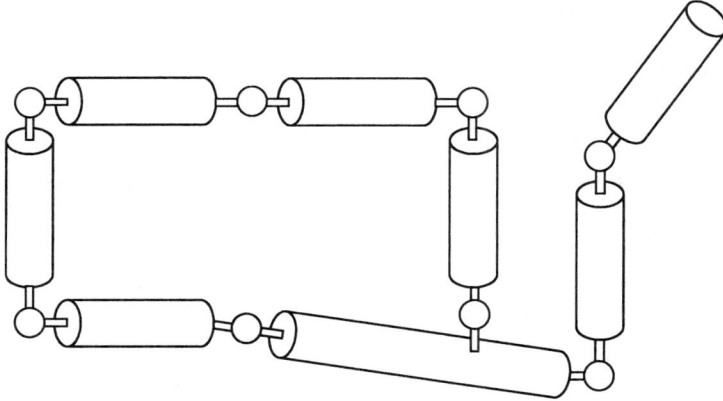

FIGURE 5.5. An articulated body.

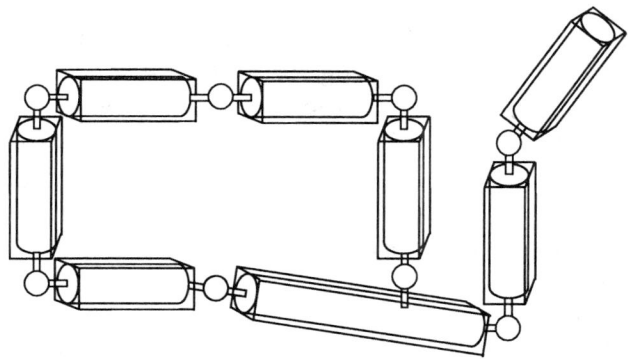

FIGURE 5.6. First-level hierarchical tree representation contains the individual representations of each link of the articulated body shown in Figure 5.5.

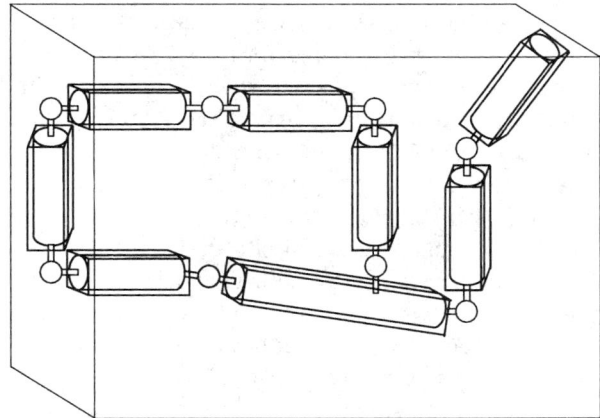

FIGURE 5.7. Root node of the hierarchical tree representation of each link obtained in the first level is used as a leaf of the hierarchical tree built at the second level. The root node of the tree at the second level contains the entire articulated structure, as shown here.

The collision check involving articulated rigid bodies is therefore a two-step process. Checking for collisions between two articulated bodies consists of first checking for geometric intersections between their second-level hierarchical tree representations. If no intersections are found, then the articulated bodies are not colliding. Otherwise, the intersecting leaves of their second-level representation are substituted for their corresponding first-level hierarchical tree (recall that each leaf of the second level is a hierarchical tree of itself). The algorithm proceeds checking for collisions between the hierarchical tree representations of the intersecting leaves. If no intersections are found, then the articulated bodies are not colliding. Otherwise, the intersecting leaves of the first level are substituted for the triangular faces they represent. At this point, several triangle-triangle intersection tests are carried out to detect whether there are pairs of intersecting triangular faces, one on each articulated body being checked for collision. Whenever an intersection is detected, the articulated bodies are said to be colliding and their trajectories are traced back in time to the moment before the most recent collision, that is, just before a pair of triangular faces intersects for the first time. The geometric displacement of the faces is then used to compute a collision normal and tangent plane, in the same manner already explained in Section 4.4, for rigid body-rigid body collisions.

Checking for collisions between an articulated body and a single rigid body is equivalent to checking for collisions between two articulated bodies, with one of them having just one link and zero joints. Again, we start by checking for geometric intersections between the root node of the hierarchical tree representation of the rigid body and the second-level hierarchical tree representing the articulated body. If an intersection is detected, then the leaves of the second-level tree that are intersecting the root node of the single rigid body are further expanded into their corresponding first-level hierarchical trees. The collision detection then proceeds by checking for intersections between the expanded leaves and the rigid body's hierarchical tree. The latter is analogous to intersecting two rigid bodies.

The two-level hierarchical tree representation of articulated bodies can also be used to detect self-collisions. Notice that each link in the articulated body can be viewed as a single rigid body itself. Therefore, checking for self-collisions is equivalent to checking for collisions between several single rigid bodies (i.e., the links) and the articulated body, as explained in the previous paragraph. Of course, the collision check between a link and the articulated body will always return the link itself as an intersection. We should discard this case and look for the existence of intersections between the link being considered and other links in the articulation. If the intersection of the link with itself is the only one found, then the link does not intersect any other link in the articulation. Repeating this process for each link in the articulation will efficiently detect whether self-collisions exist.

5.4 Collision Response

Responding to collisions involving articulated bodies is inevitably a multi-collision problem. Whenever a collision is detected with one of the articulated links, the resulting collision impulse propagates its reaction to all other bodies in the articulation through their joint connections. In other words, *external collisions* between the articulated rigid body and other bodies (possibly other articulated rigid bodies as well) create a sequence of *internal collisions* between the links at their connecting joints. The way this sequence of internal collisions is resolved depends on the mathematical model used to describe the collisions.

Consider the situation illustrated by Figure 5.8(a), where two articulated rigid bodies are colliding at point $\vec{p}_i$. The collision normal and tangent plane at the collision point $\vec{p}_i$ are computed from the bodies' relative geometric displacement just before their collision, using the same techniques discussed in Section 4.8, page 204, for the case of rigid body-rigid body collisions.

The external-collision impulse $\vec{P}_{1,4}$ generated by this collision will affect the dynamic state of bodies B_1 and B_4, as well as all other bodies attached to them. This modification of the dynamic state of each link is represented by a set of internal-collision impulses, one for each joint of the articulation (see Figure 5.8(b)). Clearly, one way to solve this multiple-collision problem would be to propagate the impulsive reactions to the external collision as if it were a wave passing through each joint. In this case, the joint-impulsive reactions would be incrementally computed as if they happened at different instants. For example, we would first compute $\vec{P}_{1,4}$ ignoring the internal

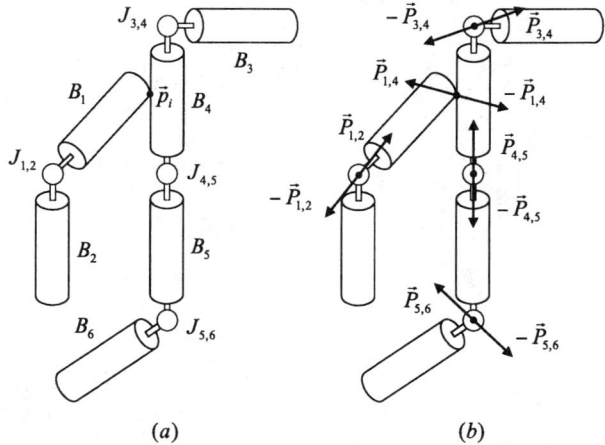

FIGURE 5.8. (a) Two articulated rigid bodies are colliding at point $\vec{p}_i$; (b) The external-collision impulse $\vec{P}_{1,4}$ between bodies B_1 and B_4 creates impulsive reactions at each joint.

dynamics of the articulation. Then we would solve for $\vec{P}_{3,4}$ and $\vec{P}_{4,5}$ on one articulation, and for $\vec{P}_{1,2}$ on the other. Finally, we would compute $\vec{P}_{5,6}$ taking into account that the other impulses have already been applied to the system (i.e., they occured at different instants).

There are two fundamental problems with this approach. The first problem is efficiency. In the particular situation shown in Figure 5.8, we would have to execute the collision-response module three consecutive times, one for computing $\vec{P}_{1,4}$, another for $\vec{P}_{3,4}$ and $\vec{P}_{4,5}$, and a third for $\vec{P}_{5,6}$. The second problem has the potential to never successfully complete the internal-collision-impulse computations, by going to an infinite-loop mode. This can potentially happen every time one of the articulated systems has cycles (see Figure 5.9).

In this book, we shall overlook the propagation time and assume that all internal collisions happen at the same time as the external collisions generating them. By so doing, the external-collision impulses will *simultaneously* affect the dynamic state of all links. Put another way, all collision impulses (external and internal) will be simultaneously computed by solving a linear system of equations involving the dynamic state of the bodies and the impulses. Notice, however, that, even though this assumption produces very good results for articulations containing a small number of links, it can significantly over simplify the impulse-propagation mechanism for articulated systems containing a large number of interconnected links.

In the following sections we shall examine in more detail how the several collision-response mechanisms discussed in Chapter 4 can be adapted to the case of articulated systems.

258   5. Articulated Rigid-Body Systems

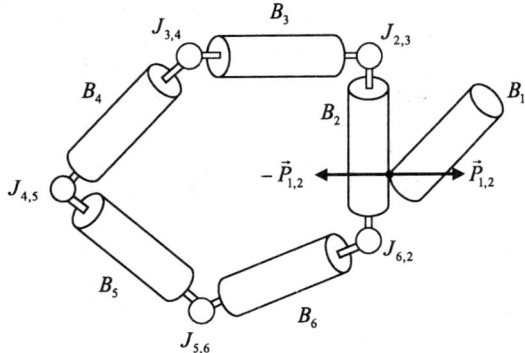

FIGURE 5.9. Hazardous situation when incrementally computing the internal-collision impulses at each joint in response to an external collision. The existence of closed loops in the articulated topology requires all internal impulses to be computed at once.

5.4.1 Computing Impulsive Forces for Single or Multiple External Collisions

The simultaneous computation of all internal and external collision impulses can be done using exactly the same multiple-rigid-body-collision techniques already developed in Section 4.8.2. Recall that the computation of multiple-collision impulses consists of building and solving a sparse linear system of equations. It was shown that the sparse-system matrix can be partitioned into block matrices of the form $\mathbf{A}_{i,q}$, $\mathbf{B}_{i,q}$, $\mathbf{C}_{i,q}$ and $\mathbf{E}_{i,q}$, where the indexes i and q refer to collision C_{iq} between bodies B_i and B_q. The rows of each block matrix are chosen from a set of four possible combinations by comparing the coefficient of friction along each tangent-plane direction with their associated critical coefficient of friction[2].

Using this multiple-rigid-body-collision response framework, each individual link will be treated as a single, independent body, and each joint will be dealt with as a collision point between its links. It remains to assure that the links stay together at their joint points, as opposed to separating because of the application of the collision impulses. In other words, we need to assure the joint constraints at each joint when computing their corresponding collision impulses.

In the case of spherical joints, the relative motion of its links is constrained to rotations only. According to equations (5.4), (5.5) and (5.6), the position, velocity and acceleration of the joint points on each link should coincide throughout the motion. This was used in Section 5.2 to derive expressions relating the joint forces and the dynamics of their links. Here

---

[2]Refer to Section 4.8.1, page 212, for a detailed discussion of the techniques used to compute the critical coefficient of friction.

we use the other two constraints, on position and velocity, to relate the impulses at each joint with the dynamic state of their corresponding links.

Let's start by analyzing the position constraint. Because the positions of the joint points on each link should coincide for the entire motion (see equation (5.4)), they cannot slide with respect to each other during a collision. In other words, the use of the critical-coefficient-of-friction value to determine whether the colliding points (i.e., joint points) are sliding or sticking during a collision is no longer necessary. There is no need to compute the value for each joint because we already know that the collision points should stick to each other during a collision.

As for the velocity constraint, the joint points should always remain with the same velocity throughout the motion (see equation (5.5)), that is, the links should not separate at their joint points after any collision. This condition can be translated into all joint collisions having a zero coefficient of restitution (inelastic collisions).

In general, the coefficient-of-restitution equation should be replaced by the velocity-constraint equation imposed by the type of joint being used. In the case of spherical joints, the velocity-constraint equation is equivalent to forcing the coefficient of restitution to be zero, but this is not always the case for other types of joints. Please refer to Section 5.5 for pointers to the literature where the velocity-constraint equations for other types of joints are derived, such as the prismatic and revolute joints.

Therefore, the four possible cases for defining the rows of each block matrix $\mathbf{A_{i,q}}$, $\mathbf{B_{i,q}}$, $\mathbf{C_{i,q}}$ and $\mathbf{E_{i,q}}$, as discussed in Section 4.8.2, are reduced to just one case, namely the case in which

$$\mathbf{A_{i,q}} = 0 \quad \mathbf{B_{i,q}} = \mathbf{I} \quad \mathbf{C_{i,q}} = -\tilde{r}_i \quad \mathbf{E_{i,q}} = \tilde{r}_q \ . \tag{5.13}$$

In summary, we build the system matrix containing all internal and external impulses by selecting the appropriate block-matrix representation for each collision. The block-matrix representation associated with all external collisions is selected following the same rules applied to the single-rigid-body-collision case, in which the colliding points can slide or stick to each other during the collision. However, the block-matrix representation associated with each internal collision is constrained to that shown in equation (5.13), where the indexes i and q refer to bodies B_i and B_q connected by joint $J_{iq}{}^3$.

---

[3] Using this formulation, a positive collision impulse $\vec{P}_{i,q}$ is applied to body B_i, and a negative collision impulse $-\vec{P}_{i,q}$ is applied to body B_q at joint J_{iq}.

5.4.2 Computing Contact Forces for Single or Multiple External Contacts

Whenever one or more links of an articulated body are in contact with each other, or with other bodies in the simulated world, the contact force at the each contact point will affect the dynamics of motion of the entire articulation. This is analogous to the multiple-rigid-body-contact situation described in Section 4.8.4 on page 234. Recall that the simultaneous computation of all contact forces involves building and solving a Linear Complementarity Problem (LCP), with the contact forces and accelerations constrained to non-negative values. Contact points that end with zero relative accelerations remain in contact after the application of the contact forces, whereas contacts with positive accelerations will break up as soon as the contact forces are applied. Contacts having zero contact forces are just touching each other and may or may not break up, depending on the value of their corresponding relative accelerations.

The idea here is to use the exact techniques already described in Section 4.8.4. The only modifications needed are those necessary to assure the spherical-joint constraints after the contact forces be applied to the system. These modifications should be implemented as part of the LCP solution method described in Appendix G. According to the results presented in Appendix G, the LCP solution method incrementally computes the contact force at contact C_i while enforcing the normal- and frictional-contact conditions at all other contacts j (with $j < i$) already considered. This requires bookkeeping the indexes of all contact points already considered such that, at any point during the computation, an index will be in one subset of each of the following three groups.

1. ZA_n or ZF_n, depending on its normal-acceleration and contact-force components along the normal direction $\vec{n}$.

2. ZA_t, $MaxF_t$ or $MinF_t$, depending on its tangential-acceleration and contact-force components along the tangent direction $\vec{t}$.

3. ZA_k, $MaxF_k$ or $MinF_k$, depending on its tangential-acceleration and contact-force components along the tangent direction $\vec{k}$.

Because the position of the joint points must coincide throughout the entire motion, it is necessary to assure that the relative acceleration at all joint contacts must always be zero. This requires modifying the LCP solution method to assign all indexes corresponding to a joint contact to groups ZA_n, ZA_t and ZA_k, and never displace them. Doing this guarantees that the relative acceleration at the joint points will be zero after all contact forces are applied, that is, that the links will remain connected at their joint points throughout the entire motion.

According to the LCP solution method discussed in Appendix G, the contact-point indexes are moved back and forth between different subsets of each of the three main groups, depending on their relative accelerations or contact forces be assigned a negative value. For instance, a contact point with index in ZA_n has a positive normal-contact-force component and zero relative normal acceleration. If in the next iteration of the LCP solution algorithm the normal-contact-force component assumes a negative value, then the system is evolved up to the point at which the normal-contact-force component assumes a zero value, and the index of such a contact point is moved from ZA_n to ZF_n in order to prevent the normal-contact-force component becoming negative. This in turn permits the relative normal acceleration to assume any positive value.

In the case of a joint contact, we need to force the relative acceleration to be always zero. This may require having a negative contact force at the joint, since we cannot move the contact index out of the zero-acceleration subsets. The following modifications are needed for computing contact forces when the contact index refers to a joint contact.

1. The joint contact force can have any value, as opposed to having only non-negative values (as is the case of external contacts). In practice, a negative joint contact force means that the local-coordinate frame used at the joint contact (i.e., the direction of the normal and tangential coordinate axes) is inverted.

2. The tangential components of the joint-contact force can also have any value, as opposed to being limited to μF_n, that is, the normal component multiplied by the coefficient of friction associated with the tangent direction being considered.

In summary, the LCP solution method discussed in Appendix G can be modified to deal with joint contact forces by forcing their indexes to be always included in the zero-acceleration subsets, and by letting the joint contact forces assume any finite value. This requires keeping track of which indexes correspond to joint contacts, and adding a conditional statement at the force-computation module to not bother if the contact force of a joint contact becomes negative.

As for implementation, numerical round-off errors can offset the joint points by small values each time the simulation loop is executed. The offset is usually not noticeable at the first few simulation time steps, but can become a problem for simulations that are executed for a substantial amount of time. In these cases, the position of the joint points computed using their corresponding links' dynamic state may no longer coincide with each other, reaching a situation in which their separation is visible on the screen. This is clearly not a "feature" that dynamic-simulation systems want to exhibit. A practical workaround to situations like this consists of adding a fictitious spring with zero resting length connecting the joint points. Whenever the joint points separate from each other owing to round-off errors, the spring force will bring the points together again. The stiffness of the spring to be used depends on the mass of the bodies connected by the joint (the heavier the bodies, the stiffer the spring should be).

5.5 Notes and Comments

In this chapter we have tried to cover a small but important subset of the several types of techniques applicable to mechanisms commonly used in dynamic simulations of articulated systems. Even though we have concentrated our analysis on rigid bodies connected by spherical joints, the same techniques can be applied to other types of joints. Different types of joints have different joint constraints to be enforced, thus requiring modifications on the actual constraint equations to be used. However, the general principle of defining the joint constraints and differentiating them twice in time to relate the joint forces and the dynamic state of the articulated system remains unchanged. The differentiation in time may be easier or more difficult to obtain depending on the joint being considered, but the principle remains the same.

As for responding to collisions, independent of the type of joint being used, its velocity constraint should always replace the coefficient-of-restitution equation. The latter is applicable only when the colliding bodies are not interconnected at the collision point. The contact-force computation, on the other hand, may require creating other types of groups on the LCP solution method described in Appendix G, in order to accommodate the joint constraints. For example, in the case of prismatic joints, the interconnected bodies are constrained to translations along the joint axis only. The relative acceleration of the bodies along the joint axis can be zero, positive or negative, whereas their relative-acceleration components on the contact plane perpendicular to the joint axis should always be zero. This in turn requires creating special-purpose subsets to keep track of the indexes of the prismatic-joint contact forces. For instance, we can think of creating a $ZAany_n$ subset for normal component accelerations that can assume any value.

The derivation of the dynamics of motion of multibody systems interconnected by spherical joints was based on the work of Hecker [Hec00a, Hec00b] and Baraff [Bar96]. The use of springs, as well as other techniques, to offset the existence of numerical round-off errors on the position of the joint points can be found in Baumgarte [Bau72] and Barzel et al. [BB88], among others.

Shabana [Sha94] presents joint-constraint formulations for other types of joints not covered in this book, such as the prismatic joint(one degree of freedom to translate along the joint axis), the revolute joint (one degree of freedom to rotate about the joint axis), and the cylindrical joint (two degrees of freedom to translate along, and rotate about, the joint axis).

Other interesting types of constraint formulations can be found in Barzel *et al.* [BB88].

Mirtich's [Mir96b] work on constrained rigid-body dynamics is focused on using revolute and prismatic joints to interconnect bodies. His formulation uses spatial-operator algebra (see also Rodriguez *et al.* [RJKD92]), and extends Featherstone [Fea83, Fea87] to cope with tree-like linkage structures; it further presented some control-systems techniques to kinematically control[4] the motion of articulated bodies. Brach [Bra91] considers chains of rigid bodies interconnected by several types of joints, but limits his analyzes to the two-dimensional case.

Finally, readers interested in the simulation of flexible bodies are referred to the work of Terzopoulos [Ter87] and Shabana [Sha98], where the work on computational dynamics is extended to the study of elastic deformable bodies.

---

[4]By kinematical control we mean that the linear and angular position, velocity and acceleration of the bodies are obtained from an animation system, possibly by interpolating their values between two consecutive animation frames.

Appendix A
Useful 3D Geometric Constructions

A.1 Introduction

In this appendix we shall cover some geometrical constructions used as building blocks to implement the several intersection tests that are part of the particle-particle, particle-rigid body and rigid body-rigid body collision-detection algorithms presented in this book. We shall also discuss in Section A.6 how the tangent plane of a collision (or contact) can be determined given the collision (or contact) normal vector.

We shall use the following notation to describe some of these tests. A point P_i in 3D space will be denoted as

$$\vec{p}_i = ((p_i)_x, (p_i)_y, (p_i)_z) \ .$$

A line segment L_s is defined by its two end points $\vec{p}_1$ and $\vec{p}_2$, and is given by

$$\vec{p} = \vec{p}_1 + k\,(\vec{p}_2 - \vec{p}_1) \ ,$$

with $0 \leq k \leq 1$.

The plane β is described by its normal $\vec{n}_\beta$ and a point $\vec{p}_\beta$ in the plane. Any point $\vec{p} \in \beta$ satisfies the equation

$$\vec{p} \cdot \vec{n}_\beta = d_\beta \ ,$$

where d_β is the plane constant obtained from

$$d_\beta = \vec{p}_\beta \cdot \vec{n}_\beta \ .$$

A.2 Projection of a Point on a Line

The projection $\vec{q}$ of point $\vec{p}$ on line L passing through points $\vec{p}_1$ and $\vec{p}_2$ is a point of L (see Figure A.1) and therefore satisfies the line equation, that is

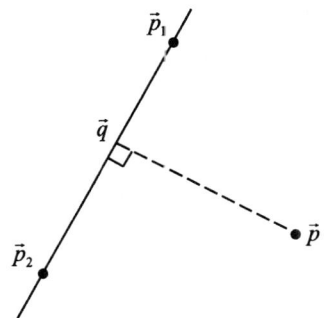

FIGURE A.1. Projecting $\vec{p}$ on line L.

$$\vec{q} = \vec{p}_1 + k\,(\vec{p}_2 - \vec{p}_1)\,, \tag{A.1}$$

where $k \in \mathbb{R}$ is a scalar variable to be determined. Notice also that the vector $(\vec{p} - \vec{q})$ is perpendicular to the line, that is

$$(\vec{p} - \vec{q}) \cdot (\vec{p}_2 - \vec{p}_1) = 0\,. \tag{A.2}$$

Substituting equation (A.1) into (A.2) and solving for k, we have

$$k = \frac{(\vec{p} - \vec{p}_1) \cdot (\vec{p}_2 - \vec{p}_1)}{(\vec{p}_2 - \vec{p}_1) \cdot (\vec{p}_2 - \vec{p}_1)}\,. \tag{A.3}$$

The projection point $\vec{q}$ is directly obtained by substituting equation (A.3) into (A.1). The distance d between $\vec{p}$ and line L is given by

$$d = |\vec{p} - \vec{q}|\,.$$

A.3 Projection of a Point on a Plane

The projection $\vec{q}$ of a point $\vec{p}$ on a plane β can be computed as follows. The projected point $\vec{q}$ satisfies the plane equation, that is

$$\vec{q} \cdot \vec{n}_\beta = d_\beta\,. \tag{A.4}$$

Moreover, the vector $(\vec{p} - \vec{q})$ is parallel to the plane normal $\vec{n}_\beta$ (see Figure A.2), that is

$$(\vec{p} - \vec{q}) = k\,\vec{n}_\beta\,, \tag{A.5}$$

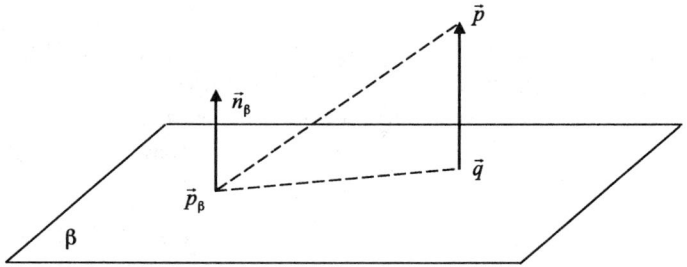

FIGURE A.2. Projecting $\vec{p}$ on plane β.

where k is a scalar that needs to be determined.

Substituting equation (A.5) into (A.4), and solving for k, we obtain

$$k = \frac{\vec{p} \cdot \vec{n}_\beta - d_\beta}{\vec{n}_\beta \cdot \vec{n}_\beta} \ . \tag{A.6}$$

Substituting equation (A.6) into (A.5) we obtain the projected point $\vec{q}$

$$\vec{q} = \vec{p} - \frac{\vec{p} \cdot \vec{n}_\beta - d_\beta}{\vec{n}_\beta \cdot \vec{n}_\beta} \vec{n}_\beta \ .$$

The distance d between point $\vec{p}$ and plane β is computed from

$$d = |\vec{p} - \vec{q}| = \left| \frac{\vec{p} \cdot \vec{n}_\beta - d_\beta}{\vec{n}_\beta \cdot \vec{n}_\beta} \right| \ .$$

A.4 Intersection of a Line Segment and a Plane

The intersection between a line segment and a plane can be the line segment itself (if the line is in the plane), a point or an empty set. Let L be the line segment connecting points $\vec{p}_1$ and $\vec{p}_2$, and let β be the plane with which we want to compute the intersection.

The intersection will be the line itself whenever

$$(\vec{p}_2 - \vec{p}_1) \cdot \vec{n}_\beta = 0 \ ,$$

that is, the line is perpendicular to the plane normal, and

$$\vec{p}_1 \cdot \vec{n}_\beta = d_\beta = \vec{p}_2 \cdot \vec{n}_\beta \ ,$$

that is, its points belong to the plane as well.

Now, suppose the line is such that

$$(\vec{p}_2 - \vec{p}_1) \cdot \vec{n}_\beta \neq 0 \ .$$

The idea is then to first intersect the infinite line supporting the line segment with the plane (let's call $\vec{g}$ this intersection point), and then check whether $\vec{g}$ lies between the end points defining the segment. If it does, then $\vec{g}$ is the actual intersection. Otherwise, the line segment does not intersect the plane. Let's first compute the intersection point $\vec{g}$. We know $\vec{g}$ lies on the infinite line supporting the line segment, that is

$$\vec{g} = \vec{p}_1 + k_g (\vec{p}_2 - \vec{p}_1), \qquad (A.7)$$

where $k_g \in \mathbb{R}$ is a scalar variable to be determined. We also know that the intersection point belongs to the plane, that is

$$\vec{g} \cdot \vec{n}_\beta = d_\beta. \qquad (A.8)$$

Substituting equation (A.8) into (A.7), and solving for k_g, we obtain

$$k_g = \frac{d_\beta - \vec{p}_1 \cdot \vec{n}_\beta}{(\vec{p}_2 - \vec{p}_1) \cdot \vec{n}_\beta}.$$

If $0 \leq k_g \leq 1$, then the intersection point $\vec{g}$ lies between the end points of the line segment, and the line segment does intersect the plane. Otherwise, the intersection point lies outside the segment and the plane does not intersect the line segment.

A.5 Closest Point between a Line and a Line Segment

Given a line L passing through points $\vec{h}_1$ and $\vec{h}_2$, and a line segment L_s defined by its two end points $\vec{p}_1$ and $\vec{p}_2$, we want to determine the point $\vec{p} \in L_s$ that is closer to line L than any other point in L_s. This is illustrated in Figure A.3.

The closest point $\vec{p}$ belongs to line segment L_s, that is

$$\vec{p} = \vec{p}_1 + k_p (\vec{p}_2 - \vec{p}_1), \qquad (A.9)$$

where $0 \leq k_p \leq 1$ is a scalar to be determined. Alternatively, we can rewrite equation (A.9) as

$$k_p = \frac{(\vec{p} - \vec{p}_1)}{(\vec{p}_2 - \vec{p}_1)}. \qquad (A.10)$$

Let $\vec{q}_1$ and $\vec{q}_2$ be the projections of points $\vec{p}_1$ and $\vec{p}_2$ on line[1] L. We compute d_1 and d_2 as the distance between points $\vec{p}_1$ and $\vec{p}_2$ and their associated projection points.

---

[1]The projection of a point on a line was already discussed in Section A.2.

A.5 Closest Point between a Line and a Line Segment   269

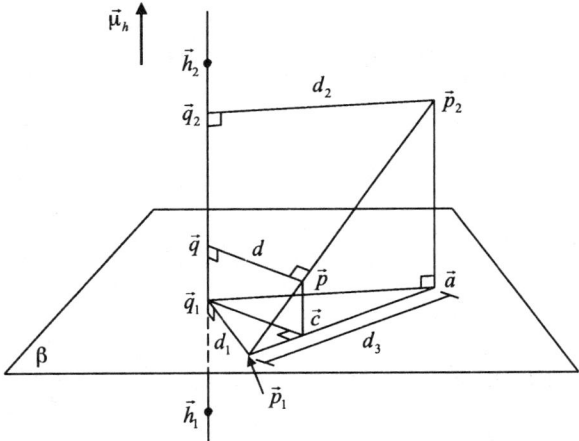

FIGURE A.3. Computing the closest point between a line and a line segment.

Now, consider the auxiliary plane β perpendicular to line L and passing through $\vec{p}_1$ (see Figure A.3). Let $\vec{a}$ be the projection of point $\vec{p}_2$ on plane β, computed according to Section A.3. We compute the distance d_3 between $\vec{a}$ and $\vec{p}_1$.

Let $\vec{c}$ represent the projection of point $\vec{p}$ (still to be determined) on plane β. By construction, triangles $(\vec{p}_1, \vec{c}, \vec{p})$ and $(\vec{p}_1, \vec{a}, \vec{p}_2)$ are similar, that is

$$\frac{(\vec{c} - \vec{p}_1)}{(\vec{a} - \vec{p}_1)} = \frac{(\vec{p} - \vec{p}_1)}{(\vec{p}_2 - \vec{p}_1)} = k_p , \qquad (A.11)$$

where the last equality was obtained using equation (A.10). Also, notice that triangles $(\vec{a}, \vec{q}_1, \vec{c})$ and $(\vec{p}_1, \vec{c}, \vec{q}_1)$ are rectangular on $\vec{c}$, and

$$|\vec{q}_1 - \vec{a}| = |\vec{q}_2 - \vec{p}_2| = d_2 .$$

Applying the Pythagorean theorem to each triangle, we obtain

$$\begin{aligned} d_1^2 &= |\vec{c} - \vec{q}_1|^2 + |\vec{c} - \vec{p}_1|^2 , \\ d_2^2 &= |\vec{c} - \vec{q}_1|^2 + |\vec{a} - \vec{c}|^2 . \end{aligned}$$

Eliminating $|\vec{c} - \vec{q}_1|^2$ from the above equations, we have

$$d_1^2 - d_2^2 = |\vec{c} - \vec{p}_1|^2 - |\vec{a} - \vec{c}|^2 . \qquad (A.12)$$

Factoring out the right-hand side of equation (A.12):

$$d_1^2 - d_2^2 = (|\vec{c} - \vec{p}_1| + |\vec{a} - \vec{c}|)(|\vec{c} - \vec{p}_1| - |\vec{a} - \vec{c}|) . \qquad (A.13)$$

From Figure A.3, we immediately have

$$|\vec{c}-\vec{p}_1|+|\vec{a}-\vec{c}| = d_3 \,. \tag{A.14}$$

Substituting equation (A.14) into (A.13):

$$|\vec{c}-\vec{p}_1|-|\vec{a}-\vec{c}| = \frac{(d_1^2 - d_2^2)}{d_3} \,. \tag{A.15}$$

Solving equations (A.15) and (A.14) for $|\vec{c}-\vec{p}_1|$, we obtain

$$|\vec{c}-\vec{p}_1| = \frac{(d_1^2 - d_2^2 + d_3^2)}{2d_3} \,. \tag{A.16}$$

Finally, substituting equations (A.16) and (A.14) into (A.11) we compute the scalar k_p as

$$k_p = \frac{(d_1^2 - d_2^2 + d_3^2)}{2d_3^2} \,.$$

If $0 < k_p < 1$, then the closest point $\vec{p}$ lies inside the line segment L_s and is obtained directly from equation (A.9). If $k_p \leq 0$, then we set $\vec{p} = \vec{p}_1$. Otherwise, $k_p \geq 1$ and we set $\vec{p} = \vec{p}_2$.

A.6 Computing the Collision- or Contact-Local Frame from the Collision- or Contact-Normal Vector

The relative displacement of the colliding particles or rigid bodies is used to determine the collision normal at the collision point. The actual computation of the collision normal is slightly different depending on whether we are considering particle-particle, particle-rigid body or rigid body-rigid body collisions. Assume that the collision- or contact-normal vector is given by $\vec{n}$, and that the tangent plane is defined by two vectors $\vec{t}$ and $\vec{k}$, mutually perpendicular and perpendicular to $\vec{n}$ as well. Together, they form the local-coordinate system, usually referred to as the *collision frame*, with origin at the collision (or contact) point. The computation of the tangent planes done after the collision normal is determined, and depends strictly on it.

All impulsive and contact forces are computed with respect to their local collision frames. Assuming $\vec{n}$ is known, the question is how we can determine the other vectors $\vec{t}$ and $\vec{k}$? The answer is simple: there are several ways we can generate the other two vectors from the normal vector. In this book, however, we shall use the following approach.

Let the collision normal be given as $\vec{n} = (n_x, n_y, n_z)$. The vector $\vec{t} = (t_x, t_y, t_z)$ is perpendicular to $\vec{n}$, that is

$$\vec{t} \cdot \vec{n} = t_x\, n_x + t_y\, n_y + t_z\, n_z = 0\ . \tag{A.17}$$

Clearly, we have just one equation and three variables, namely t_x, t_y and t_z. The idea is then to establish some rules to assign values to the variables so that equation (A.17) can be satisfied. To that end, we undertake the following steps. Compute the absolute value of each component of the normal vector $\vec{n}$, and compare them such that:

1. If $|n_x| < |n_y|$ and $|n_x| < |n_z|$, then set the auxiliary vector $\vec{a}$ to

$$\vec{a} = (0, -n_z, n_y)\ .$$

2. If $|n_y| < |n_x|$ and $|n_y| < |n_z|$, then set the auxiliary vector $\vec{a}$ to

$$\vec{a} = (-n_z, 0, n_x)\ .$$

3. If $|n_z| < |n_x|$ and $|n_z| < |n_z|$, then set the auxiliary vector $\vec{a}$ to

$$\vec{a} = (-n_y, n_x, 0)\ .$$

The tangent vector $\vec{t}$ will then be given by

$$\vec{t} = \frac{\vec{a}}{|\vec{a}|}\ .$$

The other vector $\vec{k}$ is immediately obtained as

$$\vec{k} = \vec{n} \times \vec{t},$$

since it is perpendicular to both $\vec{n}$ and $\vec{t}$.

A.7 Representing Cross-Products as Matrix-Vector multiplication

Sometimes it is useful to represent a cross-product as a matrix-vector multiplication. Consider the cross-product between vectors $\vec{a} = (a_x, a_y, a_z)$ and $\vec{b} = (b_x, b_y, b_z)$, namely

$$\vec{a} \times \vec{b} = \begin{pmatrix} a_y\, b_z - b_y\, a_z \\ -a_x\, b_z + b_x\, a_z \\ a_x\, b_y - b_x\, a_y \end{pmatrix}\ . \tag{A.18}$$

Now, let's define the matrix $\tilde{\mathbf{a}}$ obtained from vector $\vec{a}$ as

$$\tilde{\mathbf{a}} = \begin{pmatrix} 0 & -a_z & a_y \\ a_z & 0 & -a_x \\ -a_y & a_x & 0 \end{pmatrix}. \qquad (A.19)$$

If we multiply $\vec{b}$ by the matrix $\tilde{\mathbf{a}}$ we obtain

$$\tilde{\mathbf{a}}\vec{b} = \begin{pmatrix} 0 & -a_z & a_y \\ a_z & 0 & -a_x \\ -a_y & a_x & 0 \end{pmatrix} \begin{pmatrix} b_x \\ b_y \\ b_z \end{pmatrix} = \begin{pmatrix} a_y b_z - b_y a_z \\ -a_x b_z + b_x a_z \\ a_x b_y - b_x a_y \end{pmatrix}. \qquad (A.20)$$

Comparing equations (A.18) and (A.20), we immediately conclude that

$$\vec{a} \times \vec{b} = \tilde{\mathbf{a}}\vec{b}. \qquad (A.21)$$

Equation (A.21) is known as the matrix-vector representation of the cross-product between vectors $\vec{a}$ and $\vec{b}$, where the matrix $\tilde{\mathbf{a}}$ is constructed from vector $\vec{a}$ as indicated in equation (A.19).

A.8 Suggested Readings

Most of the geometric constructions presented in this appendix are standard and can be found in almost all computer graphics books. In this appendix, we decided to use the same notation and solution methods presented by Glassner [Gla90], excepting both the computation of the closest point between a line and a line segment (obtained from Karabassi et al. [KPTB99]), and the determination of the tangent plane given the normal vector (obtained from Hughes et al. [HM99]).

Appendix B
Numerical Solution of Ordinary Differential Equations of Motion

B.1 Introduction

The simulation engine is constantly required to compute the dynamic state of all rigid bodies and particles in the scene[1], owing to the net torque and net force being exerted on them. In the case of rigid bodies, this computation requires numerically solving four first-order ordinary differential equations (ODEs) of motion: two coupled equations for the linear momentum and linear position of the body, and two coupled equations for the angular momentum and angular position of the body. The four ODEs of motion are:

$$\frac{d\vec{x}(t)}{dt} = \vec{v}(t) = f_x(t, \vec{x})$$
$$\frac{d\vec{L}(t)}{dt} = \vec{F}(t) = f_L(t, \vec{L})$$
$$\frac{d\mathbf{R}(t)}{dt} = \vec{\omega}(t)\,\mathbf{R}(t) = f_R(t, \mathbf{R})$$
$$\frac{d\vec{H}(t)}{dt} = \vec{\tau}(t) = f_H(t, \vec{H})$$

(B.1)

---

[1] In this appendix, the word scene refers to the simulated world containing all bodies being simulated.

where $\vec{x}(t)$, $\vec{v}(t)$, $\vec{L}(t)$, $\vec{\omega}(t)$ and $\vec{H}(t)$ are the linear position, linear velocity, linear momentum, angular velocity and angular momentum, respectively, of the body being moved. The optional function-style representation of the time derivatives in equation (B.1) is used to simplify the notation and make the equations in this Appendix more readable.

Recall from Chapter 4 that the linear and angular momentum are computed as

$$\vec{L}(t) = m\,\vec{v}(t)$$
$$\vec{H}(t) = \mathbf{I}(t)\,\vec{\omega}(t),$$

with m and $\mathbf{I}(t)$ being the mass and inertia tensor of the body at time t. The inertia tensor of the body is computed relative to the world (fixed) frame, and is given by

$$\mathbf{I}(t) = \mathbf{R}(t)\,\mathbf{I}_b\,\mathbf{R}^{-1}(t),$$

where $\mathbf{I}_b$ is the (constant) inertia tensor relative to the body's frame[2]. The remaining variables in equation (B.1) are the rotation matrix $\mathbf{R}(t)$ representing the angular position of the body at time t, the net force $\vec{F}(t)$ acting on the center of mass of the body, and the net torque $\vec{\tau}(t)$ computed from

$$\vec{\tau}(t) = \sum_{i=1}^{n}(\vec{x}_i(t) - \vec{x}_{cm}(t)) \times \vec{F}_i(t),$$

where n is the total number of external forces acting on the body at time t, $\vec{F}_i(t)$ is the ith external force, $\vec{x}_i$ is the point on the body at which force $\vec{F}_i(t)$ is being applied, and $\vec{x}_{cm}(t)$ is the current position of the body's center of mass.

In the case of particle-systems simulation, the angular-momentum and angular-position equations in (B.1) are not applicable, and the ODEs of motion are reduced to the two coupled equations involving the linear momentum and linear position of the particle. Here, we shall represent the dynamic equations as a single state vector $\vec{Y}(t)$, given by

$$\vec{Y}(t) = \begin{pmatrix} \vec{x}(t) \\ \vec{L}(t) \end{pmatrix} \qquad \frac{d\vec{Y}(t)}{dt} = \begin{pmatrix} \vec{v}(t) \\ \vec{F}(t) \end{pmatrix} \qquad \text{(B.2)}$$

for particles, or

---

[2] See Appendix D for details on how to compute the inertia tensor $\mathbf{I}_b$ in body-frame coordinates.

$$\vec{Y}(t) = \begin{pmatrix} \vec{x}(t) \\ \mathbf{R}(t) \\ \vec{L}(t) \\ \vec{H}(t) \end{pmatrix} \quad \frac{d\vec{Y}(t)}{dt} = \begin{pmatrix} \vec{v}(t) \\ \vec{\omega}(t)\,\mathbf{R}(t) \\ \vec{F}(t) \\ \vec{\tau}(t) \end{pmatrix} \quad (B.3)$$

for rigid bodies. Most of the formulas presented in this appendix are expressed in terms of this single state vector representation, also referred to as the dynamic-state vector of a particle or rigid body in Chapters 3 and 4, respectively. Since the ODEs for particles are a subset of the ODEs for rigid bodies, we shall focus on the solution equations for the latter.

The numerical integration of the ODEs of motion consists of starting at an initial configuration where all positions and momenta are known, and gradually increase the independent variable t (time) through finite steps h (time step), computing the *approximate* value of the linear and angular positions and momenta of the body being moved that best match the Taylor series expansion of their *exact* solution. The Taylor series expansion of the exact solution $\vec{Y}(t)$ at $t = (t_i + h)$, when the values of $\vec{Y}(t_i)$ (the initial condition) and h are known, is given by the infinite sum

$$\vec{Y}(t_i + h) = \sum_{n=0}^{\infty} \frac{h^n}{n!} \frac{\partial^n \vec{Y}(t_i)}{\partial t^n} = \vec{Y}(t_i) + h\frac{\partial \vec{Y}(t_i)}{\partial t} + \frac{h^2}{2!}\frac{\partial^2 \vec{Y}(t_i)}{\partial t^2} + \ldots. \quad (B.4)$$

The degree to which the approximate solution matches the exact solution depends on how close it matches the Taylor series expansion of the exact solution. In other words, the approximate solution is usually a truncation of the Taylor series expansion and the omitted terms reflect the error between the exact and approximate solutions. The point at which the truncation takes place depends on the integration method being used.

In general, the numerical solution of equations (B.2) and (B.3) for time $t = (t_i + h)$ requires the complete knowledge of the state of the system at time $t = t_i$. Starting at the very first time step $t = t_0$, the simulation engine knows the initial state of the system, that is, it knows the linear and angular positions and momenta of each body in the scene. In most cases, the initial linear and angular momenta are set to zero, but nothing prevents us from assigning any finite value to them. At each subsequent time step, the simulation engine computes the net force and torque acting on each body in the scene, and numerically solves equations (B.2) and (B.3) for these values.

It is important to recognize that, independent of the numerical method being used, the simulation engine will always be required to compute the net torque and net force acting on each body at the time interval being considered. Simple numerical methods usually require just one computation at the beginning of each time step. However, as will be shown in Section B.3, there are numerical methods that require computing the net force and net torque acting on the body not only at the beginning of the time interval, but also at some intermediate time values along the current time step. These methods usually combine such intermediate information to obtain the approximate solution for the entire time step. In these cases, the simulation engine needs to temporarily position the bodies at these intermediate time values before it can compute the net torque and net force acting on them.

It is also clear from equation (B.4) that the choice of time step h directly affects the efficiency and stability of the numerical method being used. Too big a step, and the approximate solution may no longer resemble the exact solution; too short a step, and the efficiency can be unnecessarily dragged to unbearable levels. This is true for all methods presented in this appendix, and the problem of estimating the error and determining the right time step h to be used in order to keep the error below a threshold value is discussed in detail in Section B.4. For now, let's consider each integration method assuming that the time step h is adequate to keep the numerical error under control.

B.2 Euler Method

The Euler method is by far the simplest, and least accurate, of all methods presented in this appendix. Nonetheless, it is important to understand it because other methods are constructed on top of the basic ideas presented here.

B.2.1 Explicit Euler

The explicit Euler method, also known as the *forward Euler* method, approximates the Taylor series expansion up to its first order, that is, it approximates the infinite sum given in equation (B.4) by a straight line. The approximate solution $\vec{Y}(t_i + h)$ is then obtained from

$$\begin{aligned}
\vec{Y}(t_i + h) &= \vec{Y}(t_i) + h \frac{d\vec{Y}(t_i)}{dt} + O(h^2) \\
&= \vec{Y}(t_i) + h\, f_Y(t_i, \vec{Y}) + O(h^2) ,
\end{aligned} \qquad (B.5)$$

with $O(h^2)$ representing the order of the error obtained from truncating equation (B.4) at its second term. Figure B.1 illustrates the basic idea behind the explicit Euler method. The slope of the curve is computed once at time t_i, and is assumed to be constant for the entire time step (remember that this is a straight-line approximation!).

The explicit Euler method can be used to numerically integrate our ODEs of motion given in equations (B.2) and (B.3) for any body that is being moved in the scene, as follows. We have the initial time t_0 and the time

B.2 Euler Method

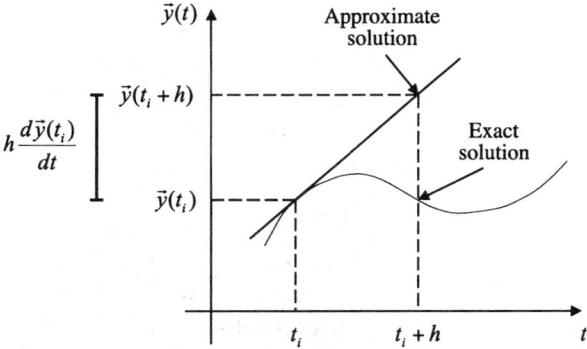

FIGURE B.1. The exact solution is approximated by a straight line segment. The slope of the line is determined at time t_i and is assumed to be constant over the entire time step h.

step h being used. We know the body's linear and angular positions and momenta at time t_0, namely

$$\vec{Y}(t_0) = \begin{pmatrix} \vec{x}(t_0) \\ \mathbf{R}(t_0) \\ \vec{L}(t_0) \\ \vec{H}(t_0) \end{pmatrix}, \qquad (B.6)$$

where

$$\vec{L}(t_0) = m\vec{v}(t_0),$$
$$\vec{H}(t_0) = \mathbf{I}(t_0)\vec{\omega}(t_0).$$

According to equation (B.5), the time derivative of the dynamic state of the body is computed once at time t_0. This computation consists of first determining the net force $\vec{F}(t_0)$ and net torque $\vec{\tau}(t_0)$ acting on the center of mass of the body at time t_0, and then substituting this information into equation (B.3) to get

$$\frac{d\vec{Y}(t_0)}{dt} = \begin{pmatrix} \vec{v}(t_0) \\ \vec{\omega}(t_0)\mathbf{R}(t_0) \\ \vec{F}(t_0) \\ \vec{\tau}(t_0) \end{pmatrix}. \qquad (B.7)$$

Therefore, the numerical solution using the explicit Euler method is given by

$$\vec{Y}(t_0+h) = \vec{Y}(t_0) + h\frac{d\vec{Y}(t_0)}{dt} = \begin{pmatrix} \vec{x}(t_0) + h\vec{v}(t_0) \\ \mathbf{R}(t_0) + h\vec{\omega}(t_0)\mathbf{R}(t_0) \\ m\vec{v}(t_0) + h\vec{F}(t_0) \\ \mathbf{I}(t_0)\vec{\omega}(t_0) + h\vec{\tau}(t_0) \end{pmatrix}.$$

B.2.2 Implicit Euler

Depending on the type of external forces acting on the body, the net torque and net force computations can sometimes introduce a linear, or even a non-linear, relation between the accelerations and the positions and velocities of the bodies in the scene. This dependence introduces linear (or non-linear) coefficients in the ODEs of motion that can considerably complicate their numerical integration. For example, if two bodies are connected by a spring, the net torque and net force acting on them will have a contribution from the spring force, which in turn depends on the relative displacement of the bodies. The same spring can be pulling the bodies together in one time step, and pushing the bodies apart in another, subsequent, time step. In this case, the position dependence of the spring force introduces linear coefficients on the ODEs of motion.

In the more general case where there is dependence between the net force, and positions and velocities of the body, equation (B.1) can be rewritten as

$$\begin{aligned}\frac{d\vec{x}(t)}{dt} &= -(\vec{c}_x(t))^t\, \vec{v}(t) \\ \frac{d\vec{L}(t)}{dt} &= -(\vec{c}_L(t))^t\, \vec{F}(t) \\ \frac{d\mathbf{R}(t)}{dt} &= -(\vec{c}_R(t))^t\, \vec{\omega}(t)\, \mathbf{R}(t) \\ \frac{d\vec{H}(t)}{dt} &= -(\vec{c}_H(t))^t\, \vec{\tau}(t) ,\end{aligned} \qquad (B.8)$$

where $(\vec{c}_x(t))^t$, $(\vec{c}_L(t))^t$, $(\vec{c}_R(t))^t$ and $(\vec{c}_H(t))^t$ are positive variables that can be either linear or non-linear. In the following paragraphs, we shall restrict our discussion to the case where all $\vec{c}$ are linear coefficients with a constant value in the time interval being considered[3]. All interaction forces covered in this book fall into this category.

A practical issue that is of special concern when such coefficients exist is the numerical stability of the solution method being applied. Clearly, the numerical stability of equations (B.8) is closely related to the time step h being used. If the time step is too big, the numerical solution may significantly differ from the exact solution, and, as the integration evolves, may no longer follow the exact solution. In extreme cases, the numerical integration becomes unstable, oscillating with increasing amplitudes and moving further away from the exact solution.

---

[3] They may have different constant values for different time intervals, but their value is constant within the same time interval.

The maximum time step h that can be used in the numerical integration of equations (B.8), still producing a stable result, is directly related to the magnitudes of the coefficients $\vec{c}(t)$. If their magnitudes differ by a significant degree (i.e., one is orders of magnitude greater than the others), the maximum time step h will be limited by the inverse of the largest magnitude value. Notice that this limitation is for purposes of stability rather than accuracy. As will be explained in more detail in Section B.4, even though the numerical error analysis may indicate that we can safely increase the current time step h being used, the stability analysis may say that we should not increase the time step if we want to keep the numerical integration stable. Since the maximum time step h is limited by the inverse of the largest magnitude value, there may be cases when the maximum time step h needed to be used to guarantee numerical stability is so small that efficiency is severely impaired, and the simulation seems to be not moving forward in time anymore. When situations like that occur, we say that equations (B.8) are *stiff*.

The implicit Euler method, also known as the *backwards Euler* method, is generally used when the ODEs of motion form a set of *stiff* equations. This method gives us a way of using larger time steps h when we have stiff equations, at the expense of being less accurate; this is a good trade-off considering that we were not moving forward using the smaller time steps anyway. Therefore, in the implicit Euler method, we care more for stability than accuracy[4]. The basic idea is to use a similar Euler approximation technique to compute $\vec{Y}(t_0 + h)$ as in equation (B.5). The difference here is that, instead of computing the time derivative at the beginning of the time interval, we compute it at the end of the time interval, that is

$$\vec{Y}(t_i + h) = \vec{Y}(t_i) + h \frac{d\vec{Y}(t_i + h)}{dt} . \tag{B.9}$$

From equations (B.8), we know that

$$\frac{d\vec{Y}(t_i + h)}{dt} = -(\vec{c}_Y)^t \vec{Y}(t_i + h) ,$$

where

$$(\vec{c}_Y)^t = (\vec{c}_x(t), \vec{c}_L(t), \vec{c}_R(t), \vec{c}_H(t))^t .$$

Substituting this information into equation (B.9), we have

$$(1 + h\,(\vec{c}_Y)^t)\,\vec{Y}(t_i + h) = \vec{Y}(t_i) , \tag{B.10}$$

which we can solve for $\vec{Y}(t_i + h)$. Even though the derivation of equation (B.10) considered only the single-variable case, we can in fact combine

---

[4] Stability analysis of this method indicates that the numerical solution is stable for *all* time-step sizes.

all linear and angular equations associated with a rigid body into a linear system of the form

$$(\mathbf{I} + h\,\mathbf{C})\,\vec{Y}(t_i + h) = \vec{Y}(t_i)\,, \tag{B.11}$$

where $\mathbf{C}$ is the positive definite coefficient matrix. In practice, the linear system in equation (B.11) is often sparse, depending on the type of force interactions between the bodies in the scene. This means that, instead of using a general $O(n^3)$ linear equation solver to determine the solution of a n dimensional system, we can take advantage of specialized sparse-matrix solvers that can compute a solution for equation (B.11) in $O(n)$.

B.3 Runge-Kutta Method

The Runge-Kutta method extends the basic idea of the explicit Euler method of computing the time derivative at the beginning of the time interval by computing intermediate values of the time derivatives throughout the time interval, and combining these values to match the Taylor series expansion up to some truncation term. The coefficients (or weights) of each of the terms being combined are carefully chosen to cancel out as many low-order derivative terms of the infinite sum in equation (B.4) as possible, leaving the higher-order derivatives not canceled out as part of the truncation error. The number of intermediate values used depends on the order of the Runge-Kutta method being applied.

B.3.1 Second-Order Runge-Kutta Method

The second-order Runge-Kutta method, also known as *midpoint* method, combines the information of two "Euler like" steps to approximate the Taylor series expansion up to its third term. The approximate solution $\vec{Y}(t_i + h)$ is then determined from

$$\vec{k}_1 = h\,f_Y(t_i, \vec{Y})\,, \tag{B.12}$$

$$\vec{k}_2 = h\,f_Y(t_i + \frac{h}{2}, \vec{Y} + \frac{\vec{k}_1}{2})\,, \tag{B.13}$$

$$\vec{Y}(t_0 + h) = \vec{Y}(t_0) + \vec{k}_2 + O(h^3)\,. \tag{B.14}$$

Figure B.2 illustrates the basic idea behind this method. The time derivative is evaluated once at time t_0 to get a first estimative of the slope of the curve. This estimative is used to determine the midpoint of the curve at time $t = (t_0 + h/2)$. Another time derivative is then evaluated to determine the slope of the curve at the midpoint, which is then used as a linear approximation of the slope of the curve for the entire time step, as is illustrated in Figure B.2(b). The numerical integration of the ODEs of motion using the second-order Runge-Kutta method can be determined from the following procedure.

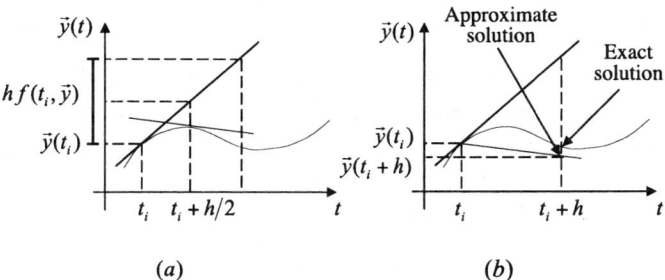

FIGURE B.2. (a) A Euler-like step is computed using the information at time t_i to determine the first estimative $f_y(t_0, \vec{y})$ of the slope of the curve. The result is then used to compute the slope of the curve at its midpoint, i.e. $f_y(t_0 + h/2, \vec{y} + \vec{k}_1/2)$. (b) The slope of the midpoint approximates the slope of the curve for the entire time step. Notice that the straight line approximating the curve is parallel to the tangent to the curve at the midpoint shown in (a), that is, has the same slope as $f_y(t_0 + h/2, \vec{y} + \vec{k}_1/2)$.

As in the explicit Euler method, we have the initial time t_0 and the time step h. We also know the body's linear and angular positions and momenta at time t_0, namely

$$\vec{Y}(t_0) = \begin{pmatrix} \vec{x}(t_0) \\ \mathbf{R}(t_0) \\ \vec{L}(t_0) \\ \vec{H}(t_0) \end{pmatrix}, \qquad (B.15)$$

where

$$\vec{L}(t_0) = m\vec{v}(t_0),$$
$$\vec{H}(t_0) = \mathbf{I}(t_0)\vec{\omega}(t_0). \qquad (B.16)$$

Again, we can compute the net force $\vec{F}(t_0)$ and net torque $\vec{\tau}(t_0)$ acting on the body at time t_0. Substituting these values into equation (B.12), we can compute the $\vec{k}_1$'s for the linear and angular positions and momenta of the body. Using the function-style notation for the time derivatives, we have:

$$\vec{k}_1 = \begin{pmatrix} \vec{k}_{1x} \\ \vec{k}_{1R} \\ \vec{k}_{1L} \\ \vec{k}_{1H} \end{pmatrix} = \begin{pmatrix} h\, f_x(t_0, \vec{x}) \\ h\, f_R(t_0, \mathbf{R}) \\ h\, f_L(t_0, \vec{L}) \\ h\, f_H(t_0, \vec{H}) \end{pmatrix} = \begin{pmatrix} h\, \vec{v}(t_0) \\ h\, \vec{\omega}(t_0)\, \mathbf{R}(t_0) \\ h\, \vec{F}(t_0) \\ h\, \vec{\tau}(t_0) \end{pmatrix}. \quad \text{(B.17)}$$

Now, according to equation (B.13), we need to use these estimates to evaluate the linear and angular accelerations of the body at the midpoint. We need to compute:

$$\vec{k}_2 = \begin{pmatrix} h\, f_x(t_0 + h/2, \vec{x} + \vec{k}_{1x}/2) \\ h\, f_R(t_0 + h/2, \mathbf{R} + \vec{k}_{1R}/2) \\ h\, f_L(t_0 + h/2, \vec{L} + \vec{k}_{1L}/2) \\ h\, f_H(t_0 + h/2, \vec{H} + \vec{k}_{1H}/2) \end{pmatrix}. \quad \text{(B.18)}$$

In plain English, equation (B.18) says that, at the midpoint, the body will be positioned at $(\vec{x} + \vec{k}_{1x}/2)$, oriented by $(\mathbf{R} + \vec{k}_{1R}/2)$, with linear and angular momenta given by $(\vec{L} + \vec{k}_{1L}/2)$ and $(\vec{H} + \vec{k}_{1H}/2)$, respectively. In other words, the second arguments of the f functions in equation (B.18) are in fact the initial conditions at the midpoint, namely:

$$\vec{Y}(t_0 + \frac{h}{2}) = \begin{pmatrix} \vec{x}(t_0 + h/2) \\ \mathbf{R}(t_0 + h/2) \\ \vec{L}(t_0 + h/2) \\ \vec{H}(t_0 + h/2) \end{pmatrix} = \begin{pmatrix} \vec{x}(t_0) + (h/2)\, \vec{v}(t_0) \\ \mathbf{R}(t_0) + (h/2)\, \vec{\omega}(t_0)\, \mathbf{R}(t_0) \\ \vec{L}(t_0) + (h/2)\, \vec{F}(t_0) \\ \vec{H}(t_0) + (h/2)\, \vec{\tau}(t_0) \end{pmatrix}.$$

We still need to determine the net force and net torque acting on the body at the midpoint. This can be done by first positioning the body at $\vec{x}(t_0 + h/2)$ with orientation $\mathbf{R}(t_0 + h/2)$, then setting its linear and angular momenta to $\vec{L}(t_0 + h/2)$ and $\vec{H}(t_0 + h/2)$, respectively. Now that the body is correctly positioned and oriented at the midpoint, we can compute the net force $\vec{F}(t_0 + h/2)$ and net torque $\vec{\tau}(t_0 + h/2)$ acting on it. Substituting this information into equation (B.18) and expanding the function-style notation, we have:

$$\vec{k}_2 = \begin{pmatrix} h\, \vec{v}(t_0 + h/2) \\ h\, \vec{\omega}(t_0 + h/2)\, \mathbf{R}(t_0 + h/2) \\ h\, \vec{F}(t_0 + h/2) \\ h\, \vec{\tau}(t_0 + h/2) \end{pmatrix} = \begin{pmatrix} h\, \frac{\vec{L}(t_0 + h/2)}{m} \\ h\, \frac{\vec{H}(t_0 + h/2)}{\mathbf{I}(t_0 + h/2)} \\ h\, \vec{F}(t_0 + h/2) \\ h\, \vec{\tau}(t_0 + h/2) \end{pmatrix}. \quad \text{(B.19)}$$

We can then substitute equations (B.15) and (B.19) back into equation (B.14) to obtain the approximate linear and angular positions and momenta of the body using the second-order Runge-Kutta method.

B.3.2 Forth-Order Runge-Kutta Method

The forth-order Runge-Kutta method combines the information of four "Euler like" steps to approximate the Taylor-series expansion up to its fifth term. The approximate solution $\vec{Y}(t_0 + h)$ is then determined from:

$$\vec{k}_1 = h f_Y(t_0, \vec{Y}), \qquad (B.20)$$

$$\vec{k}_2 = h f_Y(t_0 + \frac{h}{2}, \vec{Y} + \frac{\vec{k}_1}{2}), \qquad (B.21)$$

$$\vec{k}_3 = h f_Y(t_0 + \frac{h}{2}, \vec{Y} + \frac{\vec{k}_2}{2}), \qquad (B.22)$$

$$\vec{k}_4 = h f_Y(t_0 + h, \vec{Y} + \vec{k}_3), \qquad (B.23)$$

$$\vec{Y}(t_0 + h) = \vec{Y}(t_0) + \frac{\vec{k}_1}{6} + \frac{\vec{k}_2}{3} + \frac{\vec{k}_3}{3} + \frac{\vec{k}_4}{6} + O(h^5). \qquad (B.24)$$

Figure B.3 illustrates how this method extends the second-order Runge-Kutta method by computing one more time derivative at the mid-point, and another time derivative at the endpoint of the time interval being considered. The (1/3) and (1/6) coefficients used in equation (B.24) were chosen to cancel out the lower order time derivative terms, making the approximate solution differ by $O(h^5)$ from the exact solution.

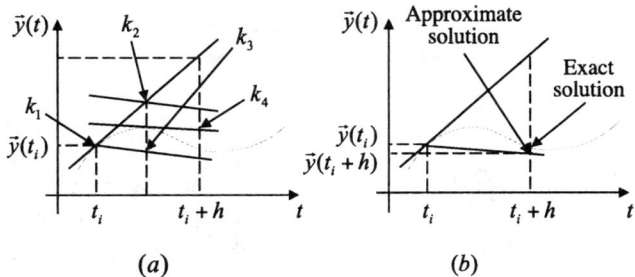

(a)   (b)

FIGURE B.3. (a) Four time derivatives are computed: one at the beginning, two at the middle, and one at the end of the time interval. The label of the points indicate which $\vec{k}$ was used to compute it. (b) The combination of all intermediate values results in a linear approximation that differs only by $O(h^5)$ from the exact solution.

The computation of the time derivatives at each of the three intermediate values requires positioning, orienting and setting the dynamic state of the body according to the second arguments of the f functions in equations (B.20) to (B.24). Only then we can compute the net force and torque acting on the body at the time interval being considered.

The time-derivative vectors $\vec{k}_1$ and $\vec{k}_2$ are computed the same way as the second-order Runge-Kutta method. We can continue applying the same techniques we used for computing $\vec{k}_2$ to obtain the time-derivative vectors $\vec{k}_3$ and $\vec{k}_4$ given in equations (B.22) and (B.23), respectively. Let's first consider the computation of $\vec{k}_3$, namely:

$$\vec{k}_3 = \begin{pmatrix} h\, f_x(t_0 + h/2, \vec{x} + \vec{k}_{2x}/2) \\ h\, f_R(t_0 + h/2, \mathbf{R} + \vec{k}_{2R}/2) \\ h\, f_L(t_0 + h/2, \vec{L} + \vec{k}_{2L}/2) \\ h\, f_H(t_0 + h/2, \vec{H} + \vec{k}_{2H}/2) \end{pmatrix}. \quad (B.25)$$

Equation (B.25) says that, at time $t = (t_0 + h/2)$, the body will be positioned at $(\vec{x} + \vec{k}_{2x}/2)$, oriented by $(\mathbf{R} + \vec{k}_{2R}/2)$, with linear and angular momenta given by $(\vec{L} + \vec{k}_{2L}/2)$ and $(\vec{H} + \vec{k}_{2H}/2)$, respectively. In other words, the second arguments of the f functions in equation (B.25) are the initial conditions at time $t = (t_0 + h/2)$, namely[5]:

$$\vec{Y}^*(t_0 + h/2) = \begin{pmatrix} \vec{x}^*(t_0 + h/2) \\ \mathbf{R}^*(t_0 + h/2) \\ \vec{L}^*(t_0 + h/2) \\ \vec{H}^*(t_0 + h/2) \end{pmatrix} = \begin{pmatrix} \vec{x}(t_0) + (h/2)\frac{\vec{L}(t_0+h/2)}{m} \\ \mathbf{R}(t_0) + (h/2)\frac{\vec{H}(t_0+h/2)}{\mathbf{I}(t_0+h/2)} \\ \vec{L}(t_0) + (h/2)\vec{F}(t_0 + h/2) \\ \vec{H}(t_0) + (h/2)\vec{\tau}(t_0 + h/2) \end{pmatrix}.$$

The net force $\vec{F}^*(t_0 + h/2)$ and net torque $\vec{\tau}^*(t_0 + h/2)$ are computed after the body is correctly positioned at $\vec{x}^*(t_0 + h/2)$ with orientation $\mathbf{R}^*(t_0 + h/2)$, and its linear and angular momenta are set to $\vec{L}^*(t_0 + h/2)$ and $\vec{H}^*(t_0 + h/2)$, respectively. Substituting this information into equation (B.25) and expanding the function-style notation we get:

$$\vec{k}_3 = \begin{pmatrix} h\,\vec{v}^*(t_0 + h/2) \\ h\,\vec{\omega}^*(t_0 + h/2)\mathbf{R}^*(t_0 + h/2) \\ h\,\vec{F}^*(t_0 + h/2) \\ h\,\vec{\tau}^*(t_0 + h/2) \end{pmatrix} = \begin{pmatrix} h\frac{\vec{L}^*(t_0+h/2)}{m} \\ h\frac{\vec{H}^*(t_0+h/2)}{\mathbf{I}^*(t_0+h/2)} \\ h\,\vec{F}^*(t_0 + h/2) \\ h\,\vec{\tau}^*(t_0 + h/2) \end{pmatrix}, \quad (B.26)$$

where

$$\mathbf{I}^*(t_0 + h/2) = \mathbf{R}^*(t_0 + h/2)\,\mathbf{I}_b\,(\mathbf{R}^*(t_0 + h/2))^{-1}.$$

The time-derivative vector:

$$\vec{k}_4 = \begin{pmatrix} h\, f_x(t_0 + h, \vec{x} + \vec{k}_{3x}/2) \\ h\, f_R(t_0 + h, \mathbf{R} + \vec{k}_{3R}/2) \\ h\, f_L(t_0 + h, \vec{L} + \vec{k}_{3L}/2) \\ h\, f_H(t_0 + h, \vec{H} + \vec{k}_{3H}/2) \end{pmatrix} \quad (B.27)$$

---

[5]The superscript $*$ is used to differentiate the $\vec{k}_3$ time-derivative estimate from the $\vec{k}_2$ estimate, since both refer to the same time $t = (t_0 + h/2)$.

is computed by first setting the dynamic state of the body to:

$$\vec{Y}(t_0 + h) = \begin{pmatrix} \vec{x}(t_0 + h) \\ \mathbf{R}(t_0 + h) \\ \vec{L}(t_0 + h) \\ \vec{H}(t_0 + h) \end{pmatrix} = \begin{pmatrix} \vec{x}(t_0) + h \frac{\vec{L}^*(t_0+h/2)}{m} \\ \mathbf{R}(t_0) + h \frac{\vec{H}^*(t_0+h/2)}{\mathbf{I}^*(t_0+h/2)} \\ \vec{L}(t_0) + h \vec{F}^*(t_0 + h/2) \\ \vec{H}(t_0) + h \vec{\tau}^*(t_0 + h/2) \end{pmatrix},$$

then determining the net force $\vec{F}(t_0 + h)$ and net torque $\vec{\tau}(t_0 + h)$ acting on the body at time $t = (t + h)$, and lastly substituting this information into equation (B.27) to obtain:

$$\vec{k}_4 = \begin{pmatrix} h\vec{v}(t_0 + h) \\ h\vec{\omega}(t_0 + h)\mathbf{R}(t_0 + h) \\ h\vec{F}(t_0 + h) \\ h\vec{\tau}(t_0 + h) \end{pmatrix} = \begin{pmatrix} h\frac{\vec{L}(t_0+h)}{m} \\ h\frac{\vec{H}(t_0+h)}{\mathbf{I}(t_0+h)} \\ h\vec{F}(t_0 + h) \\ h\vec{\tau}(t_0 + h) \end{pmatrix}. \qquad (B.28)$$

By combining all $\vec{k}$'s computed in equations (B.17), (B.19), (B.6) and (B.28) into equation (B.24), we have the approximate linear and angular positions and momenta of the body using the forth-order Runge-Kutta method.

B.4 Using Adaptive Time-Step Sizes to Speed Computations

As mentioned earlier, the choice of the time step h directly affects the efficiency and stability of the numerical method being used. If the time step h is too big, then the approximate solution may significantly differ from the exact solution, and the numerical integration is meaningless. On the other hand, if the time step h is too small, then the approximate solution may follow more closely the exact solution, but at the expense of moving forward slowly in time, possibly slower than necessary. The best choice of h depends on the equations being solved, and on their initial conditions. It is clear that, as the system evolves, the initial conditions for each time step change, and so the choice of h to be used.

Ideally, we should be able to pick the right time step h depending on the system being solved at each time step. The right time step would be one allowing the numerical integrator to move forward in time as fast as it can, still keeping the error difference between the approximate and exact solutions less than a desired threshold value. We should be able to increase or decrease the value of the time step at run-time if the error is smaller or larger than the maximum-allowed threshold value, respectively.

We already know from the previous sections that the error between the approximate and exact solutions for both explicit Euler and Runge-Kutta methods can always be expressed as being proportional to a power of h. What we still need to know is how to compute this error and take advantage of this relation with h to adaptively adjust its value as the simulation evolves.

One technique commonly used to estimate the truncation error is called *step doubling*. As its name implies, the idea is to compute the linear and angular positions and momenta using the step size h, and then compute them again using two steps of size $(h/2)$. Even though the final time step size is h in both cases, the result obtained using the two steps of size $(h/2)$ is more accurate than that using the single step size h. Their difference can therefore be used as an estimate of the truncation error, which in turn is proportional to a power of h.

To illustrate this idea, let $\vec{Y}(t_0+h)$ and $\vec{Y}(t_0+h/2+h/2)$ be the approximate values computed through the numerical integration of equation (B.1), using one time step h and two time steps $(h/2)$, respectively. The difference between these value is given by

$$\triangle_Y = \vec{Y}(t_0 + h/2 + h/2) - \vec{Y}(t_0 + h) \,.$$

We already know this difference is proportional to a power of h, that is

$$|\triangle_Y| \approx h^p \,, \tag{B.29}$$

where the value of p depends on the numerical method being applied. Table B.1 shows the values of p for the numerical methods presented in this appendix.

Integration method	Value of p
Explicit Euler	2
Implicit Euler	N/A
Second-Order Runge-Kutta	3
Fourth-Order Runge-Kutta	5

TABLE B.1. The value of p to be used depending on the numerical method applied.

So far, we have computed the difference $\triangle_Y(t_0+h)$, and have determined its relation with the time step h being used. Now, we need to ascertain how to use this information to adjust the time step h so that the error will be less than a desired threshold value.

Let $\triangle_d$ be the desired (user-definable) threshold value, and let h_d be its associated time step, that is, the time step that should be used to give an error equal to $\triangle_d$, we know that

$$|\triangle_d| \approx h_d^p. \tag{B.30}$$

Since equations (B.29) and (B.30) refer to the same ODEs, their constant of proportionality is the same. Therefore, if we divide one by the other we can cancel out their constant of proportionality, obtaining

$$\left|\frac{\triangle_Y}{\triangle_d}\right| = \left(\frac{h}{h_d}\right)^p.$$

We have the difference $\triangle_y$, the time step h being used, and the desired threshold error value $\triangle_d$. Solving for the unknown h_d, gives

$$h_d = h \left|\frac{\triangle_d}{\triangle_y}\right|^{\frac{1}{p}}. \tag{B.31}$$

Equation (B.31) shows how we can adjust the time step h using the step-doubling technique. If the computed error $\triangle_y$ is greater than the desired threshold error value $\triangle_d$, then $|\triangle_d/\triangle_y| < 1$ and h_d will be less than h, that is, the current time step h should be decreased to h_d. In this case, the integrator needs to "undo" its computations and start all over again for the new (reduced) time-step value. On the other hand, if the computed error $\triangle_y$ is smaller than the desired threshold error value $\triangle_d$, then $|\triangle_d/\triangle_y| > 1$ and h_d will be greater than h, that is, the next time step h should be increased to h_d.

In practice, because reducing the value of h is such an expensive operation (we need to redo all computations for the new value), it is important to exercise caution when we have room to increase the value of h. Keep in mind that equation (B.31) gives an estimate of how much we can increase or decrease the time step h depending on the error difference found. If we have room to increase, but increase too much, then in the next time step we may find the error is greater than the threshold error value, and we shall be forced to reduce the time step, undo the current operation and start the computations all over again. So, the gain of increasing the time step to move faster in time immediately turns out to be a considerable loss. Therefore, it is strongly advisable to increase the time step only by a percentage of the actual value computed using equation (B.31). For instance, instead of increasing h to h_d, we could increase it to $(0.8 h_d)$, leaving a 20% safety margin.

B.5 Suggested Readings

There are several other numerical methods in the literature that are applicable to our ODEs of motion, but are not so popular as the Euler and Runge-Kutta methods covered in this appendix. For example, Press *et* al. [PTVF96] present other methods such as the Bulirsch-Stoer and the fifth-order Runge-Kutta that are as effective as those presented here. In the case of the fifth-order Runge-Kutta method, the estimation of the truncation error can be done without using step doubling. This comes from the fact that the fifth-order Runge-Kutta has an embedded fourth-order Runge-Kutta on it, such that it can use the same time step to evaluate the results of two Runge-Kutta methods and compare them to estimate the integration error. Sharp *et al.* [SV94] present a general formulation for determining pairs of embedded Runge-Kutta integrators.

A more in-depth explanation of the implicit Euler method, and how the coefficients relate to the time step being used, can also be found in Press *et al.* [PTVF96], as well as in Baraff and Witkin's SIGGRAPH 98 course notes [BW98b].

Appendix C
Quaternions

C.1 Introduction

Quaternions are mathematical structures from algebraic geometry that are widely used in computer graphics as an alternate way of representing 3D rotations of objects and their orientation in a scene. Quaternions use a four-dimensional notation to represent 3×3 rotation matrices that is more efficient to manipulate, and more robust with respect to numerical round-off errors observed when combining rotation matrices.

The four-dimensional space of quaternions is composed of a real axis, and three orthogonal axes $\vec{i}, \vec{j}, \vec{k}$ known as *principal imaginaries*. We can think of the principal imaginary axis as an extension of complex numbers, since they have the same basic complex number characteristic of satisfying

$$\vec{i}^2 = \vec{j}^2 = \vec{k}^2 = -1 .$$

Being four-dimensional structures, quaternions can be represented as a quadruplet of real numbers, namely

$$q = s + x\vec{i} + y\vec{j} + z\vec{k} = s + \vec{v} , \tag{C.1}$$

consisting of a real part s and a pure imaginary part $\vec{v}$ given by

$$\vec{v} = \begin{pmatrix} x \\ y \\ z \end{pmatrix} = x\vec{i} + y\vec{j} + z\vec{k} .$$

C.2 Basic Quaternion Operations

Quaternions can be combined and manipulated in a way similar to complex numbers, where the real and imaginary parts are dealt with separately. Given two quaternions $q_1 = s_1 + \vec{v_1}$ and $q_2 = s_2 + \vec{v_2}$, the following basic rules define the most common set of operations used with quaternions.

C.2.1 Addition

The addition of two quaternions is similar to the addition of two complex numbers, and is given by

$$q_1 + q_2 = (s_1 + s_2) + (\vec{v_1} + \vec{v_2}),$$

where $(s_1 + s_2)$ is the real part of the result (notice that all summands are scalar) and $(\vec{v_1} + \vec{v_2})$ is the imaginary part of the result (notice that all summands are vectors). The result of the addition is a quaternion.

C.2.2 Dot product

The dot product of two quaternions is equivalent to the addition of the dot products of their real (scalar) and imaginary (vector) parts, namely

$$q_1 \cdot q_2 = s_1 s_2 + \vec{v_1} \cdot \vec{v_2},$$

where $\vec{v_1} \cdot \vec{v_2}$ is the usual dot product of two vectors[1] in $\mathbb{R}^3$. Notice that the result of a dot product is a scalar.

C.2.3 Multiplication

The multiplication of two quaternions can be computed as if we were multiplying two complex numbers, with the added complexity of having the imaginary part being a vector. In this case, the multiplication of the imaginary axes $\vec{i}, \vec{j}$ and $\vec{k}$ obeys the following basic rules:

$$\begin{aligned} \vec{j}\vec{i} = -\vec{k},\ \vec{k}\vec{j} = -\vec{i},\ \vec{i}\vec{k} &= -\vec{j} \\ \vec{k}\vec{i} = +\vec{j},\ \vec{i}\vec{j} = +\vec{k},\ \vec{j}\vec{k} &= +\vec{i} \\ \vec{i}\vec{j}\vec{k} &= -1. \end{aligned} \quad (C.2)$$

We can alternatively visualize the multiplication of one imaginary axis by another as a 3D rotation (see Figure C.1(a)). For example, the multiplication on the right by $\vec{k}$ can be visualized by applying the right-hand rule for rotating axis $\vec{i}$ and $\vec{j}$ in Figure C.1(b).

---

[1] Unless otherwise stated, whenever we mention $\mathbb{R}^n$, we are referring to the n-dimensional Euclidean space and its associated properties.

C.2 Basic Quaternion Operations

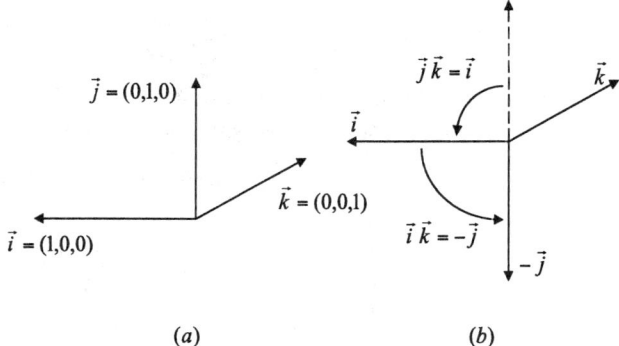

FIGURE C.1. (a) The three orthogonal principal imaginary axes viewed as a canonical basis of a 3D Euclidean space; (b) Multiplication on the right by $\vec{k}$ causes a clockwise 90 degree rotation in four-dimensional space around the $\vec{k}$ axis, rotating the $\vec{i}$ axis into the $-\vec{j}$ axis (from $\vec{i}\vec{k} = -\vec{j}$ in (C.2)), and rotating the $\vec{j}$ axis into the $\vec{i}$ axis (from $\vec{j}\vec{k} = \vec{i}$ in (C.2)). This can also be verified by applying the right-hand rule for rotating vectors represented in the right-hand coordinate system.

In order to compute an algebraic expression for the multiplication of two quaternions, we have to expand their imaginary parts and use equation (C.2) to compute the multiplication of the imaginary axes among themselves, that is

$$\begin{aligned}
q_1 q_2 &= (s_1 + \vec{v_1})(s_2 + \vec{v_2}) \\
&= (s_1 + v_{1x}\vec{i} + v_{1y}\vec{j} + v_{1z}\vec{k})(s_2 + v_{2x}\vec{i} + v_{2y}\vec{j} + v_{2z}\vec{k}) \\
&= (s_1 s_2 - v_{1x} v_{2x} - v_{1y} v_{2y} - v_{1z} v_{2z}) + \\
&\quad s_1(v_{2x}\vec{i} + v_{2y}\vec{j} + v_{2z}\vec{k}) + s_2(v_{1x}\vec{i} + v_{1y}\vec{j} + v_{1z}\vec{k}) + \\
&\quad (v_{1y} v_{2z} - v_{1z} v_{2y})\vec{i} + (v_{1z} v_{2x} - v_{1x} v_{2z})\vec{j} + \\
&\quad (v_{1x} v_{2y} - v_{1y} v_{2x})\vec{k} \\
&= (s_1 s_2 - \vec{v_1} \cdot \vec{v_2}) + (s_1 \vec{v_2} + s_2 \vec{v_1} + \vec{v_1} \times \vec{v_2}).
\end{aligned}$$

Notice that the multiplication of quaternions is not commutative. This should not come as a surprise, since multiplication of quaternions is equivalent to a 3D rotation, which itself does not enjoy the commutative property with respect to the multiplication operation.

C.2.4 Conjugate

The conjugate operation is exactly the same as its counterpart in complex number theory. We just need to negate the imaginary part to obtain the conjugate of a quaternion, that is

$$\bar{q}_1 = \overline{s_1 + \vec{v}_1} = s_1 - \vec{v}_1 \,. \tag{C.3}$$

Another basic property of the conjugate operation is that the conjugate of the multiplication of two quaternions is equivalent to the multiplication of the conjugate of each quaternion, that is

$$\begin{aligned}
\bar{q}_1\,\bar{q}_2 &= (s_1 - \vec{v}_1)(s_2 - \vec{v}_2) \\
&= (s_1\,s_2 - \vec{v}_1 \cdot \vec{v}_2) + (-s_1\,\vec{v}_2 - s_2\,\vec{v}_1 + (-\vec{v}_1) \times (-\vec{v}_2)) \\
&= (s_1\,s_2 - \vec{v}_1 \cdot \vec{v}_2) - (s_1\,\vec{v}_2 + s_2\,\vec{v}_1 + \vec{v}_2 \times \vec{v}_1) \\
&= \overline{q_1 q_2}\,.
\end{aligned}$$

C.2.5 Module

The module, also known as magnitude, is computed as the dot product between the quaternion and its conjugate, and is given by

$$|q_1|^2 = q_1 \cdot \bar{q}_1 = \bar{q}_1 \cdot q_1 = s_1^2 + \vec{v}_1 \cdot \vec{v}_1 = s_1^2 + v_{1x}^2 + v_{1y}^2 + v_{1z}^2 \,.$$

The module of the multiplication of two quaternions is the same as the multiplication of the modules of each quaternion, since

$$\begin{aligned}
|q_1\,q_2|^2 &= (q_1\,q_2)(\overline{q_1\,q_2}) = (q_1\,q_2)(\bar{q}_2\,\bar{q}_1) \\
&= q_1\,(q_2\,\bar{q}_2)\,\bar{q}_1 = q_1\,|q_2|^2\,\bar{q}_1 \\
&= |q_2|^2\,q_1\,\bar{q}_1 = |q_1|^2\,|q_2|^2\,.
\end{aligned}$$

C.2.6 Inverse

The inverse of a quaternion can be directly derived from the expression used to compute its module, resulting in

$$q_1^{-1} = \frac{\bar{q}_1}{|q_1|^2}\,. \tag{C.4}$$

C.3 Unit Quaternions

Another commonly used representation for a quaternion $q = s + \vec{v}$ is to write it as $q = s + a\,\vec{u}$, where the imaginary part $\vec{u}$ is a unitary vector (i.e., $|\vec{u}| = 1$.) We can compute a and $\vec{u}$ from $\vec{v}$ as follows:

$$\vec{u} = \begin{pmatrix} x/a \\ y/a \\ z/a \end{pmatrix} = \frac{x}{|\vec{v}|}\vec{i} + \frac{y}{|\vec{v}|}\vec{j} + \frac{z}{|\vec{v}|}\vec{k}\,.$$

A quaternion with zero real part (i.e., $s = 0$) is called a *pure quaternion*. Any arbitrary vector $\vec{v} \in \mathbb{R}^3$ can be represented as a pure quaternion:

$$q = 0 + \vec{v} \tag{C.5}$$

using the notation in (C.1). A quaternion with module equal to one is called a *unit quaternion*, also known as *unit-magnitude quaternion*. A pure unit quaternion is then a quaternion with zero real part and module equal to one. Since any arbitrary vector in $\mathbb{R}^3$ can be represented as a pure quaternion, any unitary vector in $\mathbb{R}^3$ (i.e., any normalized vector) can be represented as a pure unit quaternion.

The family of all unit quaternions (i.e., all q such that $|q| = 1$) form an hyper sphere of radius 1 in the four-dimensional space of quaternions. Because a unit quaternion q always satisfies the condition $|q| = 1$, from equation (C.4), we can directly conclude that:

$$q^{-1} = \frac{\bar{q}}{|q|^2} = \bar{q}, \tag{C.6}$$

that is, the inverse of a unit quaternion is equal to its conjugate, which in turn can be computed by negating its imaginary part (see equation (C.3)).

C.4 Rotation-Matrix Representation Using Unit Quaternions

Unit quaternions play an important role in computer graphics because they can be used as an equivalent representation of 3×3 rotation matrices. In this section, we shall briefly go over all transformations required to switch back and forth between unit quaternions and rotation-matrix representations for right-handed coordinate systems.

The 3×3 rotation-matrix representation of a rotation of an angle of θ degrees about the unit-magnitude axis $\vec{u}$ that passes through the origin is given by:

$$\mathbf{R} = \begin{pmatrix} t\,u_x^2 + \cos\theta & t\,u_x u_y + u_z \sin\theta & t\,u_x u_z - u_y \sin\theta \\ t\,u_x u_y - u_z \sin\theta & t\,u_y^2 + \cos\theta & t\,u_y u_z + u_x \sin\theta \\ t\,u_x u_z + u_y \sin\theta & t\,u_y u_z - u_x \sin\theta & t\,u_z^2 + \cos\theta \end{pmatrix}, \tag{C.7}$$

where u_x, u_y and u_z are the components of the unit-magnitude vector $\vec{u}$ and $t = (1 - \cos\theta)$. This rotation can, in turn, be represented by the unit quaternion

$$q = \cos\frac{\theta}{2} + \sin\frac{\theta}{2}\vec{u}. \tag{C.8}$$

Conversely, given a unit quaternion $q = s + \vec{v}$, the rotation axis $\vec{u}$ and the rotation angle θ represented by the quaternion can be computed as

$$\cos\theta = 2s^2 - 1$$

$$\sin\theta = 2s\sqrt{1-s^2}$$

(C.9)

and

$$\vec{u} = \frac{\vec{v}}{\sqrt{1-s^2}}.$$

(C.10)

If we are using the 3×3 matrix representation, the rotation axis $\vec{u}$ and rotation angle θ can be determined from:

$$\cos\theta = \frac{R_{xx} + R_{yy} + R_{zz} - 1}{2}$$

(C.11)

and

$$\begin{aligned} u_x &= \frac{R_{yz} - R_{zy}}{2\sin\theta} \\ u_y &= \frac{R_{zx} - R_{xz}}{2\sin\theta} \\ u_z &= \frac{R_{xy} - R_{yx}}{2\sin\theta}, \end{aligned}$$

(C.12)

with the constraint that $\sin\theta \neq 0$. If this constraint is not met, then the rotation axis is undetermined.

Lastly, given a unit quaternion $q = s + \vec{v}$, its equivalent 3×3 rotation matrix representation can be directly computed by substituting equations (C.9) and (C.10) into equation (C.7). The matrix representation of the unit quaternion is then:

$$\mathbf{R} = 2\begin{pmatrix} s^2 + v_x^2 - \frac{1}{2} & v_x v_y - s v_z & v_x v_z + s v_y \\ v_x v_y + s v_z & s^2 + v_y^2 - \frac{1}{2} & v_y v_z - s v_x \\ v_x v_z - s v_y & v_y v_z + s v_x & s^2 + v_z^2 - \frac{1}{2} \end{pmatrix},$$

(C.13)

where v_x, v_y and v_z are the components of the imaginary part $\vec{v}$, of the unit quaternion q.

At this point, we can easily switch back and forth between the quaternion and the 3×3 matrix representation of a rotation of θ degrees about the unit-magnitude axis $\vec{u}$ using equations (C.7) to (C.13). The only remaining question is how to rotate an arbitrary vector $\vec{p} \in \mathbb{R}^3$ using these representations. If we are using the 3×3 matrix representation, then we can rotate $\vec{p}$ simply by computing

$$\vec{p_r} = \mathbf{R}\,\vec{p}\,, \qquad (C.14)$$

where $\mathbf{R}$ is the rotation matrix and $\vec{p_r}$ is the rotated vector.

If we are using the quaternion representation, the vector $\vec{p} \in \mathbb{R}^3$ can be represented by the pure quaternion $q_p = 0 + \vec{p}$ (see equation (C.5)), and the rotation can then be computed as a quaternion multiplication given by

$$\begin{aligned} q_{p_r} &= q\,q_p\,\bar{q} = (s + \vec{u})\,(0 + \vec{p})\,(s - \vec{u}) \\ &= 0 + ((s^2 - \vec{u}\cdot\vec{u})\,\vec{p} + 2\,(\vec{p}\cdot\vec{u})\,\vec{u} + 2\,s\,(\vec{u}\times\vec{p})) \\ &= 0 + \vec{p_r}\,, \end{aligned} \qquad (C.15)$$

where q_{p_r} is the pure quaternion representing the rotated vector $\vec{p_r}$.

C.5 Advantages of Using Unit Quaternions

There are several advantages of using quaternion representation instead of the 3 × 3 matrix representation for rotations. Here we will focus on the most important ones.

The immediate advantage is that quaternions encode rotations by four real numbers, whereas the representation of these transformations as 3 × 3 matrices requires nine. This may save significant space for complex scenes with a large number of objects being simulated.

Besides the extra space needed, another problem of requiring more parameters to encode rotation matrices is that they are also prone to "drifting" when multiplied by one another. The drifting is caused by the fact that the "sin" and "cos" computations for a rotation round an arbitrary axis introduce round-off errors into the nine elements needed to encode the 3 × 3 rotation matrix. When multiplied by another rotation matrix, the result matrix may no longer be a rotation matrix. The round-off errors can be such that the determinant of the result matrix differs from one by a small amount. If the determinant is not equal to one, then the result matrix is no longer orthogonal, and therefore not a rotation matrix. Subsequent multiplications by other rotation matrices will rapidly increase the error to a noticeable level, such that the object being "rotated" will in fact be rotated *and* distorted or arbitrarily scaled when rendered in the scene.

A common way of solving the drifting problem for the matrix representation of rotations is to use the Gram-Schmidt orthogonalization algorithm for keeping the result matrix orthogonal. The idea is to every so often check whether the determinant of the result matrix differs from one by an amount greater than a tolerance level, and if so apply the algorithm to decompose the result matrix into an orthogonal matrix times an upper triangular matrix. This decomposition is known as the *QR* decomposition. The result matrix can then be substituted for the orthogonal matrix, and the upper triangular matrix can be used to estimate the round-off error introduced so far. The practical problem of using this scheme to correct inevitable round-off errors when combining rotation matrices is that it is time consuming, and may not be appropriate for use in simulation engines with stringent operational requirements.

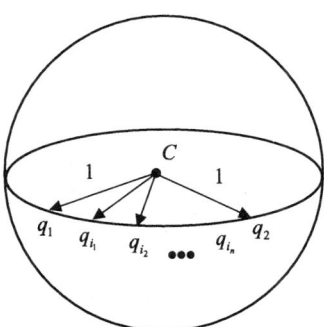

FIGURE C.2. The hyper sphere of radii 1 in the four-dimensional space of quaternions. The intermediate interpolated quaternions $q_{i_1}, \ldots, q_{i_n}$ lie on the smallest great hyper arc of the great hyper circle that contains q_1 and q_2 (the quaternions being interpolated).

With the unit quaternion representation, the drifting problem arising from the "sin" and "cos" computations and the round-off errors introduced when two unit quaternions are multiplied still exists. In these cases, the result quaternion may no longer have module equal to one, and therefore no longer represents a rotation matrix. However, with the quaternion representation, the drift problem can be easily fixed by re-normalizing the result quaternion. If we think of the unit quaternion as being constrained to the surface of a four-dimensional hyper sphere of radius one, then the drift can be visualized as the quaternion being moved in or out of the surface owing to round-off errors. The re-normalization is equivalent to projecting the quaternion back to another location on the surface close to the correct location the quaternion was supposed to be if drift had not occured.

Lastly, one of the main reasons most graphics packages use quaternions is the ability to easily interpolate between unit quaternions representing an object orientation in two consecutive animation frames. Again, if we think of the unit quaternion as being constrained to the surface of a four-dimensional hyper sphere of radius one, the intermediate interpolated unit quaternions will lie on the smallest great hyper arc that connects the unit quaternions being interpolated (see Figure C.2).

The same is not true if we are using rotation matrix representation. It is usually very hard to interpolate rotations between two orientations using the rotation matrix representation, and still get a smooth transition (i.e., one without jerkiness) between the orientations as the scene is animated.

C.6 Suggested Readings

Quaternions were originally developed as a way of rotating a 3D vector by multiplying it by another 3D vector. The method of specifying rotations and orientations of coordinate systems via unit quaternions was formally introduced to the computer graphics industry by Shoemake [Sho85]. There you can find a more detailed explanation of why the rotation of a vector $\vec{p} \in \mathbb{R}^3$ by the unit quaternion q can be computed using equation (C.15), as well as the derivation of a more sophisticated method to interpolate a set of quaternions using Bézier curves. For readers interested in quaternion visualization techniques, we strongly recommend the work of Hart et al. [HFK94], wherein is presented an alternate exponential notation for representing unit quaternions.

The derivation of equation (C.7), that is, the determination of the 3×3 rotation-matrix representation given the angle and axis of rotation, can be found in Craig [Cra89], Pique [Piq90] and in several other books in the fields of mechanics, robotics or computer graphics. An in-depth explanation of general 3D transformations, and how to convert between right-handed and left-handed coordinate systems, is given in Foley et al. [FvDFH96].

Lastly, the Gram-Schmidt algorithm used for correcting the drifting problem observed when multiplying several rotation matrices can be found in Strang [Str91], Golub et al. [GL96] and Horn et al. [HJ91].

Appendix D
Rigid-Body Mass Properties

D.1 Introduction

The way rigid bodies interact with each other in a dynamic simulation depends a great deal on their mass distribution. As was shown in Appendix B, the rigid body's total mass, center of mass and inertia tensor directly affect the computation of the net force and net torque acting on it, which in turn are used to solve the differential equations of motion. These quantities, namely total mass, center of mass and inertia tensor, are commonly known as the body's mass properties, and depend on the rigid body's shape and density.

Several algorithms have been proposed so far in the computer graphics and simulation and modeling literature to compute the mass properties of a given rigid-body object. They usually fall into one of the two high-level classes of algorithms. The first class of algorithms is frequently used in computer solid modeling. Algorithms of this class decompose the original solid-modeling representation of the object into small cells, and compute the approximate value of the object's mass properties by summing the mass properties of each cell. The degree to which the approximate value matches the exact value depends on the granularity of the cell decomposition. However, independent of how fine-grained the decomposition is, the mass properties computed using this class of algorithms are always an approximation of their exact value because of inevitable mismatches between the object's volume and the volume occupied by the cells in the decomposition.

The second class of algorithms assumes the boundary representation of the object is given, that is, the object's polygonal faces, vertices and their neighboring information, and compute the object's mass properties directly from it. Some of these algorithms are specifically designed for the case in which the object's faces are triangles. In this case, the triangular faces are connected to the origin of the coordinate system, forming several tetrahedra. The mass properties of the object are computed by combining the mass properties of each tetrahedron. The drawback to using this method is that some of the tetrahedra can be thin and tall, introducing numerical errors into the computations that can degrade the final result.

In this appendix, we present Mirtich's algorithm for computing the *exact* values of the mass properties of rigid body objects from their boundary representation. The objects are assumed to be composed by a set of homogeneous polyhedra, each with its own constant density value. The mass properties are computed by incrementally simplifying the initial volume integrals to surface integrals over the faces of the object, then to surface integrals over the faces projected to one of the coordinate planes, and finally to line integrals along the edges of each projected face, which can then be computed directly from its vertices. Despite its complex derivation, the algorithm presented here is surprisingly fast, varying linearly with respect to the total number of faces, edges and vertices of the object.

D.2 Mirtich's Algorithm

The computation of the mass properties involves computing the center of mass, total mass and inertia tensor of the object, from the center of mass, total mass and inertia tensor of each of its homogeneous polyhedra. Assuming the object is represented by n_p homogeneous polyhedra, its total mass can be directly determined from

$$M = \sum_{i=1}^{n_p} \rho_i V_i , \qquad (D.1)$$

where ρ_i and V_i are the density and volume of polyhedron i, respectively. The density values are assumed to be given, so that the volume of each polyhedron can be computed from its shape as

$$V_i = \int dV . \qquad (D.2)$$

Assuming, for the time being, that we know how to compute the volume integral in (D.2), we can determine the volume of each polyhedron, and use this information in (D.1) to compute the total mass of the object. Knowing the total mass of the object, we can compute the coordinates of its center of mass C as

$$C_x = \frac{1}{M}\sum_{i=1}^{n_p}\int_{V_i} x\,dM = \frac{1}{M}\sum_{i=1}^{n_p}\rho_i\int_{V_i} x\,dV$$
$$C_y = \frac{1}{M}\sum_{i=1}^{n_p}\int_{V_i} y\,dM = \frac{1}{M}\sum_{i=1}^{n_p}\rho_i\int_{V_i} y\,dV \qquad (D.3)$$
$$C_z = \frac{1}{M}\sum_{i=1}^{n_p}\int_{V_i} z\,dM = \frac{1}{M}\sum_{i=1}^{n_p}\rho_i\int_{V_i} z\,dV \,.$$

In equations (D.3), the homogeneity assumption of the polyhedra is used to convert the mass integrals into volume integrals, since $dM = \rho_i\,dV$ for each polyhedron i.

Having derived formulas to compute the total mass and center of mass of the object, the remaining mass property that needs to be determined is the inertia tensor. The inertia tensor is a 3×3 matrix that contains the moments and products of inertia about the center of mass of the object. In other words, it express how the mass of the object is distributed relative to its center of mass.

The computation of the inertia tensor involves determining the nine elements of its 3×3 matrix representation, given by

$$\mathbf{I} = \begin{pmatrix} I_{xx} & -I_{xy} & -I_{xz} \\ -I_{yx} & I_{yy} & -I_{yz} \\ -I_{zx} & -I_{zy} & I_{zz} \end{pmatrix}, \qquad (D.4)$$

where I_{xx}, I_{yy} and I_{zz} are the moments of inertia about axes x, y and z, respectively, and I_{xy}, I_{yx}, I_{xz}, I_{zx}, I_{zy} and I_{yz} are the products of inertia between the axes. A well known property of inertia tensors found in the mechanical engineering literature is that the inertia tensor $\mathbf{I}$ is a real symmetric matrix, that is

$$I_{xy} = I_{yx} \qquad I_{xz} = I_{zx} \qquad I_{zy} = I_{yz}\,,$$

so we only need to worry about six out of the nine elements in D.4. Another useful property of inertia tensors is that it is always possible to find a body-frame coordinate system in which all products of inertia are zero. In this case, $\mathbf{I}$ is diagonal and we just need to keep track of its three diagonal elements.

It is also important to notice that, if the mass distribution of the object doesn't change over time, its inertia tensor relative to its body frame is constant. However, it is constant only in the body-frame coordinate system, but not in the world-frame coordinate system because, as the simulation evolves, the object changes position and orientation and its mass distribution relative to the world frame changes as well. If $\mathbf{I}$ is the inertia tensor relative to the body frame, and $\mathbf{R}$ is the rotation that takes the body frame to the world frame, then the inertia tensor $\mathbf{I^w}$ relative to the world frame is given by

$$\mathbf{I^w} = \mathbf{R}\mathbf{I}\mathbf{R}^{-1} = \mathbf{R}\mathbf{I}\mathbf{R}^t. \tag{D.5}$$

The moments and products of inertia of the object relative to the world frame can be individually determined from the following set of equations.

$$\begin{aligned}
I_{xx}^w &= \sum_{i=1}^{n_p} \rho_i \int_{V_i} (y^2 + z^2) \, dV \\
I_{yy}^w &= \sum_{i=1}^{n_p} \rho_i \int_{V_i} (z^2 + x^2) \, dV \\
I_{zz}^w &= \sum_{i=1}^{n_p} \rho_i \int_{V_i} (x^2 + y^2) \, dV \\
I_{xy}^w = I_{yx}^w &= \sum_{i=1}^{n_p} \rho_i \int_{V_i} x \, y \, dV \\
I_{yz}^w = I_{zy}^w &= \sum_{i=1}^{n_p} \rho_i \int_{V_i} y \, z \, dV \\
I_{zx}^w = I_{xz}^w &= \sum_{i=1}^{n_p} \rho_i \int_{V_i} z \, x \, dV
\end{aligned} \tag{D.6}$$

We can use the *parallel axis* theorem from mechanical engineering to compute the inertia tensor relative to a body frame parallel to the world frame, but with its origin translated to the object's center of mass. The new moments and products of inertia for this new frame set forth below.

$$\begin{aligned}
I_{xx} &= I_{xx}^w - M(C_y^2 + C_z^2) \\
I_{yy} &= I_{yy}^w - M(C_z^2 + C_x^2) \\
I_{zz} &= I_{zz}^w - M(C_x^2 + C_y^2) \\
I_{xy} = I_{yx} &= I_{xy}^w - M\,C_x\,C_y \\
I_{yz} = I_{zy} &= I_{yz}^w - M\,C_y\,C_z \\
I_{zx} = I_{xz} &= I_{zx}^w - M\,C_z\,C_x
\end{aligned} \tag{D.7}$$

Equations (D.7) give us a way of computing the constant inertia tensor relative to a body frame that is parallel to the world frame, but with its origin at the object's center of mass.

A closer look at equations (D.2), (D.3), (D.6) and (D.7), quickly reveals that, in order to compute the object's mass properties, we need to be able to evaluate the following volume integrals defined over each polyhedron of the object

$$
\begin{aligned}
T_x &= \int_f x\,dV & T_{x^2} &= \int_f x^2\,dV \\
T_y &= \int_f y\,dV & T_{y^2} &= \int_f y^2\,dV \\
T_z &= \int_f z\,dV & T_{z^2} &= \int_f z^2\,dV \\
T_1 &= \int_f dV & T_{xy} &= \int_f x\,y\,dV \\
T_{yz} &= \int_f y\,z\,dV & T_{zx} &= \int_f z\,x\,dV
\end{aligned}
\qquad (D.8)
$$

The basic idea to solve the volume integrals in (D.8) is to gradually reduce their complexity from volume integrals to surface integrals, then from surface integrals to projected surface integrals, then from projected surface integrals to line integrals, and lastly to evaluate the line integrals from the object's vertex coordinates. These reductions are achieved using well known theorems from advanced calculus.

D.2.1 Volume-Integral to Surface-Integral Reduction

The very first step is to reduce the volume integrals in (D.8) to surface integrals over the faces of the object. The volume-to-surface reduction is achieved through the use of the *divergence* theorem. The divergence theorem states that, given a bounded volume V in space and its outward normal (i.e., the normal that points from the inside to the outside of V), then for any continuous vector field $\vec{F}$ defined on V, we have

$$\int_V \nabla \cdot \vec{F}\,dV = \int_{\partial V} \vec{F} \cdot \vec{n}\,dA, \qquad (D.9)$$

where ∂V is the boundary of V and ∇ is the divergence operator given by

$$\nabla \cdot \vec{F} = \frac{\partial \vec{F}}{\partial x} + \frac{\partial \vec{F}}{\partial y} + \frac{\partial \vec{F}}{\partial z}. \qquad (D.10)$$

Equation (D.9) explicitly shows the way to convert the volume integrals to surface integrals. The volume integrals in (D.8) can be reduced to surface integrals by choosing a continuous force field for each of them. A guess can certainly be used, but we prefer one that simplifies the surface-integral computation. For example, let's examine what would be a suitable choice of force field for the volume integral

$$T_x = \int_V x \, dV .$$

We have to find a force field $\vec{F}$ that satisfies

$$\nabla \cdot \vec{F} = \frac{\partial \vec{F}}{\partial x} + \frac{\partial \vec{F}}{\partial y} + \frac{\partial \vec{F}}{\partial z} = x .$$

There are many force fields that we can choose from for this given case, but a suitable guess would be one that makes the right-hand side of equation (D.9) as straightforward as possible to compute. In this context, we shall pick a force field that turns the dot product of the right-hand side of equation (D.9) into a simple scalar multiplication, or

$$\vec{F} = \left(\frac{x^2}{2}, 0, 0\right)^t .$$

Substituting this into equation (D.9) gives

$$T_x = \int_V x \, dV = \int_{\partial V} \vec{F} \cdot \vec{n} \, dA = \sum_{f \in \partial V} \int_f \left(\frac{n_x x^2}{2}\right) dA , \qquad (D.11)$$

where the surface integrals are computed for each face of the object. Because in our case each polygonal face has constant normal, we can pull the normal component out of the integral in (D.11), further simplifying the expression to

$$T_x = \sum_{f \in \partial V} \frac{n_x}{2} \int_f x^2 \, dA .$$

The same procedure can be applied for all volume integrals in (D.8), and the result is summarized in Table D.1. This table shows the appropriate force-field choice and the equivalent surface integral for each volume integral in (D.8).

Having computed the surface integral associated with each volume integral in (D.8), we are ready to proceed to the next step of the integration, which consists of reducing the surface integrals to line integrals. However, before we do that, we shall "standardize" this reduction by first projecting each face of the polyhedron onto one of the coordinate planes.

D.2.2 Surface-Integral to Projected-Surface-Integral Reduction

The surface to projected-surface reduction is achieved by projecting each face of the polyhedron onto one of the coordinate planes xy, yz or zx. The

D.2 Mirtich's Algorithm

Index i	Volume Integral T_i	Force Field $\vec{F}_i$	Equivalent Surface Integral
1	$\int_V 1\, dV$	$(x,0,0)^t$	$\sum_{f \in \partial V} n_x \int_f x\, dA$
x	$\int_V x\, dV$	$(\frac{x^2}{2},0,0)^t$	$\sum_{f \in \partial V} \frac{n_x}{2} \int_f x^2\, dA$
y	$\int_V y\, dV$	$(0,\frac{y^2}{2},0)^t$	$\sum_{f \in \partial V} \frac{n_y}{2} \int_f y^2\, dA$
z	$\int_V z\, dV$	$(0,0,\frac{z^2}{2})^t$	$\sum_{f \in \partial V} \frac{n_z}{2} \int_f z^2\, dA$
x^2	$\int_V x^2\, dV$	$(\frac{x^3}{3},0,0)^t$	$\sum_{f \in \partial V} \frac{n_x}{3} \int_f x^3\, dA$
y^2	$\int_V y^2\, dV$	$(0,\frac{y^3}{3},0)^t$	$\sum_{f \in \partial V} \frac{n_y}{3} \int_f y^3\, dA$
z^2	$\int_V z^2\, dV$	$(0,0,\frac{z^3}{3})^t$	$\sum_{f \in \partial V} \frac{n_z}{3} \int_f z^3\, dA$
xy	$\int_V xy\, dV$	$(\frac{x^2 y}{2},0,0)^t$	$\sum_{f \in \partial V} \frac{n_x}{2} \int_f x^2 y\, dA$
yz	$\int_V yz\, dV$	$(0,\frac{y^2 z}{2},0)^t$	$\sum_{f \in \partial V} \frac{n_y}{2} \int_f y^2 z\, dA$
zx	$\int_V zx\, dV$	$(0,0,\frac{z^2 x}{2})^t$	$\sum_{f \in \partial V} \frac{n_z}{2} \int_f z^2 x\, dA$

TABLE D.1. Volume- to surface-integral reduction for each volume integral in (D.8). Even though there are many possible choices of force fields that can be used, the chosen ones significantly simplify the surface-integral computations.

choice of the coordinate plane to which the face will be projected depends on the relative orientation of the face with respect to the coordinate plane, that is, it depends on the values of the components n_x, n_y and n_z of the face normal.

The surface integral over the face f can be related to the surface integral over its projection f_p as follows. Let the plane equation of face $f \in \partial V$ be given by

$$n_x \vec{x} + n_y \vec{y} + n_z \vec{z} + d = 0 ,$$

where the scalar constant d can be obtained from

$$d = -\vec{n} \cdot \vec{p}$$

for any point $\vec{p} \in f$. Then, the surface integral over the face f can be computed from the surface integral over the projected face f_p as

$$\int_f g(\alpha, \beta, \gamma) \, dA = \frac{1}{|n_\gamma|} \int_{f_p} g(\alpha, \beta, h(\alpha, \beta)) \, d\alpha \, d\beta \,, \tag{D.12}$$

where $g(\alpha, \beta, \gamma)$ is any polynomial function of α, β and γ, and $h(\alpha, \beta)$ is given by

$$h(\alpha, \beta) = -\frac{1}{n_\gamma}(n_\alpha \alpha + n_\beta \beta + d) \,. \tag{D.13}$$

The use of α, β and γ instead of x, y and z in equations (D.12) and (D.13) emphasizes the fact that the face-projection plane is selected at run-time depending on the choice that maximizes n_γ, that is, the choice that maximizes the area of the projected face (see figure D.1). Possible combination values for (α, β, γ) are (x, y, z), (y, z, x) and (z, x, y).

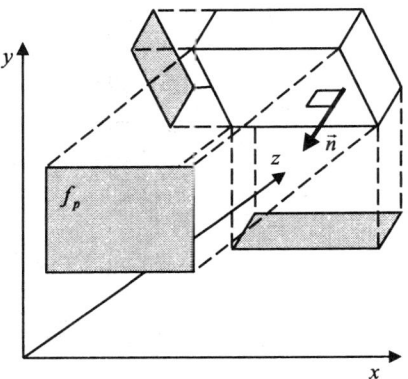

FIGURE D.1. At run-time, the largest component of the face normal is selected as n_γ and the α, β and γ axes are mapped to the x, y and z axes according to this selection. For the cubic object shown here, the largest component of the normal $\vec{n}$ of face f is n_z, and therefore γ is associated with z given the combination $(\alpha, \beta, \gamma) = (x, y, z)$.

Independent of the combination used, the computation of the surface integrals in Table D.1 can always be reduced to the computation of surface integrals of the form:

$$\begin{aligned}
&\int_f \alpha \, dA \,, \quad \int_f \alpha^2 \, dA \,, \quad \int_f \alpha^3 \, dA \,, \quad \int_f \alpha^2 \beta \, dA \\
&\int_f \beta \, dA \,, \quad \int_f \beta^2 \, dA \,, \quad \int_f \beta^3 \, dA \,, \quad \int_f \beta^2 \gamma \, dA \\
&\int_f \gamma \, dA \,, \quad \int_f \gamma^2 \, dA \,, \quad \int_f \gamma^3 \, dA \,, \quad \int_f \gamma^2 \alpha \, dA \,.
\end{aligned} \tag{D.14}$$

This convenient representation is possible because the surface integrals in Table D.1 have terms of the form α, α^2, α^3 and $\alpha^2\beta$, for any of the three possible combination values of (α, β, γ). Since the actual combination value can only be determined at run-time, we shall compute the projected surface integrals of each face as a function of α, β and γ. This we do by substituting equations (D.12) and (D.13) into each surface integral in (D.14).

The surface integrals for each face of the object as a function of its projected surface integral are then given by:

$$\int_f \alpha \, dA = \frac{1}{|n_\gamma|} P_\alpha, \tag{D.15}$$

$$\int_f \alpha^2 \, dA = \frac{1}{|n_\gamma|} P_{\alpha^2}, \tag{D.16}$$

$$\int_f \alpha^3 \, dA = \frac{1}{|n_\gamma|} P_{\alpha^3}, \tag{D.17}$$

$$\int_f \alpha^2 \beta \, dA = \frac{1}{|n_\gamma|} P_{\alpha^2 \beta}, \tag{D.18}$$

$$\int_f \beta \, dA = \frac{1}{|n_\gamma|} P_\beta, \tag{D.19}$$

$$\int_f \beta^2 \, dA = \frac{1}{|n_\gamma|} P_{\beta^2}, \tag{D.20}$$

$$\int_f \beta^3 \, dA = \frac{1}{|n_\gamma|} P_{\beta^3}, \tag{D.21}$$

$$\int_f \beta^2 \gamma \, dA = -\frac{1}{|n_\gamma| n_\gamma} (n_\alpha P_{\alpha \beta^2} + n_\beta P_{\beta^3} + d P_{\beta^2}), \tag{D.22}$$

$$\int_f \gamma \, dA = -\frac{1}{|n_\gamma| n_\gamma} (n_\alpha P_\alpha + n_\beta P_\beta + d P_1), \tag{D.23}$$

$$\int_f \gamma^2 \, dA = \frac{1}{|n_\gamma| n_\gamma^2} (n_\alpha^2 P_{\alpha^2} + 2 n_\alpha n_\beta P_{\alpha\beta} + n_\beta^2 P_{\beta^2} +$$
$$2 d n_\alpha P_\alpha + 2 d n_\beta P_\beta + d^2 P_1), \tag{D.24}$$

$$\int_f \gamma^3 \, dA = -\frac{1}{|n_\gamma| n_\gamma^3} (n_\alpha^3 P_{\alpha^3} + 3 n_\alpha^2 n_\beta P_{\alpha^2\beta} + 3 n_\alpha n_\beta^2 P_{\alpha\beta^2} +$$
$$n_\beta^3 P_{\beta^3} + 3 d n_\alpha^2 P_{\alpha^2} + 6 d n_\alpha n_\beta P_{\alpha\beta} + 3 d n_\beta^2 P_{\beta^2} +$$
$$3 d^2 n_\alpha P_\alpha + 3 d^2 n_\beta P_\beta + d^3 P_1), \tag{D.25}$$

$$\int_f \gamma^2 \alpha \, dA = \frac{1}{|n_\gamma| n_\gamma^2} (n_\alpha^2 P_{\alpha^3} + 2 n_\alpha n_\beta P_{\alpha^2\beta} + n_\beta^2 P_{\alpha\beta^2} +$$
$$2 d n_\alpha P_{\alpha^2} + 2 d n_\beta P_{\alpha\beta} + d^2 P_\alpha), \tag{D.26}$$

where the projected surface integral $P_{\alpha^u \beta^v}$ is computed as

$$P_{\alpha^u \beta^v} = \int_{f_p} \alpha^u \beta^v \, dA . \qquad (D.27)$$

Having determined the surface integrals as a function of the projected surface integrals, we can then proceed to the next step, which consists of reducing the projected surface integrals to line integrals along each edge of the projected face.

D.2.3 Projected-Surface-Integral to Line-Integral Reduction

The projected surface integral is reduced to a line integral using the *Green's theorem*. The Green's theorem can be envisage as the 2D case of the divergence theorem presented in Section D.2.1. It states that, given a planar surface f_p, a continuous force field $\vec{H}$ defined over f_p, and the outward normal $\vec{m}$ along the boundary ∂f_p of f_p, the surface integral is then equivalent to the line integral

$$\int_{f_p} \triangle \cdot \vec{H} \, dA = \oint_{\partial f_p} \vec{H} \cdot \vec{m} \, ds , \qquad (D.28)$$

where the circular line integral traverses *counterclockwise* the boundary of f_p (see Figure D.2). Again, we need to choose a suitable force field $\vec{H}$ such that the right-hand side of equation (D.28) is simplified as much as possible for the following line integral computations.

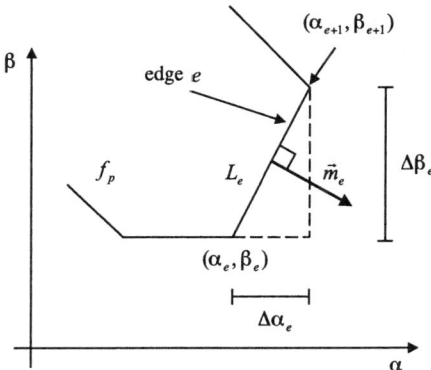

FIGURE D.2. The projected surface integral is reduced to a line integral over the edges of the projected face f_p. The line integral traverses each edge $e \in f_p$ in a counterclockwise direction.

The selection process is identical to that described in Section D.2.1. We shall pick a force field $\vec{H}$ that turns the dot product of the right-hand side of equation (D.28) into a simple scalar multiplication. We shall also break the line integral into smaller pieces along each edge of the projected face,

D.2 Mirtich's Algorithm

since each edge has a constant outward normal that can be extracted from the integration. Table D.2 shows the selected force fields for each projected surface integral that appears in equations (D.15) to (D.26).

Index i	Proj. Surf. Integral P_i	Force Field $\vec{H}_i$	Equivalent Line Integral
α	$\int_{P_i} \alpha \, dA$	$(\frac{\alpha^2}{2}, 0)^t$	$\frac{sign(n_\gamma)}{2} \sum_{e=1}^{n_e} \Delta\beta_e \int_0^1 \alpha^2(L_e\lambda) \, d\lambda$
α^2	$\int_{P_i} \alpha^2 \, dA$	$(\frac{\alpha^3}{3}, 0)^t$	$\frac{sign(n_\gamma)}{3} \sum_{e=1}^{n_e} \Delta\beta_e \int_0^1 \alpha^3(L_e\lambda) \, d\lambda$
α^3	$\int_{P_i} \alpha^3 \, dA$	$(\frac{\alpha^4}{4}, 0)^t$	$\frac{sign(n_\gamma)}{4} \sum_{e=1}^{n_e} \Delta\beta_e \int_0^1 \alpha^4(L_e\lambda) \, d\lambda$
β	$\int_{P_i} \beta \, dA$	$(0, \frac{\beta^2}{2})^t$	$-\frac{sign(n_\gamma)}{2} \sum_{e=1}^{n_e} \Delta\alpha_e \int_0^1 \beta^2(L_e\lambda) \, d\lambda$
β^2	$\int_{P_i} \beta^2 \, dA$	$(0, \frac{\beta^3}{3})^t$	$-\frac{sign(n_\gamma)}{3} \sum_{e=1}^{n_e} \Delta\alpha_e \int_0^1 \beta^3(L_e\lambda) \, d\lambda$
β^3	$\int_{P_i} \beta^3 \, dA$	$(0, \frac{\beta^4}{4})^t$	$-\frac{sign(n_\gamma)}{4} \sum_{e=1}^{n_e} \Delta\alpha_e \int_0^1 \beta^4(L_e\lambda) \, d\lambda$
$\alpha\beta$	$\int_{P_i} \alpha\beta \, dA$	$(\frac{\alpha^2\beta}{2}, 0)^t$	$\frac{sign(n_\gamma)}{2} \sum_{e=1}^{n_e} \Delta\beta_e \int_0^1 \alpha^2(L_e\lambda)\beta(L_e\lambda) \, d\lambda$
$\alpha^2\beta$	$\int_{P_i} \alpha^2\beta \, dA$	$(\frac{\alpha^3\beta}{3}, 0)^t$	$\frac{sign(n_\gamma)}{3} \sum_{e=1}^{n_e} \Delta\beta_e \int_0^1 \alpha^3(L_e\lambda)\beta(L_e\lambda) \, d\lambda$
$\alpha\beta^2$	$\int_{P_i} \alpha\beta^2 \, dA$	$(0, \frac{\alpha\beta^3}{3})^t$	$\frac{sign(n_\gamma)}{3} \sum_{e=1}^{n_e} \Delta\beta_e \int_0^1 \alpha(L_e\lambda)\beta^3(L_e\lambda) \, d\lambda$

TABLE D.2. The line integrals along each edge of the projected face f_p in terms of the α and β components selected at run-time. The change of variable $ds = L_e d\lambda$, with L_e being the length of edge e, is used to simplify the integral to vary from 0 to 1.

At this point, we have successfully reduced the volume integrals to line integrals over each edge of each projected face of each polyhedron that composes the object. Now, we need to address the last remaining part of the algorithm, which is to compute the line integrals given in Table D.2 as a function of the vertex coordinates of each edge.

D.2.4 Computing the Line Integrals from the Vertex Coordinates

The line integrals given in Table D.2 are all of the form

$$\int_0^1 \alpha^p(L_e\lambda)\,\beta^q(L_e\lambda)\,d\lambda\,,$$

where L_e is the length of the edge e being considered in the integration, and $0 \leq p \leq 4$ and $0 \leq q \leq 4$ are the coefficients of α and β, respectively.

Evaluating the line integrals from the vertex coordinates consists of computing the following expression for each edge being integrated:

$$\int_0^1 \alpha^p(L_e\lambda)\,\beta^q(L_e\lambda)\,d\lambda = \frac{1}{k_{pq}} \sum_{i=0}^p \sum_{j=0}^q \frac{\binom{p}{i}\binom{q}{j}}{\binom{p+q}{i+j}} \alpha_{e+1}^i\, \alpha_e^{p-i}\, \beta_{e+1}^j\, \beta_e^{q-j}\,,$$

where $k_{pq} = (p+q+1)$ and

$$\binom{p}{i} = \frac{p!}{i!\,(p-i)!}\,.$$

Table D.3 shows the vertex-based computation for each line integral in Table D.2. We use the vertex coordinates to compute the line integrals, and substitute the result into Table D.2 to compute the projected surface integrals. The surface integrals are computed from the projected surface integrals using equations D.15 to D.26, and their values are substituted back into Table D.1 to compute all volume integrals. Finally, the mass properties are evaluated from the volume integrals using equations (D.2), (D.3), (D.6) and (D.7).

D.3 Suggested Readings

The algorithm presented in this appendix was developed by Mirtich [Mir96a, Mir96b]. Despite its complex derivation and endless expressions relating volume to surface integrals, surface to line integrals, and line to vertex computations, the implementation is fairly straightforward. All volume integrals can be computed in a single pass through the polyhedrons' faces, edges and vertices, resulting in a fast practical algorithm.

A comprehensive survey of the various types of algorithms used to compute the mass properties of objects can be found in Lee et al. [LR82a]. The approximate cell-decomposition methods mentioned in the introduction can also be found in Lee et al. [LR82a, LR82b]. The other boundary-representation method specially designed for polyhedra with triangular faces was developed by Lien et al. [LK84].

Lastly, the derivation and analysis of the inertia tensor formulas, as well as the parallel-axis theorem, can be found in Beer et al [BJ77a], Alonso et al. [AF67], and many other mechanical engineering books that deal with the dynamics of mechanical parts.

Line Integral	Equivalent Vertex-Based Computation
$\int_0^1 \alpha^2(L_e\lambda)\,d\lambda$	$\frac{1}{3}\sum_{i=0}^{2} \alpha_{e+1}^i \alpha_e^{2-i}$
$\int_0^1 \alpha^3(L_e\lambda)\,d\lambda$	$\frac{1}{4}\sum_{i=0}^{3} \alpha_{e+1}^i \alpha_e^{3-i}$
$\int_0^1 \alpha^4(L_e\lambda)\,d\lambda$	$\frac{1}{5}\sum_{i=0}^{4} \alpha_{e+1}^i \alpha_e^{4-i}$
$\int_0^1 \beta^2(L_e\lambda)\,d\lambda$	$\frac{1}{3}\sum_{j=0}^{2} \beta_{e+1}^j \beta_e^{2-j}$
$\int_0^1 \beta^3(L_e\lambda)\,d\lambda$	$\frac{1}{4}\sum_{j=0}^{3} \beta_{e+1}^j \beta_e^{3-j}$
$\int_0^1 \beta^4(L_e\lambda)\,d\lambda$	$\frac{1}{5}\sum_{j=0}^{4} \beta_{e+1}^j \beta_e^{4-j}$
$\int_0^1 \alpha^2(L_e\lambda)\beta(L_e\lambda)\,d\lambda$	$\frac{1}{12}(\beta_{e+1}\sum_{i=0}^{2}(i+1)\alpha_{e+1}^i \alpha_e^{2-i} + \beta_e\sum_{i=0}^{2}(3-i)\alpha_{e+1}^i \alpha_e^{2-i})$
$\int_0^1 \alpha^3(L_e\lambda)\beta(L_e\lambda)\,d\lambda$	$\frac{1}{20}(\beta_{e+1}\sum_{i=0}^{3}(i+1)\alpha_{e+1}^i \alpha_e^{3-i} + \beta_e\sum_{i=0}^{3}(4-i)\alpha_{e+1}^i \alpha_e^{3-i})$
$\int_0^1 \alpha(L_e\lambda)\beta^3(L_e\lambda)\,d\lambda$	$\frac{1}{20}(\alpha_{e+1}\sum_{j=0}^{3}(j+1)\beta_{e+1}^j \beta_e^{3-j} + \alpha_e\sum_{j=0}^{3}(4-j)\beta_{e+1}^j \beta_e^{3-j})$

TABLE D.3. Vertex-based computation of each line integral in Table D.2.

Appendix E
Useful Time Derivatives

E.1 Introduction

In this appendix we shall present a detailed description of how the time derivatives of a normal vector, a rotation matrix and a quaternion are computed. These time derivatives are used extensively in Chapters 4 and 5 to describe the dynamic equations of a rigid body.

E.2 Computing the Time Derivative of a Vector Attached to a Rigid Body

The time derivative of a vector attached to a rigid body is used as a auxiliary result to most of the following sections in this appendix. Let points $\vec{p}_1(t)$ and $\vec{p}_2(t)$ of a rigid body B define a general vector $\vec{p}(t)$ given by

$$\vec{p}(t) = \vec{p}_1(t) - \vec{p}_2(t) , \qquad (E.1)$$

that is, the general vector $\vec{p}(t)$ is attached to the rigid body, so that its linear and angular velocities can be computed as a function of the rigid body's linear and angular velocities. The time derivative of the general vector $\vec{p}(t)$ is then

$$\frac{d\vec{p}(t)}{dt} = \frac{d\vec{p}_1(t)}{dt} - \frac{d\vec{p}_2(t)}{dt} . \qquad (E.2)$$

Assume that at time instant t the rigid body is moving with a linear velocity $\vec{v}(t)$ and an angular velocity $\vec{\omega}(t)$. Since $\vec{p}_1$ and $\vec{p}_2$ are points on the rigid body, the time derivatives of their positions is directly obtained as:

$$\frac{d\vec{p}_1(t)}{dt} = \vec{v}(t) + \vec{\omega}(t) \times \vec{p}_1$$

$$\frac{d\vec{p}_2(t)}{dt} = \vec{v}(t) + \vec{\omega}(t) \times \vec{p}_2$$

(E.3)

Substituting equations (E.3) into (E.2), we have

$$\frac{d\vec{p}(t)}{dt} = \vec{\omega}(t) \times \vec{p}_1 - \vec{\omega}(t) \times \vec{p}_2 = \vec{\omega}(t) \times (\vec{p}_1 - \vec{p}_2).$$

Using equation (E.1) we immediately have that the time derivative of a general vector attached to a rigid body is computed as:

$$\frac{d\vec{p}(t)}{dt} = \vec{\omega}(t) \times \vec{p}(t)$$

E.3 Computing the Time Derivative of a Contact-Normal Vector

Whenever particles or rigid bodies are in contact, the determination of the contact force necessary to prevent their interpenetration requires the computation of the time derivative of their contact normal. This computation is done slightly differently depending on whether we have a particle-particle, particle-rigid body or rigid body-rigid body contact.

E.3.1 Particle-Particle Contact

In the particle-particle case, the contact-normal direction is defined as the vector connecting the contacting particles O_1 and O_2, that is

$$\vec{n}(t) = \vec{p}_1(t) - \vec{p}_2(t).$$

The time derivative of the normal vector direction is obtained as

$$\begin{aligned}\frac{d\vec{n}(t)}{dt} &= \frac{d\vec{p}_1(t)}{dt} - \frac{d\vec{p}_2(t)}{dt}, \\ &= \vec{v}_1 - \vec{v}_2\end{aligned}$$

(E.4)

E.3 Computing the Time Derivative of a Contact-Normal Vector

where $\vec{p}_1$ and $\vec{p}_2$ are the velocities of particles O_1 and O_2, respectively. The actual derivative of the normal vector is obtained by normalizing equation (E.4), that is

$$\frac{d\vec{n}(t)}{dt} = \frac{(\vec{v}_1 - \vec{v}_2)}{|\vec{v}_1 - \vec{v}_2|} .$$

E.3.2 Rigid Body-Rigid Body Contact

There are two possible ways the contact-normal direction can be computed for a rigid body-rigid body contact. If the contact is between a vertex or an edge of rigid body B_1 with a face of rigid body B_2, then the contact-normal direction is given by the face-normal direction, that is

$$\vec{n}(t) = \vec{a}(t) \times \vec{b}(t) , \qquad (E.5)$$

where $\vec{a}(t)$ and $\vec{b}(t)$ are two edges of the face used to compute the face normal. However, if the contact is between an edge of B_1 with an edge of B_2, then the contact-normal direction is given by

$$\vec{n}(t) = \vec{e}_1(t) \times \vec{e}_2(t) , \qquad (E.6)$$

where $\vec{e}_1(t)$ is the edge of B_1 and $\vec{e}_2(t)$ is the edge of B_2.

Let's start by examining the time derivative of equation (E.6), namely

$$\frac{d\vec{n}(t)}{dt} = \frac{d\vec{e}_1(t)}{dt} \times \vec{e}_2(t) + \vec{e}_1(t) \times \frac{d\vec{e}_2(t)}{dt} . \qquad (E.7)$$

If we think of edge $\vec{e}_1(t)$ as a general vector attached to rigid body B_1, then using the results of Section E.2 we obtain

$$\frac{d\vec{e}_1(t)}{dt} = \vec{\omega}_1(t) \times \vec{e}_1(t) , \qquad (E.8)$$

where $\vec{\omega}_1(t)$ is the angular velocity of rigid body B_1. Conversely, we have

$$\frac{d\vec{e}_2(t)}{dt} = \vec{\omega}_2(t) \times \vec{e}_2(t) , \qquad (E.9)$$

where $\vec{\omega}_2(t)$ is the angular velocity of rigid body B_2. Substituting equations (E.8) and (E.9) into equation (E.7) we get

$$\frac{d\vec{n}(t)}{dt} = (\vec{\omega}_1(t) \times \vec{e}_1(t)) \times \vec{e}_2(t) + \vec{e}_1(t) \times (\vec{\omega}_2(t) \times \vec{e}_2(t)) . \qquad (E.10)$$

Using the general cross-product relation

$$\vec{a} \times (\vec{b} \times \vec{c}) = -(\vec{b} \times \vec{c}) \times \vec{a}$$

in the first cross-product term of equation (E.10), we obtain

$$\frac{d\vec{n}(t)}{dt} = \vec{e}_1(t) \times (\vec{\omega}_2(t) \times \vec{e}_2(t)) - \vec{e}_2(t) \times (\vec{\omega}_1(t) \times \vec{e}_1(t)) \,. \tag{E.11}$$

Using yet another general cross-product relation

$$\vec{a} \times (\vec{b} \times \vec{c}) = (\vec{a} \cdot \vec{c})\,\vec{b} - (\vec{a} \cdot \vec{b})\,\vec{c}\,, \tag{E.12}$$

we can substitute both cross-product terms of equation (E.11) by

$$\begin{aligned}\frac{d\vec{n}(t)}{dt} &= (\vec{e}_1(t) \cdot \vec{e}_2(t))\,\vec{\omega}_2(t) - (\vec{e}_1(t) \cdot \vec{\omega}_2(t))\,\vec{e}_2(t) - \\ &\quad ((\vec{e}_2(t) \cdot \vec{e}_1(t))\,\vec{\omega}_1(t) - (\vec{e}_2(t) \cdot \vec{\omega}_1(t))\,\vec{e}_1(t)) \,.\end{aligned}$$

Merging similar terms, we have that the time derivative of the contact-normal direction for edge-edge contact is given by

$$\frac{d\vec{n}(t)}{dt} = (\vec{e}_1(t)\cdot\vec{e}_2(t))\,(\vec{\omega}_2(t)-\vec{\omega}_1(t))+(\vec{e}_2(t)\cdot\vec{\omega}_1(t))\,\vec{e}_1(t)-(\vec{e}_1(t)\cdot\vec{\omega}_1(t))\,\vec{e}_2(t)\,. \tag{E.13}$$

The derivations for vertex-face contact, represented by equation (E.5), are almost the same as the derivations for edge-edge contact, represented by equation (E.5), that is

$$\frac{d\vec{n}(t)}{dt} = (\vec{a}(t) \cdot \vec{b}(t))\,(\vec{\omega}_b(t) - \vec{\omega}_a(t)) + (\vec{b}(t) \cdot \vec{\omega}_a(t))\,\vec{a}(t) - (\vec{a}(t) \cdot \vec{\omega}_a(t))\,\vec{b}(t) \,. \tag{E.14}$$

The only difference is that, since the normal in the vertex-face contact is computed as the cross-product of two edges belonging to the same face (i.e., the same rigid body), their angular velocities are the same. In other words

$$\vec{\omega}_a(t) = \vec{\omega}_b(t) = \vec{\omega}(t) \,, \tag{E.15}$$

where $\vec{\omega}(t)$ is the angular velocity of the rigid body the face belongs to. Substituting equation (E.15) into equation (E.14) we get

$$\begin{aligned}\frac{d\vec{n}(t)}{dt} &= (\vec{a}(t) \cdot \vec{b}(t))\,(\vec{\omega}(t) - \vec{\omega}(t)) + (\vec{b}(t) \cdot \vec{\omega}(t))\,\vec{a}(t) - (\vec{a}(t) \cdot \vec{\omega}(t))\,\vec{b}(t)) \\ &= (\vec{b}(t) \cdot \vec{\omega}(t))\,\vec{a}(t) - (\vec{a}(t) \cdot \vec{\omega}(t))\,\vec{b}(t) \,.\end{aligned} \tag{E.16}$$

Using the general cross-product relation described in equation (E.12) we obtain

$$\frac{d\vec{n}(t)}{dt} = \vec{\omega}(t)\,(\vec{a}(t) \times \vec{b}(t)) = \vec{\omega}(t) \times \vec{n}(t)\ . \tag{E.17}$$

which is compatible with the expression obtained for the time derivative of a general vector attached to a rigid body.

The actual time derivative of the contact normal for both situations is then obtained by normalizing equations (E.13) and (E.17), that is, by computing

$$\left(\frac{d\vec{n}(t)}{dt}\right) \bigg/ \left|\frac{d\vec{n}(t)}{dt}\right|\ .$$

E.4 Computing the Time Derivative of the Tangent Plane

Following the convention described in Section A.6 of Appendix A, the tangent plane direction $\vec{t}(t)$ is directly obtained from the normal vector direction $\vec{n}(t) = (n_x(t), n_y(t), n_z(t))$ by setting to zero its component with the smallest absolute value, swapping the remaining two components and multiplying one of them by -1. We use an auxiliary vector $\vec{a}$ as follows.

1. If $|n_x| < |n_y|$ and $|n_x| < |n_z|$, then set the auxiliary vector $\vec{a}$ to

$$\vec{a} = (0, -n_z, n_y)\ . \tag{E.18}$$

2. If $|n_y| < |n_x|$ and $|n_y| < |n_z|$, then set the auxiliary vector $\vec{a}$ to

$$\vec{a} = (-n_z, 0, n_x)\ . \tag{E.19}$$

3. If $|n_z| < |n_x|$ and $|n_z| < |n_z|$, then set the auxiliary vector $\vec{a}$ to:

$$\vec{a} = (-n_y, n_x, 0)\ . \tag{E.20}$$

The tangent vector $\vec{t}$ is then set to

$$\vec{t} = \frac{\vec{a}}{|\vec{a}|}\ .$$

The time derivative of the tangent vector $\vec{t}$ is computed as a function of the time derivative of the auxiliar vector, namely

$$\frac{d\vec{t}}{dt} = \left(\frac{d\vec{a}}{dt}\right) \bigg/ \left|\frac{d\vec{a}}{dt}\right|\ ,$$

where:

318   Appendix E. Useful Time Derivatives

1. If $\vec{a} = (0, -n_z, n_y)$, then

$$\frac{d\vec{a}}{dt} = \left(0, -\frac{dn_z(t)}{dt}, \frac{dn_y(t)}{dt}\right).$$

2. If $\vec{a} = (-n_z, 0, n_x)$, then

$$\frac{d\vec{a}}{dt} = \left(-\frac{dn_z(t)}{dt}, 0, \frac{dn_x(t)}{dt}\right).$$

3. If $\vec{a} = (-n_y, n_x, 0)$, then

$$\frac{d\vec{a}}{dt} = \left(-\frac{dn_y(t)}{dt}, \frac{dn_x(t)}{dt}, 0\right).$$

The tangent-plane direction $\vec{k}(t)$ is computed as the cross-product of $\vec{n}(t)$ and $\vec{t}(t)$, that is

$$\vec{k} = \vec{n} \times \vec{t}.$$

Therefore, its time derivative can be obtained as

$$\frac{d\vec{k}}{dt} = \frac{d\vec{n}}{dt} \times \vec{t} + \vec{n} \times \frac{d\vec{t}}{dt}$$

after the result is normalized.

E.5 Computing the Time Derivative of a Rotation Matrix

A rotation matrix $\mathbf{R}(t)$ can be viewed as a transformation between a coordinate frame[1] $\mathcal{F}_1$ and the canonical coordinate frame $\mathcal{F}_0$, with the origin of $\mathcal{F}_1$ being coincident with the origin of $\mathcal{F}_0$. Let $\mathcal{F}_1$ be defined by the coordinate vectors $\vec{x}_1$, $\vec{y}_1$ and $\vec{z}_1$, and let the canonical vectors be $\vec{x}_0 = (1, 0, 0)^t$, $\vec{y}_0 = (0, 1, 0)^t$ and $\vec{z}_0 = (0, 0, 1)^t$. Then, a point $\vec{p}_1$ in the coordinate frame $\mathcal{F}_1$ is transformed to a point $\vec{p}_0$ in the canonical frame $\mathcal{F}_0$ by applying the rotation matrix

$$\vec{p}_0 = \mathbf{R}(t)\,\vec{p}_1.$$

Viewed as a "change of basis" transformation, it is natural to describe the rotation matrix $\mathbf{R}(t)$ using its column-vector representation:

---

[1] In the rigid-body case, the coordinate frame $\mathcal{F}_1$ is the body frame.

E.5 Computing the Time Derivative of a Rotation Matrix

$$\mathbf{R}(t) = \begin{pmatrix} \vec{c}_1(t) & \vec{c}_2(t) & \vec{c}_3(t) \end{pmatrix}, \quad (E.21)$$

where the column vectors $\vec{c}_1$, $\vec{c}_2$ and $\vec{c}_3$ represent the coordinate axes $\vec{x}_1$, $\vec{y}_1$ and $\vec{z}_1$ expressed in canonical-frame coordinates.

The time derivative of the rotation matrix is then computed as

$$\frac{d\mathbf{R}(t)}{dt} = \begin{pmatrix} d\vec{c}_1(t)/dt & d\vec{c}_2(t)/dt & d\vec{c}_3(t)/dt \end{pmatrix}. \quad (E.22)$$

Let $\vec{\omega}(t)$ be the angular velocity of frame $\mathcal{F}_1$ expressed in canonical-frame coordinates. Using the results obtained in Section E.2, we have that the time derivative of each column vector can be obtained as

$$\begin{aligned}
\frac{d\vec{c}_1(t)}{dt} &= \vec{\omega}(t) \times \vec{c}_1(t) \\
\frac{d\vec{c}_2(t)}{dt} &= \vec{\omega}(t) \times \vec{c}_2(t) \\
\frac{d\vec{c}_3(t)}{dt} &= \vec{\omega}(t) \times \vec{c}_3(t).
\end{aligned} \quad (E.23)$$

Substituting equations (E.23) into (E.22), we have

$$\frac{d\mathbf{R}(t)}{dt} = \begin{pmatrix} \vec{\omega}(t) \times \vec{c}_1(t) & \vec{\omega}(t) \times \vec{c}_2(t) & \vec{\omega}(t) \times \vec{c}_3(t) \end{pmatrix}.$$

Using the matrix-vector representation of a cross-product as described in Section A.7, we get

$$\frac{d\mathbf{R}(t)}{dt} = \begin{pmatrix} \tilde{\omega}(t)\,\vec{c}_1(t) & \tilde{\omega}(t)\,\vec{c}_2(t) & \tilde{\omega}(t)\,\vec{c}_3(t) \end{pmatrix}, \quad (E.24)$$

where $\vec{\omega} = (\omega_x, \omega_y, \omega_z)$ and

$$\tilde{\omega}(t) = \begin{pmatrix} 0 & -\omega_z & \omega_y \\ \omega_z & 0 & -\omega_x \\ -\omega_y & \omega_x & 0 \end{pmatrix}.$$

Equation (E.24) can then be written as

$$\frac{d\mathbf{R}(t)}{dt} = \tilde{\omega}(t) \begin{pmatrix} \vec{c}_1(t) & \vec{c}_2(t) & \vec{c}_3(t) \end{pmatrix}.$$

Thus, the time derivative of a rotation matrix $\mathbf{R}(t)$ is given by

$$\frac{d\mathbf{R}(t)}{dt} = \tilde{\omega}(t)\,\mathbf{R}(t)\,. \tag{E.25}$$

E.6 Computing the Time Derivative of a Unit Quaternion

The time derivative of a unit quaternion $q = s + \vec{v}$ will be computed using the results already obtained for the time derivative of a rotation matrix. Recall from Appendix C that the rotation-matrix representation of a unit quaternion is given by

$$\mathbf{R} = 2\begin{pmatrix} s^2 + v_x^2 - \frac{1}{2} & v_x\,v_y - s\,v_z & v_x\,v_z + s\,v_y \\ v_x\,v_y + s\,v_z & s^2 + v_y^2 - \frac{1}{2} & v_y\,v_z - s\,v_x \\ v_x\,v_z - s\,v_y & v_y\,v_z + s\,v_x & s^2 + v_z^2 - \frac{1}{2} \end{pmatrix}, \tag{E.26}$$

where v_x, v_y and v_z are the components of the imaginary part $\vec{v}$ of the unit quaternion q. The time derivative of this rotation matrix is then[2]

$$\frac{d\mathbf{R}}{dt} = 2\begin{pmatrix} 2(s\,\dot{s} + v_x\,\dot{v}_x) & \dot{v}_x\,v_y + v_x\,\dot{v}_y - \dot{s}\,v_z - s\,\dot{v}_z \\ \dot{v}_x\,v_y + v_x\,\dot{v}_y + \dot{s}\,v_z + s\,\dot{v}_z & 2(s\,\dot{s} + v_y\,\dot{v}_y) \\ \dot{v}_x\,v_z + v_x\,\dot{v}_z - \dot{s}\,v_y - s\,\dot{v}_y & \dot{v}_y\,v_z + v_y\,\dot{v}_z + \dot{s}\,v_x + s\,\dot{v}_x \end{pmatrix}$$

$$\begin{pmatrix} \dot{v}_x\,v_z + v_x\,\dot{v}_z + \dot{s}\,v_y - s\,\dot{v}_y \\ \dot{v}_y\,v_z + v_y\,\dot{v}_z - \dot{s}\,v_x - s\,\dot{v}_x \\ 2\,(s\,\dot{s} + v_z\,\dot{v}_z) \end{pmatrix}. \tag{E.27}$$

As mentioned in Section E.5, the rotation matrix $\mathbf{R}(t)$ can be viewed as a transformation between the body frame and the canonical frame. Let $\vec{\omega}(t)$ be the angular velocity of the body frame expressed in canonical-frame coordinates. Using equation (E.25), we have that the time derivative of the rotation matrix is computed as

$$\frac{d\mathbf{R}(t)}{dt} = \tilde{\omega}(t)\,\mathbf{R}(t)\,, \tag{E.28}$$

where $\vec{\omega} = (\omega_x, \omega_y, \omega_z)$ and

$$\tilde{\omega}(t) = \begin{pmatrix} 0 & -\omega_z & \omega_y \\ \omega_z & 0 & -\omega_x \\ -\omega_y & \omega_x & 0 \end{pmatrix}. \tag{E.29}$$

Right multiplying both sides of equation (E.28) by $\mathbf{R}^{-1}(t) = \mathbf{R}^t(t)$ gives

---

[2] To simplify the notation, we shall use $\dot{a}$ to represent the time derivative $\frac{da}{dt}$.

E.6 Computing the Time Derivative of a Unit Quaternion

$$\tilde{\omega}(t) = \frac{d\mathbf{R}(t)}{dt}\mathbf{R}^t(t) . \tag{E.30}$$

Substituting equations (E.26), (E.27) and (E.29) into equation (E.30), we obtain a linear system that can be solved for ω_x, ω_y and ω_z as follows.

Inspecting equation (E.29), we have that ω_x is obtained by multiplying the third row of $d\mathbf{R}(t)/dt$ by the second column of $\mathbf{R}^t(t)$, that is

$$\frac{\omega_x}{4} = (\dot{v}_x v_z + \dot{v}_z v_x - \dot{v}_y s - \dot{s} v_y)(v_x v_y + v_z s) +$$
$$(\dot{v}_y v_z + \dot{v}_z v_y + \dot{v}_x s + \dot{s} v_x)(s^2 + v_y^2 - \frac{1}{2}) +$$
$$2(\dot{s} s + \dot{v}_z v_z)(\dot{v}_y v_z - \dot{v}_x s) .$$

Grouping the terms with common derivatives we get

$$\frac{\omega_x}{4} = \dot{v}_x (v_x v_y v_z + s(v_z^2 + s^2 + v_y^2 - \frac{1}{2})) +$$
$$\dot{v}_y (v_z (v_y^2 - \frac{1}{2}) - v_x v_y v_z) +$$
$$\dot{v}_z (v_y (v_x^2 + v_y^2 + v_z^2 + s^2 - \frac{1}{2}) + v_y v_z^2 - v_x v_z s) +$$
$$\dot{s} (s v_y v_z - v_x (s^2 + \frac{1}{2})) . \tag{E.31}$$

Since $q = s + \vec{v}$ is a unit quaternion, it must satisfy

$$v_x^2 + v_y^2 + v_z^2 + s^2 = 1 \tag{E.32}$$
$$\dot{v}_x v_x + \dot{v}_y v_y + \dot{v}_z v_z + \dot{s} s = 0 . \tag{E.33}$$

Using equation (E.32) into (E.31) we have

$$\frac{\omega_x}{4} = \dot{v}_x (v_x v_y v_z + s(\overbrace{v_z^2 + s^2 + v_y^2}^{1-v_x^2} - \frac{1}{2})) +$$
$$\dot{v}_y (v_z (v_y^2 - \frac{1}{2}) - v_x v_y v_z) +$$
$$\dot{v}_z (v_y (\overbrace{v_x^2 + v_y^2 + v_z^2 + s^2}^{1} - \frac{1}{2}) + v_y v_z^2 - v_x v_z s) +$$
$$\dot{s} (s v_y v_z - v_x (s^2 + \frac{1}{2}))$$
$$= \dot{v}_x (v_x v_y v_z + s(\frac{1}{2} - v_x^2)) +$$

$$\dot{v}_y \left(v_z \left(v_y^2 - \frac{1}{2}\right) - v_x v_y v_z\right) +$$
$$\dot{v}_z \left(v_y \left(\frac{1}{2} + v_z^2\right) + v_y v_z^2 - v_x v_z s\right) +$$
$$\dot{s} \left(s v_y v_z - v_x \left(s^2 + \frac{1}{2}\right)\right).$$

Regrouping the terms and using equation (E.33) we have

$$\frac{\omega_x}{4} = (\overbrace{\dot{v}_x v_x + \dot{s} s}^{-\dot{v}_y v_y - \dot{v}_z v_z}) v_y v_z - (\overbrace{\dot{v}_y v_y + \dot{v}_z v_z}^{-\dot{v}_x v_x - \dot{s} s}) v_x s +$$
$$\dot{v}_x s \left(\frac{1}{2} - v_x^2\right) + \dot{v}_y v_z \left(v_y^2 - \frac{1}{2}\right) +$$
$$\dot{v}_z v_y \left(\frac{1}{2} + v_z^2\right) - \dot{s} v_x \left(\frac{1}{2} + s^2\right)$$
$$= \frac{s}{2} \dot{v}_x - \frac{v_z}{2} \dot{v}_y + \frac{v_y}{2} \dot{v}_z - \frac{v_x}{2} \dot{s},$$

that is

$$\omega_x = 2 s \dot{v}_x - 2 v_z \dot{v}_y + 2 v_y \dot{v}_z - 2 v_x \dot{s}. \qquad (E.34)$$

Inspecting equation (E.29) one more time, we have that ω_y is obtained by multiplying the first row of $d\mathbf{R}(t)/dt$ by the third column of $\mathbf{R}^t(t)$, that is

$$\frac{\omega_y}{4} = 2(\dot{s} s + \dot{v}_x v_x)(v_z v_x - v_y s) +$$
$$(\dot{v}_y v_x + \dot{v}_x v_y - \dot{v}_z s - \dot{s} v_z)(v_z v_y + v_x s) +$$
$$(\dot{v}_z v_x + \dot{v}_x v_z + \dot{v}_y s + \dot{s} v_y)(s^2 + v_z^2 - \frac{1}{2}).$$

Doing groupings and substitutions similar to those we employed when computing ω_x, we obtain

$$\omega_y = 2 v_z \dot{v}_x + 2 s \dot{v}_y - 2 v_x \dot{v}_z - 2 v_y \dot{s}. \qquad (E.35)$$

Finally, inspecting equation (E.29) one last time, we have that ω_z is obtained by multiplying the second row of $d\mathbf{R}(t)/dt$ by the first column of $\mathbf{R}^t(t)$, that is

$$\frac{\omega_z}{4} = (\dot{v}_x v_y + \dot{v}_y v_x + \dot{v}_z s + \dot{s} v_z)(s^2 + v_x^2 - \frac{1}{2}) +$$
$$2(\dot{s} s + \dot{v}_y v_y)(v_x v_y - s v_z) +$$
$$(\dot{v}_z v_y + \dot{v}_y v_z - \dot{v}_x s - \dot{s} v_x)(v_x v_z + v_y s).$$

Again, doing groupings and substitutions similar to those we employed when computing ω_x, we obtain

$$\omega_z = -2\, v_y\, \dot{v}_x + 2\, v_x\, \dot{v}_y + 2\, s\, \dot{v}_z - 2\, v_z\, \dot{s}\ . \tag{E.36}$$

Equations (E.34), (E.35) and (E.36) form a linear system:

$$\begin{pmatrix} \omega_x \\ \omega_y \\ \omega_z \\ 0 \end{pmatrix} = 2 \begin{pmatrix} s & -v_z & v_y & -v_x \\ v_z & s & -v_x & -v_y \\ -v_y & v_x & s & -v_z \\ v_x & v_y & v_z & s \end{pmatrix} \begin{pmatrix} \dot{v}_x \\ \dot{v}_y \\ \dot{v}_z \\ \dot{s} \end{pmatrix}, \tag{E.37}$$

where equation (E.33) was used in the last row to make the matrix square. Because the determinant of this matrix is -1, the matrix is always invertible. Inverting the system in equation (E.37), we obtain

$$\begin{pmatrix} \dot{v}_x \\ \dot{v}_y \\ \dot{v}_z \\ \dot{s} \end{pmatrix} = \frac{1}{2} \begin{pmatrix} s & v_z & -v_y & v_x \\ -v_z & s & v_x & v_y \\ v_y & -v_x & s & v_z \\ -v_x & -v_y & -v_z & s \end{pmatrix} \begin{pmatrix} \omega_x \\ \omega_y \\ \omega_z \\ 0 \end{pmatrix}, \tag{E.38}$$

which gives the time derivative of the unit quaternion as a function of the angular-velocity components.

E.7 Suggested Readings

Most of the derivations in this appendix were either directly obtained, or inspired, from Baraff et al [BW98a] and Mirtich's [Mir96b] work. The main difference in the derivations presented here from their work is in the computation of the time derivative of a unit quaternion. Mirtich assumed the angular velocity of a rigid body is expressed in the body-frame coordinates, instead of being given in the canonical-frame coordinates as used in this book. For that reason, the results obtained in Section E.6 are different from Mirtich's final equations.

Baraff et al., on the other hand, presented a totally different approach to compute the time derivative of a unit quaternion. They represented the angular velocity $\vec{\omega}(t)$ as a rotation about the $(\vec{\omega}(t)/|\vec{\omega}(t)|)$ axis with magnitude $|\vec{\omega}(t)|$. A unit quaternion $q(t)$ can then be built from this rotation axis and rotation angle as

$$q(t) = \cos(\frac{|\vec{\omega}(t)|\, t}{2}) + \sin(\frac{|\vec{\omega}(t)|\, t}{2}) \frac{\vec{\omega}(t)}{|\vec{\omega}(t)|}\ .$$

The time derivative of the quaternion is then, after some manipulation, given by

$$\frac{d\,q(t)}{dt} = \frac{1}{2}\,q_\omega(t)\,q(t)\ ,$$

where $q_\omega(t) = 0 + \vec{\omega}(t)$ is the pure quaternion representing the angular velocity.

Lastly, the general cross-product relations were obtained from Gardshteyn et al. [GR80].

Appendix F
Convex Decomposition of 3D Polyhedra

F.1 Introduction

Most algorithms presented in this book are specially tailored for convex objects. The assumption that the objects being manipulated are convex guarantees faster solutions and much more efficient implementations that take full advantage of the nice properties of convex polyhedra. Nevertheless, most interesting dynamic-simulation scenarios contain at least one non-convex object, making it necessary to pre-process non-convex objects into a set of convex polyhedra before applying most of the algorithms described in this book.

The general convex decomposition problem of partitioning a non-convex 3D polyhedron into a minimum number of convex parts is a rather complex one, known to be NP-hard. There is a significant amount of work in the computational-geometry literature establishing worst-case bounds on the time complexity of convex decomposition algorithms, as well as lower bounds on the total number of convex polyhedra found in the decomposition. However, in the context of dynamic simulations, we are more concerned with the quality of the convex decomposition than the actual number of convex parts it comprises. We want a convex decomposition of "good" quality, tolerant to numerical round-off errors that may be introduced during the computations. We want a convex partition that is suitable for computing hierarchical decompositions of the object (such as those presented in Chapter 2), for checking geometrical intersections between objects to detect the existence of collisions, and for accurately computing the normal vector at the collision point between two objects.

In this appendix, we shall present a simplified version of Joe's algorithm, a convex-decomposition algorithm developed in the context of mesh generation for finite element analysis of complex 3D shapes. The algorithm decomposes non-convex 3D polyhedra into a "good" quality set of convex parts. By "good" quality we mean that the algorithm avoids long and skinny convex parts, as well as unnecessarily short edges and narrow subregions in the convex decomposition. In the case of finite-element analysis, this means that the tetrahedral mesh created from the convex decomposition contains tetrahedra of about the same size and shape. In the context of dynamic simulation, this means that the convex decomposition can be effectively used to produce high-quality hierarchical decompositions of the non-convex object.

There are a few restrictions, though, to the type of 3D polyhedra the algorithm can handle. The original algorithm is restricted to decomposing *simple* non-convex polyhedra, that is, polyhedra that satisfy the following five conditions.

1. The polyhedron has no interior holes.

2. The faces of the polyhedron may have holes. These holes can either stop somewhere inside the polyhedron, or go through the polyhedron from one face to another.

3. The polyhedron is not self-intersecting.

4. Each edge is incident on exactly two faces.

5. The faces surrounding each vertex form a simple circuit.

The last condition states that, for each vertex of the non-convex polyhedron, if we construct a double linked-list of neighbor faces in which a face is linked to another if they share a common edge that has the given vertex as one of its extreme points, then all faces are reachable from any other face in the list. This condition is necessary to avoid special cases wherein the polyhedron can be separated into two parts that share only a single vertex (see Figure F.1).

The simplified version presented in this appendix restricts even further the simple-polyhedron assumption stated in the original algorithm. Besides being simple, the polyhedra being decomposed must also satisfy the following two conditions.

6. The faces of the polyhedron have no holes.

7. The faces are convex polygons.

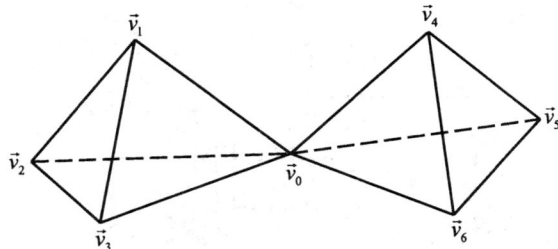

FIGURE F.1. An example of a special case that is not handled by the algorithm: a polyhedron made of two tetrahedra connected by a single vertex v_0.

Even though condition 6 limits the polyhedron faces to be simple polygons themselves, this does not prevent the non-convex polyhedron from having exterior holes. Figure F.2 shows the difference between a face having a hole, and the polyhedron itself having an exterior hole. Exterior holes are allowed, provided that they don't go through the interior of any face.

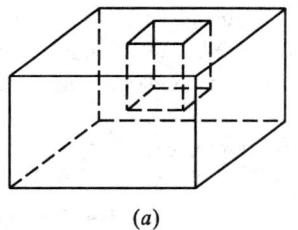 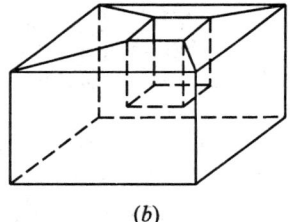

(a) (b)

FIGURE F.2. (a) The face has an interior hole that stops somewhere inside the polyhedron; (b) The faces have no interior holes. Nonetheless, the polyhedron itself has an exterior hole with the same depth as (a).

These extra assumptions have no effect on the main body of Joe's algorithm itself. However, they will be useful to restrict the types of non-convex polyhedra that can be handled, simplify the occurrence of special cases, and reduce the complexity of a software implementation. An example of special cases that are avoided in the simplified version is the creation of double-occuring faces after the non-convex polyhedron is split by a cut plane (see Figure F.3).

F.2 Joe's Algorithm

The algorithm that computes the convex decomposition of a 3D polyhedron that satisfies conditions 1 to 7 above is simple to state, but much more complicated to implement. The basic idea is to go through the list of edges of the polyhedron and compute the dihedral angle associated with each edge (see Section 2.2.4 of Chapter 2). The dihedral angle is the internal

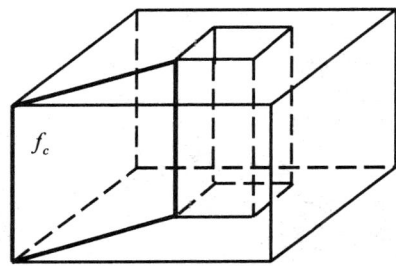

FIGURE F.3. Double-occuring faces may appear in the original Joe's algorithm, where faces are allowed to have internal holes. In this case, the cut face f_c sub-divides the original non-convex polyhedron in a simply connected polyhedron with face f_c occuring twice.

angle formed by the faces that share the edge. If the dihedral angle is less than or equal to π, then the polyhedron is convex at the edge. Otherwise, the polyhedron is non-convex at the edge, and the edge is said to be a *reflex* edge. It is clear that the polyhedron will be non-convex if and only if it has at least one reflex edge.

Having determined all reflex edges of the non-convex polyhedron, the algorithm proceeds by recursively resolving each reflex edge. The process of resolving reflex edges consists of splitting their dihedral angle into sub-angles less than or equal to π. The way the dihedral angle is split depends on the conditions that need to be met to obtain a "good" quality convex partition. In our case, we want to avoid small dihedral angles, short edges and narrow subregions when resolving a reflex edge. In other words, the selection of a cut plane that split the dihedral angle of a reflex edge is a process of accepting (or rejecting) cut-plane candidates that are computed from a set of conditions that need to be satisfied to obtain our "good" quality convex partition.

Once a cut plane that satisfies the conditions is obtained, a cut face associated with the cut plane is traced. The cut face is computed by intersecting the cut plane with the faces of the polyhedron, such that the interior of the cut face lies in the interior of the polyhedron. This procedure will usually split the polyhedron into two polyhedra, and the convex-decomposition algorithm is recursively applied for each of the two polyhedra[1] until the algorithm terminates either with a valid convex decomposition, or with one or more reflex edges that could not be resolved for the given conditions.

---

[1] This is true provided the cut face turns out to be a simple polygon. However, there are cases where the cut face can be either a multiply connected polygon with the reflex edge lying on the outer boundary or the inner boundary of a hole, or a simply connected, but non-simple, polygon.

F.2.1 Determining Candidate Cut Planes

Candidate cut planes are determined for each reflex edge being resolved, according to the set of conditions that we want to meet.

8. Sufficiently large dihedral angles.

9. Not too narrow subregions.

10. Edges with reasonable length.

In order to quantify conditions 8 to 10, we use the following variables.

11. The minimum acceptable internal dihedral angle, denoted by θ_{acc}.

12. The minimum acceptable *relative* distance between the cut plane and other vertices of the polyhedron not on the plane, denoted by d_{acc}. Notice that the actual distance depends on the size of the polyhedron being decomposed and is computed as $d_i = d_{acc} \bar{e}_i$, where $\bar{e}_i$ is the average length of the edges of polyhedron P_i.

The minimum acceptable dihedral angle variable in 11 is used to address condition 8 in avoiding long and skinny convex parts, whereas the minimum acceptable relative distance variable in 12 is used to address conditions 9 and 10 in avoiding both short edges and narrow convex regions.

Another important variable to be considered is the total number of candidate cut planes n_c considered for each reflex edge. This variable is used to limit the amount of execution time spent trying to resolve each reflex edge, and restrict the set of candidate cut planes to those that are most likely to be accepted. Therefore, for each reflex edge, up to n_c candidate cut planes are computed and tested against the desired conditions 8 to 10.

The actual list of candidate cut planes is constructed for each reflex edge as follows. Let e_r be a reflex edge of the non-convex polyhedron P_i. Let faces f_1 and f_2 be incident on edge e_r, and let $\theta_e > \pi$ be its associated dihedral angle.

Naturally, the first two choices of candidate cut planes are those forming a dihedral angle of $\theta = \theta_e - \pi$ with faces f_1 and f_2, respectively (see Figure F.4). These choices are added to the list of candidate cut planes only if their dihedral angle θ satisfies the minimum acceptable dihedral angle condition 11, that is, only if

$$\theta \geq \theta_{acc} \, . \tag{F.1}$$

The next choices of candidate cut planes are those that contain the reflex edge e_r and another polyhedron edge sharing a vertex with e_r. Again, these choices will be added to the list of candidate cut planes only if their corresponding dihedral angles satisfy equation (F.1).

330   Appendix F. Convex Decomposition of 3D Polyhedra

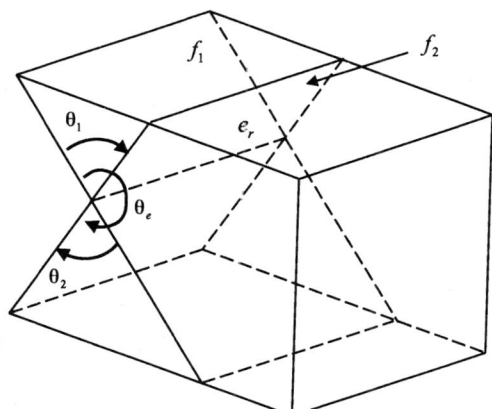

FIGURE F.4. First two choices of candidate cut planes that form an angle of $\theta_e - \pi$ with each of the faces f_1 and f_2 incident on the reflex edge e_r. The angles θ_1 and θ_2 must be tested against the minimum dihedral angle condition of equation (F.1) before they can be added to the list of candidate cut planes.

If the list of candidate cut planes obtained so far has not yet achieved the limit of n_c cut planes, then we continue choosing candidate cut planes that bisect the internal dihedral angle θ_e forming angles θ with face f_1, that satisfy equation (F.1) and are of the form

$$\theta = \rho \theta_e ,$$

with $\rho \in \{0.25, 0.5, 0.75, 0.375, 0.625\}$. The actual number of candidate cut planes added in this last step depends on the value of θ_e, and on the number of candidate cut planes selected.

Having determined the list of candidate cut planes for the reflex edge e_r, we are ready to compute a cut face associated with each candidate cut plane. As each edge of the cut face is being traced, conditions 8 to 10 are constantly checked to verify whether the candidate cut plane should be rejected.

A candidate cut plane is rejected if one of the following occurs.

13. The minimum dihedral-angle condition is not satisfied. In this case, we have found an edge of the cut face that forms an internal dihedral angle θ that does not satisfy equation (F.1).

14. The minimum acceptable relative-distance condition is not satisfied. In this case, there exists at least one vertex of the polyhedron that does not lie on the cut plane and is distant from the plane by an amount less than or equal to the minimum acceptable distance $d = d_{acc} \bar{e}$, where $\bar{e}$ is the average length of the edges of the polyhedron.

15. The cut face is not a simple polygon, that is, it contains one or more holes.

If none of the candidate cut planes in the list satisfies the above criteria, then the reflex edge e_r cannot be resolved at this time, and the algorithm proceeds to another reflex edge, leaving the unresolved one to be addressed again later. The rationale behind this strategy is that, after resolving other reflex edges, the sub-polyhedron containing the reflex edge e_r may be smaller and easier to decompose, or the reflex edge e_r may have been subdivided by other cut faces into two or more reflex sub-edges of less complexity.

F.2.2 Computing the Cut Face Associated with a Cut Plane

The most complicated part of the algorithm is undoubtedly computing the edges of the cut face associated with a candidate cut plane. The idea is to intersect the cut plane with the faces of the polyhedron being decomposed and construct the cut face from the intersection edges, such that the cut face lies in the interior of the polyhedron.

Starting with the reflex edge e_r, we trace the cut face, one edge at a time, moving along the direction corresponding to the normal vector of the cut plane, such that the interior of the polyhedron being cut is to the left of the boundary of the cut face. In most cases, this direction is counterclockwise, but it can be clockwise as well, and we should keep track of the relative direction of the faces in each sub-polyhedra as they are cut by the cut planes.

One advantage of tracing the cut face one edge at a time is that the internal dihedral angle of each new edge can be immediately computed and tested using equation (F.1). The candidate cut plane is rejected if the internal dihedral angle of the new edge is less than the minimum acceptable value. Also, as new edges of the cut face are traced, if a duplicate vertex other than the first vertex is found, then the cut face is at least simply connected, and the candidate cut plane is rejected as well.

The cut face associated with a candidate cut plane is determined as follows. Let P be the non-convex polyhedron to which we are applying the convex decomposition algorithm. Let f_c be the cut face associated with the cut plane α passing through e_r, and let $\vec{v}_0, \vec{v}_1, \ldots, \vec{v}_i$ be the cut-face vertices traced so far. Assume that the last-computed edge $(\vec{v}_{i-1}, \vec{v}_i)$ of the cut face f_c lies on face $f_k \in P$. The next vertex v_{i+1} of the cut face is computed according to one of the following two possible situations.

In the first situation considered, the last-computed vertex v_i lies in the interior of an edge $e_k \in P$ (see Figure F.5).

In this case, the direction $\vec{d}_{next}$ of the next edge $(\vec{v}_i, \vec{v}_{i+1})$ of the cut face f_c is unique, and can be determined from the cross-product of the normal vector $\vec{n}_\alpha$ of the cut plane α, and the normal vector $\vec{n}_f$ of the polyhedron face f_j (the other face incident on e_r). Since the next vertex $\vec{v}_{i+1}$ of the cut face f_c lies on an edge of the polyhedron face f_j, and we know the direction

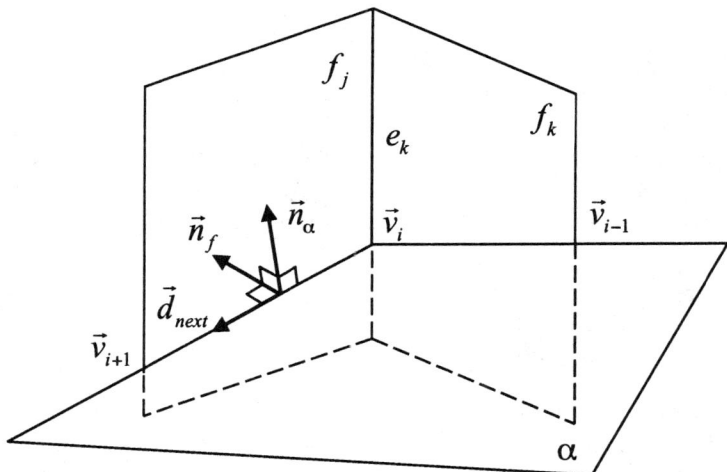

FIGURE F.5. Case when the last-computed vertex v_i lies in the interior of a polyhedron edge e_k. Here, there is only one way to compute the direction of the next edge $(\vec{v}_i, \vec{v}_{i+1})$ of the cut face f_c.

$\vec{d}_{next}$ of the edge $(\vec{v}_i, \vec{v}_{i+1})$, the next vertex $\vec{v}_{i+1}$ can be found as the first intersection of the ray $(\vec{v}_i + t\,\vec{d}_{next})$, for $t > 0$, with the edges of f_j that are also intersected by the cut plane[2].

In the second possible situation, the last-computed vertex v_i is a vertex of an edge $e_k \in P$ (see Figure F.6).

Hence, the direction $\vec{d}_{next}$ of the next edge $(\vec{v}_i, \vec{v}_{i+1})$ of the cut face f_c may not be unique. In fact, if the polyhedron being decomposed has faces coplanar to the cut plane α that contain the last edge $(\vec{v}_{i-1}, \vec{v}_i)$ of the cut face f_c, then any edge of these faces can be the next edge. Therefore, the next vertex $\vec{v}_{i+1}$ may either lie in the interior of an edge of f_j (see Figure F.6), or be a vertex of the polyhedron P (see Figure F.7).

The way to resolve this ambiguity is to remember that, by construction, we are always keeping the interior of the polyhedron P to the left of the cut face f_c, that is, to the left of the directed sub-chain $(\vec{v}_{i-1}, \vec{v}_i, \vec{v}_{i+1})$ defining the two consecutive edges $(\vec{v}_{i-1}, \vec{v}_i)$ and $(\vec{v}_i, \vec{v}_{i+1})$ of f_c. Because of this, the direction we should pick to determine the next vertex $\vec{v}_{i+1}$ from the set of candidate directions must be the one that minimizes the interior angle at vertex $\vec{v}_i$. If the direction that minimizes the interior angle at vertex $\vec{v}_i$ coincides with the direction of an edge of f_j, then the next vertex $\vec{v}_{i+1}$ is the other vertex of the edge. Otherwise, the next vertex v_{i+1} lies in the interior of an edge of f_j and can be computed as before.

---

[2]In our simplified version of the algorithm, the faces of the polyhedron are assumed to be convex, and therefore there is only one edge of the convex face f_j that intersects the cut plane α.

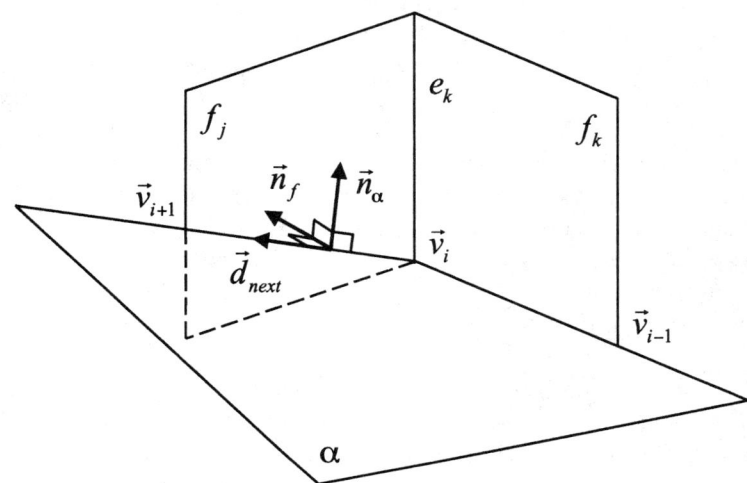

FIGURE F.6. If the polyhedron being decomposed has no coplanar faces with the cut plane α that contain the last edge $(\vec{v}_{i-1}, \vec{v}_i)$, then the direction that minimizes the internal angle at $\vec{v}_i$ is unique. In this case, the procedure used to compute the next vertex $\vec{v}_{i+1}$ is identical to that illustrated in Figure F.5.

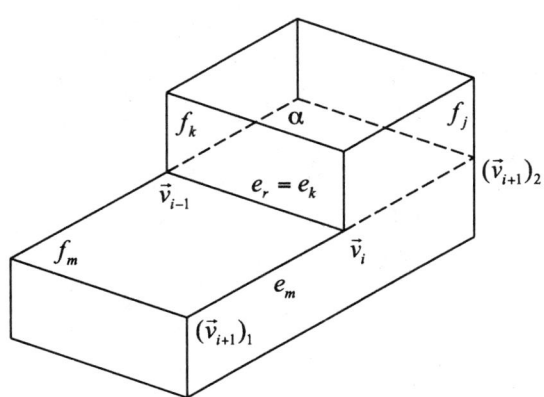

FIGURE F.7. The polyhedron face f_m is coplanar with the cut plane α. In this case, there are two possible directions to compute the next edge $(\vec{v}_i, \vec{v}_{i+1})$, one coincident with edge e_m of f_m, namely $(\vec{v}_i, (\vec{v}_{i+1})_1)$, and another along the interior of face f_j, namely $(\vec{v}_i, (\vec{v}_{i+1})_2)$. The internal angle at $\vec{v}_i$ for each of the choices is $\frac{3\pi}{2}$ and $\frac{\pi}{2}$, respectively. The selected direction should minimize the internal angle at $\vec{v}_i$, making $(\vec{v}_{i+1})_2$ our choice for the next vertex $\vec{v}_{i+1}$.

After determining all edges of the cut face f_c, we still need to carry out two more tests before accepting f_c as a valid cut face that meets all desired conditions 8 to 10.

The first test consists of checking whether the cut face is oriented counter-clockwise with respect to the normal vector of the cut plane. The orientation can be determined by summing the exterior angles at each vertex of the cut face. By construction, the correct orientation of the cut face should be counter-clockwise with respect to the normal vector of the cut plane, since the interior of the polyhedron P is always kept to the left of the edges of the cut face, as they are discovered. However, if the cut face is multiply connected, then the orientation will be clockwise instead of counter-clockwise. In this case, the cut plane associated with this cut face is rejected.

The second test consists of checking whether the cut face has interior holes. This is done as follows. Initially, we consider the set of edges of the polyhedron P that intersect the cut plane α. From this set, we remove any edge that contains a vertex of the cut face, either in its interior or as a vertex of the edge itself. Finally, for each edge left in the set, we compute its intersection with the cut plane α. If the intersection point lies inside the cut face, then the cut face has holes and its associated cut plane is rejected.

F.2.3 Termination Conditions

It is clear that there may be cases where one or more reflex edges cannot be successfully resolved for the given minimum internal-dihedral-angle and relative-distance values. A workaround for these cases is to reduce the minimum acceptable values and execute the algorithm one more time, not on the original non-convex polyhedron, but on the convex decomposition of the polyhedron found so far.

Even though this strategy considerably improves the chances of resolving all reflex edges, there may be still cases wherein some reflex edges may not be resolved. For these cases, a more sophisticated algorithm must be used, such as one that extends the original Joe's algorithm to allow the creation of simply connected or multiply connected cut faces.

F.3 Suggested Readings

The algorithm presented in this appendix is part of a tetrahedral mesh-generation algorithm for convex and non-convex polyhedra developed by Joe [Joe94, Joe91]. The original algorithm is itself a modification of Chazelle's algorithm [Cha84], which resolves each reflex edge by a cut face without taking into account the quality of the convex partition produced. Bajaj *et al.* [BD92] developed an extension to Chazelle's algorithm capable of handling internal holes on the non-convex polyhedra.

As far as triangulating the faces of the resulting convex decomposition, several algorithms can be used depending on the faces being convex or non-convex polygons. If the faces are convex, then we suggest computing the Delaunay triangulation of each face using the algorithm presented in Joe [Joe86]. Otherwise, we suggest using more sophisticated triangulation algorithms for convex and non-convex polygons, such as those described in Joe [JS86] and Keil [Kei96].

Appendix G
The Linear-Complementarity Problem

G.1 Introduction

As explained in Chapters 3 and 4, there exists a linear relation between the relative acceleration $\vec{a}_i$ between two contacting bodies and the contact force $\vec{F}_i$ at contact point C_i, expressed by

$$\vec{a}_i = \begin{pmatrix} (a_i)_n \\ (a_i)_t \\ (a_i)_k \end{pmatrix} = \begin{pmatrix} (a_{ii})_n & (a_{i(i+1)})_t & (a_{i(i+2)})_k \\ (a_{(i+1)i})_n & (a_{(i+1)(i+1)})_t & (a_{(i+1)(i+2)})_k \\ (a_{(i+2)i})_n & (a_{(i+2)(i+1)})_t & (a_{(i+2)(i+2)})_k \end{pmatrix} \begin{pmatrix} (F_i)_n \\ (F_i)_t \\ (F_i)_k \end{pmatrix} + \begin{pmatrix} (b_i)_n \\ (b_i)_t \\ (b_i)_k \end{pmatrix} = \mathbf{A}_i \vec{F}_i + \vec{b}_i , \quad \text{(G.1)}$$

where the index n indicates the component along the contact-normal direction, and the indexes t and k indicate the components along the contact-tangent plane (i.e., the plane passing through the contact point with normal vector parallel to the contact normal). The coefficients of matrix $\mathbf{A}_i$ and vector $\vec{b}_i$ are computed from the mass properties and relative geometrical displacement of the contacting objects at contact C_i. For example, coefficient $(a_{ii})_n$ relates the normal contact-force component $(F_i)_n$ with the normal relative-acceleration component $(a_i)_n$ at contact C_i. Analogously, coefficients $(a_{i(i+1)})_t$ and $(a_{i(i+2)})_k$ relate the force components $(F_i)_t$ and

$(F_i)_k$ with the acceleration[1] components $(a_i)_t$ and $(a_i)_t$ along the tangent plane. As for vector $\vec{b}_i$, it can be shown that it is a vector in the column space of $\mathbf{A}_i$, that is, there exists a non-zero vector $\vec{z}$ such that $\vec{b}_i = \mathbf{A}_i \vec{z}$. As we shall see later in this appendix, the fact that $\vec{b}$ lies in the column space of $\mathbf{A}_i$ turns out to be a fundamental result to the analysis of the existence of valid contact forces, since it guarantees that a solution to the system of equations always exists for frictionless contacts.

In the case of multiple simultaneous contacts, the accelerations and contact forces at each individual contact point are merged into a system-wide acceleration and contact-force vector

$$\vec{a} = (\vec{a}_1, \vec{a}_2, \ldots, \vec{a}_m)^t$$
$$\vec{F} = (\vec{F}_1, \vec{F}_2, \ldots, \vec{F}_m)^t ,$$
(G.2)

with m being the total number of simultaneous contacts. In the context of simultaneous contacts, equation (G.1) is replaced by

$$\vec{a} = \mathbf{A}\vec{F} + \vec{b}$$
$$= \begin{pmatrix} \mathbf{A}_{11} & \mathbf{A}_{12} & \mathbf{A}_{13} & \ldots & \mathbf{A}_{1m} \\ \mathbf{A}_{12}{}^t & \mathbf{A}_{22} & \mathbf{A}_{23} & \ldots & \mathbf{A}_{2m} \\ \mathbf{A}_{13}{}^t & \mathbf{A}_{23}{}^t & \mathbf{A}_{33} & \ldots & \mathbf{A}_{3m} \\ \ldots & \ldots & \ldots & \ldots & \ldots \\ \mathbf{A}_{1m}{}^t & \mathbf{A}_{2m}{}^t & \mathbf{A}_{3m}{}^t & \ldots & \mathbf{A}_{mm} \end{pmatrix} \vec{F} + \begin{pmatrix} b_1 \\ b_2 \\ b_3 \\ \vdots \\ b_m \end{pmatrix}$$
(G.3)

where each sub-matrix $\mathbf{A}_{ij}$ is a 3×3 matrix of the same form as matrix $\mathbf{A}_i$ in equation (G.1). For instance, sub-matrix $\mathbf{A}_{23}$ relates the contact force $\vec{F}_3 = ((F_3)_n, (F_3)_t, (F_3)_k)^t$ of contact C_3 with the acceleration $\vec{a}_2 = ((a_2)_n, (a_2)_t, (a_2)_k)^t$ of contact C_2.

According to the contact-force derivations presented in Chapters 3 and 4, the contact force at each contact point C_i is said to be valid if it satisfies the *non-interpenetration conditions* (also referred to as *normal conditions*) at C_i, namely

$$\begin{aligned} (F_i)_n (a_i)_n &= 0 \\ (a_i)_n &\geq 0 \\ (F_i)_n &\geq 0 , \end{aligned}$$
(G.4)

---

[1] Unless otherwise stated, the term acceleration used in this appendix means the relative acceleration between the contacting bodies at a contact point.

that is, either the contact-force or the acceleration components along the contact normal can be greater than zero, but never both of them simultaneously.

The actual derivation of equation (G.4), as well as how the coefficients of matrix $\mathbf{A}$ and vector $\vec{b}$ are computed for a given contact configuration, were already covered in detail in Chapters 3 and 4. Here, we shall focus on how the system of equations is actually solved.

It happens that the formulation presented in equations (G.4) and (G.3) fits the well known Linear Complementarity Problem (LCP) formulation of linear programming theory. The approach indicated depends on whether friction is being considered. There are three possible cases to be addressed: frictionless contacts, contacts with static friction only, and the more general case of contacts with dynamic friction. In the following sections, we shall present approaches to the LCP problem associated with each of these three possible cases.

G.2 Dantzig's Algorithm: The Frictionless Case

Dantzig's algorithm works by incrementally computing intermediate solutions for instances of the problem defined by equations (G.4) and (G.3), where each instance takes into account one more contact point than the previous instance. At instance i, the algorithm computes the contact force for the ith contact point without violating the non-interpenetration conditions for the $(i-1)$ contact points already resolved at instance $(i-1)$. Assuming we have m simultaneous contact points, the solution of the LCP problem is immediately obtained after solving each of the m instances.

In the frictionless case, the direction of the contact force is the same as the direction of the contact normal, that is, the contact force has no tangential components. Therefore, the system of equations is formulated using only the normal components of the contact-force and acceleration vectors given in equation (G.2). In other words, for each frictionless contact C_i we have

$$\vec{a}_i = ((a_i)_n, (a_i)_t, (a_i)_k)^t = ((a_i)_n, 0, 0)^t$$
$$\vec{F}_i = ((F_i)_n, (F_i)_t, (F_i)_k)^t = ((F_1)_n, 0, 0)^t,$$

and the system-wide vectors representing the accelerations and contact forces can then be written using the shorthand format

$$\vec{a} = ((a_1)_n, \ldots, (a_i)_n, \ldots, (a_m)_n)^t$$
$$\vec{F} = ((F_1)_n, \ldots, (F_i)_n, \ldots, (F_m)_n)^t,$$

omitting the tangential components. This significantly simplifies the computation of matrix $\mathbf{A_i}$ and vector $\vec{b}_i$ associated with contact C_i, since there

340   Appendix G. The Linear-Complementarity Problem

is no need to compute the coefficients related to the tangential components. Therefore, for the frictionless case, $\mathbf{A_i}$ and vector $\vec{b}_i$ are reduced to

$$\mathbf{A_i} = ((a_{ii})_n)$$
$$\vec{b} = ((b_i)_n),$$

that is, they become scalars.

According to equation (G.4), a solution is achieved whenever either the contact force or the relative acceleration along the contact normal of each contact point is zero. As shown in Figure G.1, a positive relative normal acceleration $(a_i)_n > 0$ indicates that the contacting bodies are moving away from each other at contact point C_i, and the contact is about to break. In this situation, the contact force $(F_i)_n$ in equation (G.4) should be set to zero to enforce

$$(F_i)_n (a_i)_n = 0 .$$

If the normal acceleration $(a_i)_n$ is zero (see Figure G.2), contact between the bodies is maintained, and any non-negative contact force $(F_i)_n$ can be used in equation (G.4). The problem arises when the normal acceleration $(a_i)_n$ is negative, as shown in Figure G.3. Here, the bodies are accelerating towards each other at the contact point, indicating that they are about to interpenetrate. A sufficiently powerful positive contact force $(F_i)_n$ should then be applied to make the negative normal acceleration become zero. Clearly, the contact points we need to worry about are those having negative normal acceleration.

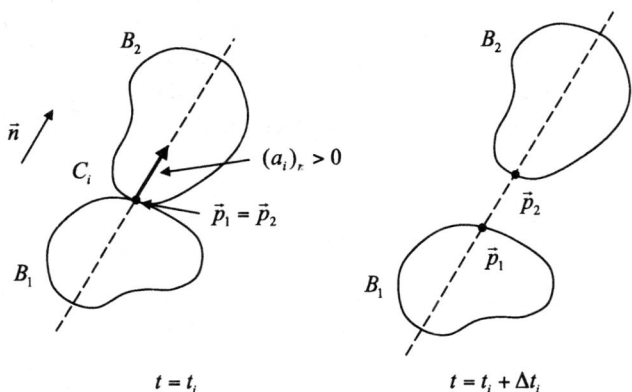

FIGURE G.1. The relative-acceleration component $(a_i)_n$ along the contact-normal direction is positive and the contact force f_i is zero (contact is about to break at $t = (t_i + \triangle t_i)$.

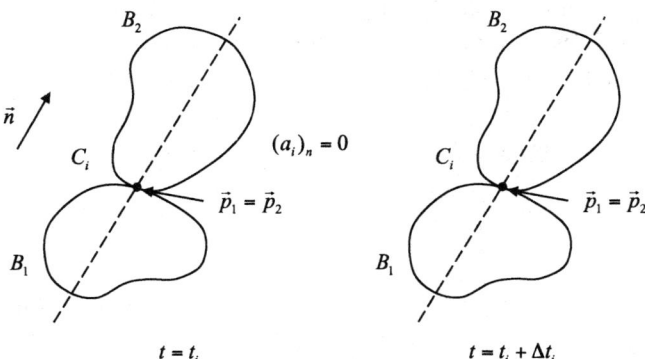

FIGURE G.2. The relative-acceleration component $(a_i)_n$ along the contact-normal direction is zero, and the contact force can be any non-negative value (bodies remain in contact).

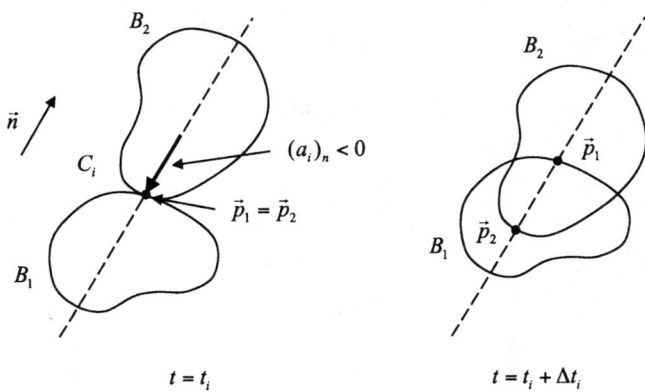

FIGURE G.3. The relative-acceleration component $(a_i)_n$ along the contact-normal direction is negative, and the contact force should be large enough to reduce it to zero to prevent interpenetration.

Appendix G. The Linear-Complementarity Problem

As mentioned before, the main idea of Dantzig's algorithm is to assure the non-interpenetration conditions at each new contact point while maintaining these conditions at the contact points already resolved. For example, at the first instance, the algorithm ignores all contact points save one[2]. This has the same effect as setting all contact forces to zero, and solving equations (G.4) and (G.3) for just one contact point, that is, solving

$$
\begin{aligned}
(a_1)_n = ((a_{11})_n (F_1)_n + (b_1)_n) &\geq 0 \\
(F_1)_n (a_1)_n = (F_1)_n ((a_{11})_n (F_1)_n + (b_1)_n) &= 0 \quad\quad (G.5)\\
(F_1)_n &\geq 0 .
\end{aligned}
$$

For this contact point, the algorithm checks whether the normal acceleration $(a_1)_n$ is positive, zero or negative, and makes the necessary adjustment to the contact force $(F_1)_n$ in the latter case so as to prevent interpenetration of the contacting bodies at the contact point. At the end of the first instance of the algorithm, the contact-force vector will have the intermediate value

$$\vec{F} = ((F_1)_n, 0, 0, \ldots, 0)^t ,$$

with

$$(F_1)_n \geq 0 .$$

At the second instance, the algorithm computes the contact force $(F_2)_n$ so that the normal acceleration $(a_2)_n$ is non-negative. This should be done without violating the non-interpenetration conditions along the normal direction of the contact point already resolved on the first instance. So, at the second instance, equations (G.4) and (G.3) are reduced to:

$$
\begin{aligned}
\begin{pmatrix} (a_1)_n \\ (a_2)_n \end{pmatrix} = \left(\begin{pmatrix} (a_{11})_n & (a_{12})_n \\ (a_{21})_n & (a_{22})_n \end{pmatrix} \begin{pmatrix} (F_1)_n \\ (F_2)_n \end{pmatrix} + \begin{pmatrix} (b_1)_n \\ (b_2)_n \end{pmatrix} \right) &\geq \begin{pmatrix} 0 \\ 0 \end{pmatrix} \\
\begin{pmatrix} (F_1)_n \\ (F_2)_n \end{pmatrix}^t \left(\begin{pmatrix} (a_{11})_n & (a_{12})_n \\ (a_{21})_n & (a_{22})_n \end{pmatrix} \begin{pmatrix} f_1 \\ f_2 \end{pmatrix} + \begin{pmatrix} (b_1)_n \\ (b_2)_n \end{pmatrix} \right) &= \begin{pmatrix} 0 \\ 0 \end{pmatrix} \\
\begin{pmatrix} (F_1)_n \\ (F_2)_n \end{pmatrix} &\geq \begin{pmatrix} 0 \\ 0 \end{pmatrix}
\end{aligned}
$$

It is clear from the above equation that, as we change $(F_2)_n$, the values of $(a_2)_n$ and $(a_1)_n$ change as well, thus requiring an update $\triangle(F_1)_n$ to the contact force $(F_1)_n$ in order to enforce the non-interpenetration conditions of equation (G.5).

---

[2] The order in which the contact points are resolved is irrelevant.

In general, at instance m, an increase of $\triangle(F_m)_n$ on the contact force $(F_m)_n$ requires an update $\triangle(F_i)_n$ on the values of the contact forces $(F_i)_n$ for all $i \in \{1, \ldots, (m-1)\}$ already computed at instance $(m-1)$. This update is absolutely essential to the maintenance of the non-interpenetration conditions at these contact points. In the following paragraphs, we shall examine in more detail how the non-interpenetration conditions are actually maintained.

Let $\triangle \vec{F}$ and $\triangle \vec{a}$ be the increment to the force and acceleration vectors computed at instance m, that is

$$\triangle \vec{F} = (\triangle(F_1)_n, \triangle(F_2)_n, \ldots, \triangle(F_m)_n)^t$$
$$\triangle \vec{a} = (\triangle(a_1)_n, \triangle(a_2)_n, \ldots, \triangle(a_m)_n).$$

The updated force and acceleration vectors are then given by

$$\vec{F}_{new} = \vec{F} + \triangle \vec{F}$$
$$\vec{a}_{new} = \vec{a} + \triangle \vec{a}.$$
(G.6)

Substituting equation (G.3) into (G.6) we get

$$\begin{aligned} \triangle \vec{a} &= \vec{a}_{new} - \vec{a} \\ &= (\mathbf{A} \vec{F}_{new} + \vec{b}) - (\mathbf{A} \vec{F} + \vec{b}) \\ &= (\mathbf{A} (\vec{F} + \triangle \vec{F}) + b) - (\mathbf{A} \vec{F} + \vec{b}) \\ &= \mathbf{A} \triangle \vec{F}. \end{aligned}$$
(G.7)

So, as we increase $(F_m)_n$ by $\triangle(F_m)_n$, some $(a_i)_n$'s and $(F_j)_n$'s will increase or decrease according to equation (G.7), depending on the values of the coefficients of matrix $\mathbf{A}$. Clearly, the problem arises when the adjustments associated with contact point C_m violate the non-interpenetration conditions for one or more contact points C_i with $i \in \{1, 2, \ldots, (m-1)\}$.

As stated in equation (G.4), the non-interpenetration conditions are achieved at the contact point C_i whenever the collision force $(F_i)_n$ is zero and $(a_i)_n > 0$ (i.e., we need to maintain the condition $(F_i)_n = 0$), or the relative acceleration $(a_i)_n$ is zero and $(F_i)_n \geq 0$ (i.e., we need to maintain the condition $(a_i)_n = 0$). Therefore, there are only two ways the non-interpenetration conditions at contact point C_i can be violated.

1. If $(F_i)_n = 0$ and an increase in $(F_m)_n$ by $\triangle(F_m)_n$ forces $(a_i)_n$ to assume a negative value.

2. If $(a_i)_n = 0$ and an increase in $(F_m)_n$ by $\triangle(F_m)_n$ forces $(F_i)_n$ to assume a negative value.

If the non-interpenetration conditions at contact point C_i were achieved with $(F_i)_n = 0$, then we need to check whether condition (1) is valid after each increase in $(F_m)_n$. On the other hand, if the non-interpenetration conditions were achieved with $(a_i)_n = 0$, then we need to check whether condition (2) is valid. In other words, we need to keep track of which case the contact point falls into (either zero contact force or zero acceleration) before we can decide the condition in need of verification. This can be efficiently implemented as follows.

At the beginning of instance m, we sub-divide the contact points into two groups. The first group, called ZA (zero acceleration), contains the indexes of all contact points C_i with $i < m$ that have $(a_i)_n = 0$. Because instance $(m-1)$ was already resolved, the contact force for each contact point in ZA is guaranteed to have $(F_i)_n \geq 0$, that is

$$ZA = \{(1 \leq i < m) : (a_i)_n = 0 \text{ and } (F_i)_n \geq 0\}.$$

The second group, called ZF (zero force), contains the indexes of all contact points C_i with $i < m$ that have $(F_i)_n = 0$. Again, since instance $(m-1)$ was already resolved, the normal acceleration for these contact points is guaranteed to be $(a_i)_n > 0$, that is

$$ZF = \{(1 \leq i < m) : (F_i)_n = 0 \text{ and } (a_i)_n > 0\}.$$

As we increase $(F_m)_n$ by $\triangle(F_m)_n$, the algorithm tries to keep $(a_i)_n = 0$ for all $i \in ZA$, and to keep $(F_j)_n = 0$ for all $j \in ZF$, while updating the contact forces and normal accelerations using equation (G.7). The idea is then to set $\triangle(a_i)_n = 0$ for all $i \in ZA$, such that $(a_i)_n$ remains the same, then set $\triangle(F_j)_n = 0$ for all $j \in ZF$, such that $(F_j)_n$ remains the same, and lastly solve equation (G.7) for the unknowns $\triangle(F_i)_n$ for $i \in ZA$. If while solving equation (G.7) we detect that some $(F_i)_n$ with $i \in ZA$ has decreased to zero, then we temporarily stop the computations and move contact point C_i from ZA to ZF. By so doing, we prevent $(F_i)_n$ from decreasing even further and end up assuming a negative value. On the other hand, if we detect that some $(a_i)_n$ with $i \in ZF$ has decreased to zero, then we temporarily stop the computations and move contact point C_i from ZF to ZA to prevent $(a_i)_n$ from assuming a negative value. In both cases, equation (G.7) is re-arranged according to the index update of the contact points in both ZA and ZF groups, and the computation continues until we have increased $(F_m)_n$ enough to make $(a_m)_n = 0$. Figure G.4 shows a finite-state machine representation of the possible group changes that contact point C_i can make while solving equation (G.7).

Let's examine in more detail how equation (G.7) is actually solved for the unknowns $\triangle(F_i)_n$ for $i \in ZA$. Because the order in which the contact points are numbered is irrelevant, let's assume that $ZA = \{1, 2, \ldots, k\}$ and

G.2 Dantzig's Algorithm: The Frictionless Case

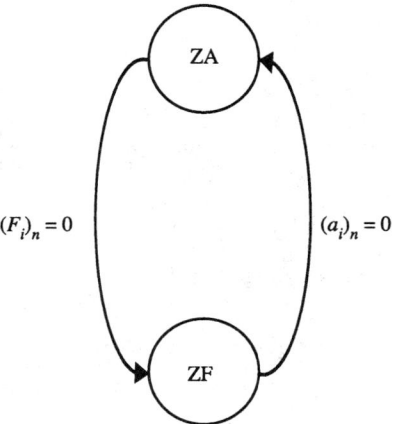

FIGURE G.4. At any instant during the solution of equation (G.7), a contact point C_i can be in one of two states, namely ZA and ZF. The arrows show the direction of the possible movements, together with the condition that needs to be satisfied to trigger the movement.

$ZF = \{(k+1), (k+2), \ldots, (m-1)\}$, and that we have $(a_m)_n < 0^3$. The matrix $\mathbf{A}$ and increment vector $\triangle \vec{F}$ can then be partitioned such that the first k columns correspond to contact points in ZA, and the remaining columns to contact points in ZF. By so doing, we have

$$\mathbf{A} = \begin{pmatrix} \mathbf{A}_{11} & \mathbf{A}_{12} & \vec{v}_1 \\ \mathbf{A}_{12}{}^t & \mathbf{A}_{22} & \vec{v}_2 \\ \vec{v}_1^t & \vec{v}_2^t & c \end{pmatrix} \qquad (G.8)$$

and

$$\triangle \vec{F} = \begin{pmatrix} \vec{x} \\ \vec{0} \\ \triangle (F_m)_n \end{pmatrix}, \qquad (G.9)$$

where $\mathbf{A}_{11} \in \mathbb{R}^{k \times k}$ and $\mathbf{A}_{22} \in \mathbb{R}^{(m-1-k) \times (m-1-k)}$ are square symmetric matrices, $\vec{v}_1 \in \mathbb{R}^k$ and $\vec{v}_2 \in \mathbb{R}^{(m-1-k)}$ are vectors, c is a scalar and $\vec{x} \in \mathbb{R}^k$ is the unknown contact-force increment we want to determine. Substituting equations (G.8) and (G.9) into equation (G.7), we obtain

$$\triangle \vec{a} = \mathbf{A} \triangle \vec{F} = \mathbf{A} \begin{pmatrix} \vec{x} \\ \vec{0} \\ \triangle (F_m)_n \end{pmatrix} = \begin{pmatrix} \mathbf{A}_{11} \vec{x} + \vec{v}_1 \triangle (F_m)_n \\ \mathbf{A}_{12}^t \vec{x} + \vec{v}_2 \triangle (F_m)_n \\ \vec{v}_1^t \vec{x} + c \triangle (F_m)_n \end{pmatrix}. \qquad (G.10)$$

We can then rearrange $\triangle \vec{a}$ in the same way we did for $\triangle \vec{F}$, that is

---

[3] If $(a_m)_n \geq 0$, then we can immediately solve instance m by setting $(F_m)_n = 0$.

$$\triangle \vec{a} = \begin{pmatrix} \vec{0} \\ \vec{w} \end{pmatrix}, \qquad (G.11)$$

where $\vec{0} \in \mathbb{R}^{k \times k}$. Combining equations (G.10) and (G.11), we have

$$\mathbf{A}_{11}\vec{x} + \vec{v}_1 \triangle(F_m)_n = \vec{0},$$

that is, we need to solve the linear system

$$\mathbf{A}_{11}\vec{x} = -\vec{v}_1 \triangle(F_m)_n \qquad (G.12)$$

for the unknown contact force increment $\vec{x}$. Since we don't know in advance which increment $\triangle(F_m)_n$ should be used in order to increase $(a_m)_n$ to zero, we initially solve for

$$\triangle(F_m)_n = 1$$

and then adjust the solution by computing the smallest positive scalar s such that, when we increase $\vec{F}$ by $s\triangle\vec{F}$, we have either $(a_m)_n = 0$, or some contact point moved between ZA and ZF. In other words, the scalar s is used to limit how much we can increase $(F_m)_n$ without violating the non-interpenetration conditions for the remaining contact points.

There are three situations to consider when computing the scalar s. If $\triangle(a_m)_n > 0$, then the normal acceleration at contact point C_m is increasing. Because we want to increase $(a_m)_n < 0$ up to the point at which it becomes zero, then the maximum allowed step size s that does not make $(a_m)_n > 0$ is

$$s = -\frac{(a_m)_n}{\triangle(a_m)_n}. \qquad (G.13)$$

The second situation occurs wherever $\triangle(F_i)_n < 0$ for $i \in ZA$, that is, the contact force at contact point C_i is decreasing. Because we want to enforce $(F_i)_n \geq 0$ for $i \in ZA$, the maximum allowed step size s that does not make $(F_i)_n$ negative is

$$s \leq -\frac{(F_i)_n}{\triangle(F_i)_n}, \forall i \in ZA \text{ with } \triangle(F_i)_n < 0. \qquad (G.14)$$

The third and last situation occurs when we have $\triangle(a_i)_n < 0$ for $i \in ZF$, that is, the normal acceleration at contact point C_i is decreasing. Because we want to enforce that $(a_i)_n > 0$ for $i \in ZA$, the maximum allowed step size s that does not make $(a_i)_n$ negative is

$$s \leq -\frac{(a_i)_n}{\triangle(a_i)_n}, \forall i \in ZF \text{ with } \triangle(a_i)_n < 0. \qquad (G.15)$$

G.2 Dantzig's Algorithm: The Frictionless Case

The goal is to select the smallest s satisfying equations (G.13), (G.14) and (G.15). Having determined s, we update $\vec{F}$ and $\vec{a}$ by

$$\vec{F}_{new} = \vec{F} + (s \triangle \vec{F}),$$
$$\vec{a}_{new} = \vec{a} + (s \triangle \vec{a}).$$

If the smallest s was obtained from equation (G.13), then $(a_m)_n = 0$ after the update, and the non-interpenetration conditions at contact point C_m are satisfied. In this case, contact point C_m is moved into ZA and the mth instance of the algorithm is completed.

However, if the smallest s was obtained from equation (G.14), then there exists a contact point C_i with $i \in ZA$ that had its force $(F_i)_n$ decreased to zero before we could reach $(a_m)_n = 0$. In this case, we need to move index i from ZA to ZF to prevent the force from becoming negative. The partition of matrix A is then updated to account for the contact point moved, and the algorithm loops back to continue increasing $(F_m)_n$ until $(a_m)_n = 0$.

A similar update happens if the smallest s was obtained from equation (G.15). In this case, there exists a contact point C_i with $i \in ZF$ that had its acceleration $(a_i)_n$ decreased to zero before we could reach $(a_m)_n = 0$. This makes it necessary to move index i from ZF to ZA to prevent the acceleration from becoming negative. Again, the partition of matrix A is rearranged to account for the contact point moved, and the algorithm loops back to continue increasing $(F_m)_n$ until $(a_m)_n = 0$.

G.2.1 Termination Conditions

There are two critical assumptions in the frictionless algorithm described in the previous section that have the potential of being invalid, meaning they can compromise the existence of a solution. The first critical assumption is the existence of a solution $\vec{x}$ to the linear system defined by equation (G.12) with $\triangle(F_m)_n = 1$, that is, the existence of a solution to

$$\mathbf{A}_{11} \vec{x} = -\vec{v}_1 . \tag{G.16}$$

Fortunately, it can be shown that the vector $\vec{v}_1$ is always in the column space of $\mathbf{A}_{11}$ for all possible combinations of indexes in ZA and ZF. Therefore, the linear system in equation (G.16) is well conditioned and a solution $\vec{x}$ is guaranteed to always exist.

The second critical assumption is that an increase in $(F_m)_n$ will also cause an increase in $(a_m)_n$ by a positive amount, such that $(a_m)_n$ will eventually reach the desired zero value. If we substitute $(s \triangle \vec{F})$ back into equation (G.10), we have that $(a_m)_n$ increases by

$$s\left((\vec{v}_1)^t \vec{x} + c\right) = s \triangle (a_m)_n .$$

In this case, it can be shown that, if **A** is positive definite, then both $((\vec{v}_1)^t \vec{x} + c)$ and step size s are always positive. This in turn means that $(a_m)_n$ will always increase by a positive amount at the end of each step, that is, the algorithm is guaranteed to terminate after a finite number of steps.

The interested reader is referred to Section G.4 for pointers to the literature containing formal proofs of these assumptions.

G.3 Baraff's Algorithm: Coping with Friction

Baraff's extension to Dantzig's algorithm addresses the friction problem in much the same way as the normal conditions were addressed in the frictionless case. Again, the idea is to incrementally compute intermediate solutions for instances of the problem defined by equations (G.4) and (G.3), where each instance takes into account one more contact point than the previous instance. At instance i, the algorithm computes the contact force for the ith contact point without violating the normal *and* friction conditions for the $(i-1)$ contact points already resolved at instance $(i-1)$. The actual friction conditions to be satisfied at each contact point depend on the contact's being static or dynamic.

Respecting friction, the contact force will have a normal component *and* a tangential component, the latter owing to the friction force acting at the contact point. Clearly, the relation between the normal and tangential contact-force components depends on the contact model adopted.

In this book, we use the Coulomb friction model to relate the tangential and normal contact-force components. More specifically, we use a directional-friction model to compute the tangential contact-force components[4]. Let $\vec{F}_i$ be the contact force associated with contact C_i, that is

$$\vec{F}_i = ((F_i)_n, (F_i)_t, (F_i)_k)^t .$$

Using the directional-friction model, the contact-force components on the tangent plane are obtained from the contact-force component along the normal direction using

$$\begin{aligned} (F_i)_t &= \mu_t (F_i)_n \\ (F_i)_k &= \mu_k (F_i)_n , \end{aligned}$$

where μ_t and μ_k are the coefficients of friction along the tangent-plane directions $\vec{t}$ and $\vec{k}$.

---

[4]This is the same model used in Chapters 3 and 4.

G.3 Baraff's Algorithm: Coping with Friction

It is important to notice that the directional-friction model is a generalization of the widely used model of relating the tangential and normal contact-force components using just one omnidirectional coefficient of friction μ, as in

$$(F_i)_{tk} = \mu \, (F_i)_n \, . \tag{G.17}$$

In equation (G.17), the term $(F_i)_{tk}$ refers to the net contact-force component on the tangent plane given by

$$(F_i)_{tk} = \sqrt{(F_i)_t^2 + (F_i)_k^2} \, .$$

For example, if friction is isotropic, that is, independent of direction, we can write

$$\mu_t = \mu \cos \phi$$
$$\mu_k = \mu \sin \phi$$

for some angle ϕ, and so

$$\begin{aligned}
(F_i)_{tk} &= \sqrt{(F_i)_t^2 + (F_i)_k^2} \\
&= \sqrt{\mu^2 \, (F_i)_n^2 \cos \phi^2 + \mu^2 \, (F_i)_n^2 \sin \phi^2} \\
&= \mu \, (F_i)_n \, ,
\end{aligned}$$

which is the same result obtained using the omnidirectional friction model of equation (G.17). The main advantage of using the directional-friction model is that the non-linear equation

$$|(F_i)_{tk}| = \sqrt{(F_i)_t^2 + (F_i)_k^2} \leq \mu \, (F_i)_n$$

that needs to be enforced when the contacting bodies are not sliding at the contact point (i.e., static friction) can be substituted for two linear equations

$$\begin{aligned}
|(F_i)_t| &\leq \mu_t \, (F_i)_n \\
|(F_i)_k| &\leq \mu_k \, (F_i)_n \, ,
\end{aligned}$$

which are equivalent to the non-linear equation if friction is isotropic, and, most important, can be independently resolved.

The way the normal and tangential contact-force components are related in the Coulomb friction model depends on the contact's being static or dynamic. A contact is said to be static if the net relative velocity along its tangent plane is zero, or less than a threshold value. Otherwise, the contact is said to be dynamic. In either case, the system-wide contact-force and relative-acceleration-vector components for m simultaneous contacts are expressed as

$$\vec{a} = ((a_1)_n, (a_1)_t, (a_1)_k, \ldots, (a_m)_n, (a_m)_t, (a_m)_k)^t \quad (G.18)$$
$$\vec{F} = ((F_1)_n, (F_1)_t, (F_1)_k, \ldots, (F_m)_n, (F_m)_t, (F_m)_k)^t, \quad (G.19)$$

that is, the tangential contact-force components are no longer omitted, as in the frictionless case (see equation (G.5)). Therefore, the frictional contact-force computation requires solving the following system (which is the same as that shown in equation (G.3), but repeated here for convenience):

$$\vec{a} = \mathbf{A}\vec{F} + \vec{b} = \begin{pmatrix} \mathbf{A}_{11} & \mathbf{A}_{12} & \mathbf{A}_{13} & \cdots & \mathbf{A}_{1m} \\ \mathbf{A}_{12}{}^t & \mathbf{A}_{22} & \mathbf{A}_{23} & \cdots & \mathbf{A}_{2m} \\ \mathbf{A}_{13}{}^t & \mathbf{A}_{23}{}^t & \mathbf{A}_{33} & \cdots & \mathbf{A}_{3m} \\ \cdots & \cdots & \cdots & \cdots & \cdots \\ \mathbf{A}_{1m}{}^t & \mathbf{A}_{2m}{}^t & \mathbf{A}_{3m}{}^t & \cdots & \mathbf{A}_{mm} \end{pmatrix} \vec{F} + \begin{pmatrix} b_1 \\ b_2 \\ b_3 \\ \vdots \\ b_m \end{pmatrix}, \quad (G.20)$$

where $\vec{a}$ and $\vec{F}$ are given by equations (G.18) and (G.19), and each submatrix $\mathbf{A}_{ij}$ is a 3×3 matrix relating the relative acceleration at contact C_i with the contact force at contact C_j. Clearly from equation (G.20), an increment of

$$\triangle \vec{F} = (\triangle(F_1)_n, \triangle(F_1)_t, \triangle(F_1)_k, \ldots, \triangle(F_m)_n, \triangle(F_m)_t, \triangle(F_m)_k)^t$$

to the contact-force components will cause a variation of

$$\vec{a} = (\triangle(a_1)_n, \triangle(a_1)_t, \triangle(a_1)_k, \ldots, \triangle(a_m)_n, \triangle(a_m)_t, \triangle(a_m)_k)^t$$

to the acceleration components obtained from

$$\triangle \vec{a} = \mathbf{A} \triangle \vec{F}. \quad (G.21)$$

The problem is that, depending on the values of the coefficients of matrix $\mathbf{A}$, an increment of $\triangle(F_m)_n$ to the normal contact-force component of contact C_m at the mth iteration can increase or decrease the contact-force and acceleration components of other contacts already resolved in one of the previous iterations. Not only that, an increment of $\triangle(F_m)_t$ or $\triangle(F_m)_k$ to the tangential-force components of contact C_m can not only affect the normal and friction conditions already established for the previous $(m-1)$ contacts, but can also increase or decrease the normal components $(F_m)_n$ and $(a_m)_n$ associated with contact C_m. Another important issue is that, because we are adopting the Coulomb friction model, the tangential

G.3 Baraff's Algorithm: Coping with Friction

contact-force components and their increments are computed as a function of the normal-force component and its increment. This in turn, requires that we first resolve the normal conditions at contact C_m, and only then resolve the friction conditions while enforcing the normal conditions just obtained.

Therefore, enforcing the normal and friction conditions at the mth iteration is a two-step process. We need to first resolve the normal conditions at contact C_m, assuming $\triangle(F_m)_t$ and $\triangle(F_m)_k$ are zero, while maintaining the normal and friction conditions for the previous $(m-1)$ contact points. We need then to adjust $\triangle(F_m)_t$ and $\triangle(F_m)_k$ to assure the friction conditions along the tangent-plane directions $\vec{t}$ and $\vec{k}$, while maintaining the normal and friction conditions for the previous $(m-1)$ contact points, as well as the normal condition for contact C_m. At the end of the mth iteration, we have assured both normal and friction conditions for all m contacts.

In the following sections, we shall apply this solution scheme to both static and dynamic friction, and study the termination conditions for each case.

G.3.1 Static-Friction Conditions

In the static-friction case, the relation between the normal and tangential contact-force components depends on the value of the relative tangential acceleration. Let $\vec{a}_i$ and $\vec{F}_i$ be, respectively, the relative acceleration and contact force at contact C_i, that is

$$\vec{a}_i = ((a_i)_n, (a_i)_n, (a_i)_n),$$
$$\vec{F}_i = ((F_i)_n, (F_i)_n, (F_i)_n).$$

If the relative tangential acceleration is zero, then the tangential components of the contact force are constrained within a range proportional to the value of the normal component. Since we are using the directional-friction model, this condition translates into assuring

$$|(F_i)_t| \leq \mu_t (F_i)_n, \text{ if } (a_i)_t = 0$$
$$|(F_i)_k| \leq \mu_k (F_i)_n, \text{ if } (a_i)_k = 0,$$
(G.22)

as opposed to assuring the commonly used condition

$$|(F_i)_t^2 + (F_i)_k^2| \leq \mu (F_i)_n$$

that arises when the anisotropic-friction model is considered. Notice that the coefficients μ_t and μ_k in equation (G.22) are the static-friction coefficients along the tangent-plane directions $\vec{t}$ and $\vec{k}$, respectively.

352   Appendix G. The Linear-Complementarity Problem

If on the other hand the relative tangential acceleration is not zero, then the tangent components of the contact force will have maximum magnitude

$$|(F_i)_t| = \mu_t (F_i)_n, \text{ if } (a_i)_t \neq 0$$
$$|(F_i)_k| = \mu_k (F_i)_n, \text{ if } (a_i)_k \neq 0$$

and opposite direction with respect to the relative tangential acceleration, that is, the contact force and relative acceleration must have opposite signs:

$$(F_i)_t (a_i)_t < 0$$
$$(F_i)_k (a_i)_k < 0.$$

Therefore, the static friction conditions that need to be assured at each contact point C_i are

$$\begin{aligned} |(F_i)_t| &\leq \mu_t (F_i)_n \\ (a_i)_t (F_i)_t &\leq 0 \\ (a_i)_t (\mu_t (F_i)_n - |(F_i)_t|) &= 0 \end{aligned} \quad (G.23)$$

and

$$\begin{aligned} |(F_i)_k| &\leq \mu_k (F_i)_n \\ (a_i)_k (F_i)_k &\leq 0 \\ (a_i)_k (\mu_k (F_i)_n - |(F_i)_k|) &= 0 \end{aligned} \quad (G.24)$$

The last condition in equations (G.23) and (G.24) assures that $(F_i)_t$ and $(F_i)_k$ will have maximum magnitude $\mu_t(F_i)_n$ and $\mu_k(F_i)_n$ whenever $(a_i)_t \neq 0$ and $(a_i)_k \neq 0$, respectively.

As mentioned, the way the static-friction conditions are assured is very similar to that used to assure the normal conditions. In the frictionless case, we created two groups of indexes, namely ZA and ZF, and used them to partition the contact-force and acceleration vectors $\Delta\vec{F}$ and $\Delta\vec{a}$ such that the first rows are filled in with contact points with index in ZA, the next rows are filled in with contact points with index in ZF, and the last row is filled in with $\Delta(F_m)_n$ and $\Delta(a_m)_n$ associated with contact C_m. The partition results in

$$\Delta\vec{F} = \begin{pmatrix} \vec{x} \\ \vec{0} \\ \Delta(F_m)_n \end{pmatrix} \quad \Delta\vec{a} = \begin{pmatrix} \vec{0} \\ \vec{y} \\ \Delta(a_m)_n \end{pmatrix}, \quad (G.25)$$

where $\vec{F}$ and $\vec{a}$ are related by

$$\triangle \vec{a} = \mathbf{A}\triangle \vec{F}. \qquad (\text{G.26})$$

We then made $\triangle(F_m)_n = 1$ and solved a sub-system of equation (G.26) of the form

$$\mathbf{A}_{11}\vec{x} = -\vec{v}_1$$

for $\vec{x}$, that is, for the force increments $\triangle(F_i)_n$ with $i \in ZA$ (see Section G.2 for details on how this sub-system is constructed from equation (G.26)). We substitute $\vec{x}$ and $\triangle(F_m)_n = 1$ back into equation (G.25) to obtain all components of $\triangle \vec{F}$, and use equation (G.26) again to obtain all components of $\triangle \vec{a}$. Finally, we compute the minimum scalar s to be used in

$$\begin{aligned} \vec{a} &= \vec{a} + s\triangle\vec{a} \\ \vec{F} &= \vec{F} + s\triangle\vec{F} \end{aligned}$$

such that, either the normal conditions are met to contact C_m, or a change in the index sets ZA or ZF is required, in which case we need to loop back and re-partition vectors $\triangle \vec{F}$ and $\triangle \vec{a}$ according to the updated groups and solve the updated system.

Following the same principles used to resolve the normal conditions for the frictionless case, we shall create eight groups to manage the indexes of the contact points. The first two groups are identical to the frictionless case, namely groups ZA_n (zero normal acceleration) and ZF_n (zero normal contact force). These groups are used to assure the normal conditions for static contacts.

The next three groups are ZA_t, $MaxF_t$ and $MinF_t$. They are used to classify the contact points with respect to their static-friction conditions along the tangent-plane direction $\vec{t}$. The ZA_t (zero acceleration) group is used to keep track of the contact points that have zero tangential acceleration along $\vec{t}$, that is

$$ZA_t = \{(1 \le i < m) : (a_i)_t = 0 \text{ and } |(F_i)_t| \le \mu_t(F_i)_n\}, \qquad (\text{G.27})$$

whereas the $MaxF_t$ (maximum friction force) and $MinF_t$ (minimum friction force) groups are used to keep track of the contact points that have non-zero tangential acceleration along $\vec{t}$, that is

$$MaxF_t = \{(1 \le i < m) : (a_i)_t < 0 \text{ and } (F_i)_t = \mu_t(F_i)_n\} \qquad (\text{G.28})$$

$$MinF_t = \{(1 \le i < m) : (a_i)_t > 0 \text{ and } (F_i)_t = -\mu_t(F_i)_n\}.$$

Notice that, by construction, the contact-force components $(F_i)_n$ are always non-negative. The last three groups are ZA_k, $MaxF_k$ and $MinF_k$. They are used to classify the contact points with respect to their static-friction conditions along the tangent-plane direction $\vec{k}$. Their descriptions are analogous to those given to groups ZA_t, $MaxF_t$ and $MinF_t$, that is:

$$\begin{aligned}
ZA_k &= \{(1 \leq i < m) : (a_i)_k = 0 \text{ and } |(F_i)_k| \leq \mu_k (F_i)_n\} \\
MaxF_k &= \{(1 \leq i < m) : (a_i)_k < 0 \text{ and } (F_i)_k = \mu_k (F_i)_n\} \\
MinF_k &= \{(1 \leq i < m) : (a_i)_k > 0 \text{ and } (F_i)_k = -\mu_k (F_i)_n\}\,.
\end{aligned} \quad (G.29)$$

Let's examine how these eight groups are used at the mth iteration to assure the normal and static-friction conditions for all other $(m-1)$ contact points already resolved at iteration $(m-1)$. At any point in the algorithm, the index i of each contact point C_i appears in three out of the eight groups above described. More specifically, we assure the normal conditions at contact C_i with $i \leq m$ by setting:

- $i \in ZA_n$ if $(a_i)_n = 0$ and $(F_i)_n \geq 0$, or
- $i \in ZF_n$ if $(f_i)_n = 0$ and $(a_i)_n \geq 0$.

We assure the static-friction conditions along the tangent-plane direction $\vec{t}$ by setting:

- $i \in ZA_t$ if $(a_i)_t = 0$ and $-\mu_t (F_i)_n \leq (F_i)_t \leq \mu_t (F_i)_n$, or
- $i \in MinF_t$ if $(F_i)_t = -\mu_t (F_i)_n$ and $(a_i)_t > 0$, or
- $i \in MaxF_t$ if $(F_i)_t = \mu_t (F_i)_n$ and $(a_i)_t < 0$.

Finally, we assure the static-friction conditions along the tangent-plane direction $\vec{k}$ by setting:

- $i \in ZA_k$ if $(a_i)_k = 0$ and $-\mu_t(F_i)_n \leq (F_i)_k \leq \mu_t(F_i)_n$, or
- $i \in MinF_k$ if $(F_i)_k = -\mu_t (F_i)_n$ and $(a_i)_k > 0$, or
- $i \in MaxF_k$ if $(F_i)_k = \mu_t (F_i)_n$ and $(a_i)_k < 0$.

The first step at the mth iteration is to assure the normal conditions at C_m. This can be done by partitioning the vectors $\triangle \vec{F}$ and $\triangle \vec{a}$ such that the first rows are filled in with contact points with index in ZA_n, the next rows are filled in with contact points with index in ZF_n, and the last row is filled in with $\triangle \vec{F}_m$ and $\triangle \vec{a}_m$ associated with contact C_m. The partition results in

G.3 Baraff's Algorithm: Coping with Friction

$$\Delta \vec{F} = \begin{pmatrix} \vec{x} \\ \vec{0} \\ \Delta(F_m)_n \\ \Delta(F_m)_t \\ \Delta(F_m)_k \end{pmatrix} \quad \Delta \vec{a} = \begin{pmatrix} \vec{0} \\ \vec{y} \\ \Delta(a_m)_n \\ \Delta(a_m)_t \\ \Delta(a_m)_k \end{pmatrix}, \quad \text{(G.30)}$$

where $\vec{F}$ and $\vec{a}$ are related by equation (G.26). Notice that $\vec{x}$ and $\vec{y}$ in equation (G.30) are of the form:

$$\vec{x} = (\Delta(F_1)_n, \Delta(F_1)_t, \Delta(F_1)_k, \ldots, \Delta(F_j)_n, \Delta(F_j)_t, \Delta(F_j)_k)^t$$
$$\vec{y} = (\Delta(a_{j+1})_n, \Delta(a_{j+1})_t, \Delta(a_{j+1})_k, \ldots,$$
$$\Delta(a_{m-1})_n, \Delta(a_{m-1})_t, \Delta(a_{m-1})_k)^t,$$

assuming $ZA_n = \{1, 2, \ldots, j\}$ and $ZF_n = \{(j+1), (j+2), \ldots, (m-1)\}$. Because we are assuring the normal conditions first, we set

$$\Delta(F_i)_t = \Delta(F_i)_k = 0 \; \forall i$$
$$\Delta(a_i)_t = \Delta(a_i)_k = 0 \; \forall i$$
$$\Delta(F_m)_n = 1$$
$$\Delta(F_m)_t = 0$$
$$\Delta(F_m)_k = 0$$

and solve a sub-system of the form of equation (G.30) for $\vec{x}$, that is, for the force increments $\Delta(F_i)_n$ with $i \in ZA_n$. The matrix $\mathbf{A}_{11}$ and vector $\vec{v}_1$ defining the sub-system are constructed exactly as described in Section G.2. Having computed $\vec{x}$, we substitute its value back into $\Delta \vec{F}$ and use equation (G.26) again to obtain all components of $\Delta \vec{a}$. Lastly, we compute the minimum scalar s to be used in

$$\vec{a} = \vec{a} + s \Delta \vec{a}$$
$$\vec{F} = \vec{F} + s \Delta \vec{F}$$

such that, either the normal conditions are met to contact C_m, or a change in the index sets ZA_n or ZF_n is required, in which case we need to loop back and re-partition vectors $\Delta \vec{F}$ and $\Delta \vec{a}$ according to the updated groups and solve the updated system. The scalar s is determined as follows.

- If $i \in ZA_n$ and $\Delta(F_i)_n < 0$, then

$$s \leq -\frac{(F_i)_n}{\Delta(F_i)_n} \quad \text{(G.31)}$$

- If $i \in ZF_n$ and $\triangle(a_i)_n < 0$, then

$$s \leq -\frac{(a_i)_n}{\triangle(a_i)_n} \qquad (G.32)$$

- As long as $(a_m)_n < 0$, we want to force it to zero in order to satisfy the normal conditions at contact C_m. So, if $\triangle(a_m)_n > 0$, then

$$s \leq -\frac{(a_m)_n}{\triangle(a_m)_n} \qquad (G.33)$$

If the minimum scalar s comes from equation (G.33), then the normal conditions are established at contact point C_m. If it comes from equation (G.32), then we need to move its associated index i from ZF_n to ZA_n. Lastly, if the scalar s comes from equation (G.31), then we need to move its associated index i from ZA_n to ZF_n. Notice that, in both cases where the index sets are modified, we need to loop back and re-partition the force and acceleration vectors until equation (G.33) is satisfied.

Having resolved the normal conditions at contact C_m, the next step consists of resolving its static-friction conditions. If the normal contact-force component $(F_m)_n$ obtained after assuring the normal conditions at contact C_m is zero, then the static-friction conditions are satisfied by setting $(F_m)_t = (F_m)_k = 0$, and we are done with the mth iteration. Also, if $m \in ZA_n$, that is if $(a_m)_n = 0$ after the normal conditions are assured, then setting $(F_m)_t = (F_m)_k = 0$ also assures the static friction conditions because

$$\begin{aligned}(F_m)_t = 0 &< \mu_t (F_m)_n \\ (F_m)_k = 0 &< \mu_k (F_m)_n\ .\end{aligned}$$

If none of the above conditions is satisfied, then, whenever $(a_m)_t$ or $(a_m)_k$ are negative, we resolve the static-friction conditions by applying a solution method identical to that used for the normal conditions. The solution method consists of solving equation (G.26) by first partitioning it in a way that the first rows correspond to zero-acceleration increments, and the next rows correspond to zero-force increments. The partition is computed after setting:

$$\begin{aligned}\triangle(F_i)_t &= \mu_t \triangle(F_i)_n, \ \forall i \in MaxF_t \\ \triangle(F_i)_t &= -\mu_t \triangle(F_i)_n, \ \forall i \in MinF_t \\ \triangle(F_i)_k &= \mu_k \triangle(F_i)_n, \ \forall i \in MaxF_k \\ \triangle(F_i)_k &= -\mu_k \triangle(F_i)_n, \ \forall i \in MinF_k \\ \triangle(a_i)_t &= 0, \ \forall i \in ZA_t\end{aligned} \qquad (G.34)$$

$$\triangle(a_i)_k = 0, \forall i \in ZA_k$$
$$\triangle(F_m)_n = 1.$$

The components $\triangle(F_m)_t$ and $\triangle(F_m)_k$ are set according to the following[5]:

$$\triangle(F_m)_t = \mu_t \overbrace{\triangle(F_m)_n}^{1} = \mu_t, \text{ if } (a_m)_t < 0$$

$$\triangle(F_m)_t = -\mu_t \overbrace{\triangle(F_m)_n}^{1} = -\mu_t, \text{ if } (a_m)_t > 0$$

(G.35)

$$\triangle(F_m)_k = \mu_k \overbrace{\triangle(F_m)_n}^{1} = \mu_k, \text{ if } (a_m)_k < 0$$

$$\triangle(F_m)_k = -\mu_k \overbrace{\triangle(F_m)_n}^{1} = -\mu_k, \text{ if } (a_m)_k > 0.$$

After the assignments described by equations (G.34) and (G.35) are completed, we partition the original system such that:

- The first rows are filled with the rows of the original system that have zero acceleration.

- The next rows are filled with the rows of the original system that have zero force.

- The last three rows correspond to the normal and tangential equations defined by contact C_m.

Notice that the original system given by equation (G.26) is laid out such that each contact corresponds to three consecutive rows of the system, one for the normal and two for the tangential contact directions. This may no longer be the case after the above-mentioned partition is carried out. In other words, the system to be solved is partitioned, not with respect to the contact points in the sense that their three corresponding equations are always kept together (i.e., laid out consecutively in the system), but with respect to which rows have zero force or acceleration assigned to them. The partition result is shown in equation (G.30).

Another fact worth mentioning is that, when $\triangle(F_i)_t = \pm\mu_t \triangle(F_i)_t$ and $i \notin ZF_n$ (i.e., $\triangle(F_i)_n \neq 0$), then we need to merge this row with the row corresponding to $\triangle(F_i)_n$. We then do a substitution of variable

$$\triangle(q_i)_n = (1 \pm \mu_t) \triangle(F_i)_n$$

---

[5] Notice that, if $(a_m)_t = 0$, then we the static-friction condition is immediately satisfied by setting $(F_m)_t = 0$. The same applies to $(F_m)_k$ if $(a_m)_k = 0$.

to solve the system for $\triangle(q_i)_n$ and use its value to compute $\triangle(F_i)_n$ and $\triangle(F_i)_t$. The same applies for $\triangle(F_i)_k$.

Having partitioned the system according to equation (G.30), we solve for the sub-system containing the rows associated with zero-acceleration increments. This will give us the complete vector $\triangle\vec{F}$, which can be substituted back again into equation (G.26) to compute $\triangle\vec{a}$. Lastly, we compute the minimum scalar s to be used in

$$\vec{a} = \vec{a} + s\triangle\vec{a}$$
$$\vec{F} = \vec{F} + s\triangle\vec{F}$$

such that, either the static-friction conditions are met for contact C_m, or a change in the index sets is required, in which case we need to loop back and re-partition vectors $\triangle\vec{F}$ and $\triangle\vec{a}$ according to the updated groups and solve the updated system. The minimum scalar $s \geq 0$ is determined as follows

- If $i \in ZA_n$ and $\triangle(F_i)_n < 0$, then

$$s \leq -\frac{(F_i)_n}{\triangle(F_i)_n} \qquad (G.36)$$

- If $i \in ZF_n$ and $\triangle(a_i)_n < 0$, then

$$s \leq -\frac{(a_i)_n}{\triangle(a_i)_n} \qquad (G.37)$$

- If $i \in ZA_t$ and $\triangle(a_i)_t \neq 0$, then we set

$$s = 0 \qquad (G.38)$$

- If $i \in MaxF_t$ and $\triangle(a_i)_t < 0$, then

$$s \leq -\frac{(a_i)_t}{\triangle(a_i)_t} \qquad (G.39)$$

- If $i \in MinF_t$ and $\triangle(a_i)_t > 0$, then

$$s \leq -\frac{(a_i)_t}{\triangle(a_i)_t} \qquad (G.40)$$

- If $i \in ZA_k$ and $\triangle(a_i)_k \neq 0$, then we set

$$s = 0 \qquad (G.41)$$

G.3 Baraff's Algorithm: Coping with Friction

- If $i \in MaxF_k$ and $\triangle(a_i)_k < 0$, then

$$s \leq -\frac{(a_i)_k}{\triangle(a_i)_k} \tag{G.42}$$

- If $i \in MinF_k$ and $\triangle(a_i)_k > 0$, then

$$s \leq -\frac{(a_i)_k}{\triangle(a_i)_k} \tag{G.43}$$

- If $i = m$, then,

  - If $(a_m)_t > 0$ and $\triangle(a_m)_t < 0$, then

  $$s \leq -\frac{(a_m)_t}{\triangle(a_m)_t} \tag{G.44}$$

  - If $(a_m)_t < 0$ and $\triangle(a_m)_t > 0$, then

  $$s \leq -\frac{(a_m)_t}{\triangle(a_m)_t} \tag{G.45}$$

  - If $(a_m)_t = 0$ and $\triangle(F_m)_t \neq 0$, then

  $$s \leq \frac{(\mu_t(F_m)_n - (F_m)_t)}{\triangle(F_m)_t} \tag{G.46}$$

  - If $(a_m)_k > 0$ and $\triangle(a_m)_k < 0$, then

  $$s \leq -\frac{(a_m)_k}{\triangle(a_m)_k} \tag{G.47}$$

  - If $(a_m)_k < 0$ and $\triangle(a_m)_k > 0$, then

  $$s \leq -\frac{(a_m)_k}{\triangle(a_m)_k} \tag{G.48}$$

  - If $(a_m)_k = 0$ and $\triangle(F_m)_k \neq 0$, then

  $$s \leq \frac{(\mu_k(F_m)_n - (F_m)_k)}{\triangle(F_m)_k} \tag{G.49}$$

If the minimum s comes from one of equations (G.44) to (G.49), then the static-friction conditions are met at least in one of the tangent directions. We can carry out a quick test to determine whether they are met on both directions, and if so the mth iteration is completed. However, if the minimum s comes from any other equation, then we need to update the groups accordingly and loop back to re-partition the system and solve it again. There are several cases to be considered when carrying out the update.

360   Appendix G. The Linear-Complementarity Problem

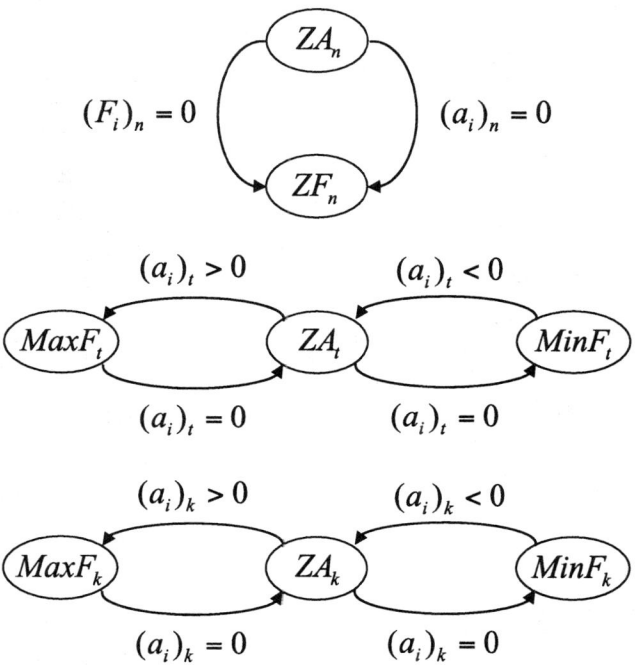

FIGURE G.5. At any instant during the solution of equation (G.26), a contact point C_i can appear in three out of eight possible states. The arrows show the direction of the possible movements, together with the condition that needs to be satisfied to trigger the movement.

- If minimum s comes from equation (G.36), then move its associated index i from ZA_n to ZF_n.

- If minimum s comes from equation (G.37), then move its associated index i from ZF_n to ZA_n.

- If minimum s comes from equation (G.38), then move its associated index i from ZA_t to $MaxF_t$ if $\triangle(a_i)_t < 0$, or $MinF_t$ otherwise.

- If minimum s comes from equation (G.39), then move its associated index i from $MaxF_t$ to ZA_t.

- If minimum s comes from equation (G.40), then move its associated index i from $MinF_t$ to ZA_t.

- If minimum s comes from equation (G.41), then move its associated index i from ZA_k to $MaxF_k$ if $\triangle(a_i)_k < 0$, or $MinF_k$ otherwise.

- If minimum s comes from equation (G.42), then move its associated index i from $MaxF_k$ to ZA_k.

- If minimum s comes from equation (G.43), then move its associated index i from $MinF_k$ to ZA_k.

Figure G.5 illustrates the possible state transitions between group indexes.

G.3.2 Dynamic Friction

Dynamic friction occurs whenever the relative tangential velocity at contact C_i differs from zero. In this case, the magnitude of the tangential contact-force components are always maximum, that is

$$\begin{aligned} (F_i)_t &= \pm\mu_t (F_i)_n \\ (F_i)_k &= \pm\mu_k (F_i)_n \,. \end{aligned} \tag{G.50}$$

The sign in equations (G.50) is chosen such that the contact force is pointing in the opposite direction of the tangential velocity, that is

$$\begin{aligned} \text{sign}\,((F_i)_t) &= -\,\text{sign}\,((\vec{v}_r)_i \cdot \vec{t}) \\ \text{sign}\,((F_i)_k) &= -\,\text{sign}\,((\vec{v}_r)_i \cdot \vec{k})\,, \end{aligned}$$

where $(\vec{v}_r)_i$ represents the relative tangential velocity at contact point i.

Since the magnitude of the dynamic contact-force components is fixed and linearly proportional to the normal contact-force component, we can merge the three rows associated with contact C_i in equation (G.26) and make the following substitution of variable

$$\triangle(q_i)_n = (1 \pm \mu_t \pm \mu_k) \triangle(F_i)_n ,$$

solve the system for $\triangle(q_i)_n$, and use its value to compute $\triangle(F_i)_n$, $\triangle(F_i)_t$ and $\triangle(F_i)_k$. In other words, we can use the same algorithm proposed in Section G.3.1 for the static-friction case to enforce the dynamic-friction conditions at contact C_m.

At the m-iteration, we make the variable substitution and compute the appropriate $\triangle \vec{F}$ that enforces the normal conditions at all contact points, including C_m. Having done so, the tangential contact-force components are computed using equations (G.50). Lastly, index m is assigned to the following groups.

- If $(a_m)_n = 0$, then add m to ZA_n.
- If $(F_m)_n = 0$, then add m to ZF_n.
- If $(a_m)_t = 0$, then add m to ZA_t.
- If $(a_m)_t \neq 0$, then add m to $MaxF_t$ if sign $(F_m)_t > 0$, or to $MinF_t$ otherwise.
- If $(a_m)_k = 0$, then add m to ZA_k.
- If $(a_m)_k \neq 0$, then add m to $MaxF_k$ if sign $(F_m)_k > 0$, or to $MinF_k$ otherwise.

G.3.3 Termination Conditions

Most of the termination-condition properties discussed for the frictionless case are still valid for the static-friction case, save the one that guarantees the existence of a positive scalar s. It can be shown that, for the frictionless case, the existence of a positive scalar $s > 0$ guarantees that the algorithm will always progress and will eventually terminate. However, in the static-friction case, there can be situations wherein the scalar s is zero.

Consider the case wherein the contact point C_i is initially assigned to ZA_n, and $(F_i)_n$ decreases to zero after some steps $s\triangle(F_i)_n$. In this case, the algorithm temporarily stops and moves index i from ZA_n to ZF_n, and adjusts the corresponding columns of matrix $\mathbf{A}$. The algorithm then proceeds to the next iteration, and we may find that $(a_i)_n$ immediately assumes a negative value. Again, the algorithm temporarily stops and moves index i from ZF_n to ZA_n, and adjusts the corresponding columns of matrix $\mathbf{A}$. Clearly, the algorithm is locked in an infinite loop with index i moving back and forth groups ZA_n and ZF_n.

Steps of size zero, such as the one just described, can be detected during program execution by keeping track of the group the index i just came from in the previous step, if any. If the group the index i is moving to is the same as the group it came from in the previous step, then the contact C_i is oscillating. The way to deal with this is to temporarily give up trying to assure both normal and static conditions at contact C_i. This has the effect of postponing the assurance of such conditions at C_i to a later time during the algorithm's execution. The rationale behind this strategy is that, when the algorithm comes back to contact C_i, more contacts have been assured and the conditions that created the loop may no longer be the current conditions of the problem. Unfortunately, the bottom line is that it can't be proved whether the algorithm will always terminate when static friction is considered.

In the dynamic-friction case, the situation gets even worse because the system matrix **A** is no longer symmetric, and is possibly indefinite. In this context, several properties that were used to guarantee the existence of a solution in the frictionless case no longer apply. Here, it is possible to have a situation in which, either the normal force component $(F_i)_n$, or the tangential force components $(F_i)_t$ and $(F_i)_k$, will indefinitely increase without driving $(a_i)_n$, $(a_i)_t$ or $(a_i)_k$ to zero. Another possible situation is that increasing the contact force components may not necessarily trigger a change in the index set of the groups ZA_n, ZF_n, ZA_t, $MaxF_t$, $MinF_t$, ZAk, $MaxF_k$ or $MinF_k$.

As for implementation, these cases correspond to finding an unbounded scalar[6] $s = \infty$. The way to deal with this is to treat as collisions all contacts C_i that are generating an unbounded scalar s, and apply the multiple-collision techniques presented in Chapters 3 and 4 to resolve them. In summary, we resolve all contact first, and treat the remaining contacts that are generating unbounded scalars s as collisions.

G.4 Suggested Readings

The determination of the existence of a solution for a general LCP problem is known to be NP-hard, whereas finding the solution itself is a NP-complete problem. However, there are some special cases of the coefficient matrix **A** in which the LCP problem becomes convex, that is, the existence and actual computation of a solution can be done in polynomial time, with worst-case exponential time complexity. Such cases are addressed in detail in Cottle et al. [CPS92].

---

[6]Before computing s, we initialize it to ∞, such that we can detect at the end of the iteration whether $s = \infty$.

In the context of frictionless contact-force computations, the matrix $\mathbf{A}$ is symmetric and positive semi-definite (PSD), and the vector $\vec{b}$ lies in the column space of $\mathbf{A}$. Fortunately, this is one of the cases when a polynomial time algorithm can be used, such as Dantzig's algorithm described in Cottle et al. [CD68]. Baraff [Bar94, Bar93] presented a modification of Dantzig's pivot-based method to cope with both static and dynamic friction. In the latter case, the matrix $\mathbf{A}$ is not symmetric and possibly indefinite, and a solution may or may not be found. If a solution is not found, then Baraff's algorithm is capable of detecting such a situation, and indicates the contacts that should be dealt with as collisions in order to resolve the inconsistencies.

Other inconsistencies may also occur whenever vector $\vec{b}$ does not lie in the column space of $\mathbf{A}$. Such situations can happen when the contact geometry represents an unsatisfiable combination of kinematic constraints of the system. A more sophisticated implementation could use this to check for inconsistencies on user-definable configurations where some objects are attached to others by joints. Since in most cases users have the freedom to attach joints to whichever objects they deem appropriate, inconsistencies may arise from an inattentive selection of joint attachments.

Usually, the linear system defined by equation (G.16) can be solved using standard Gaussian elimination techniques readily available from several books on linear algebra and matrix theory, such as Strang [Str91], Golub et al. [GL96] and Horn et al. [HJ91]. However, efficacy can be boosted if we use techniques that take into account the fact that consecutive invocations of the linear system differ by a few rows and columns from the previous invocation. Such techniques would incrementally update the LU decomposition of matrix $\mathbf{A}$ already computed in the previous step, as opposed to computing the decomposition from scratch. This has the effect of reducing the computational cost from $\mathcal{O}(n^3)$ to $\mathcal{O}(n^2)$. Gill et al. [GMSW87] discuss in detail an incremental factorization technique that maintains LU factors for general sparse matrices.

Another commonly used formulation of the contact-force computation states the problem as a quadratic programming problem (QP), that is

$$\min_f(f^t \mathbf{A} f - b^t f) \text{ subjected to } \left\{ \begin{array}{c} \mathbf{A} f \geq b \\ f \geq 0 \end{array} \right\}.$$

Lötstedt [Löt84] originally proposed this approach based on a simplification of the Coulomb friction model used to avoid the non-linear constraint:

$$|(f_i)_F| = \sqrt{(f_i)_{F_x}^2 + (f_i)_{F_y}^2} \leq \mu (f_i)_N .$$

In this case, the LCP problem is transformed into a LQP optimization problem. Gill et al. [GM78] and Lawson et al. [LH74] present some numerically stable techniques that can be used to solve the LQP problem.

Finally, the proof of the termination conditions discussed in Sections G.2.1 and G.3.3 can be found in Baraff [Bar94].

Appendix H
Software Implementation

The implementation of a rigid-body and particle-systems dynamic simulator requires a considerable amount of work to design the system architecture and write the code. A high-level architectural specification of the dynamic-simulation engine discussed in this book would, at a minimum, include the following features.

- Load and Save rigid-body boundary descriptions from common graphics files formats.

- Keep an internal representation of the objects to quickly carry out several commonly used operations, such as:

    - Given a face, retrieve all its edges and vertices.
    - Given an edge, retrieve its vertices and the faces that have the edge in common.
    - Given a vertex, retrieve all edges and faces that contain the vertex.

    Preferably, the internal data structures for vertices, edges and faces should be redundant to carry out the above operations in constant time.

- Decompose all non-convex polyhedra being simulated into a set of convex parts.

- Decompose all polygonal faces of each polyhedra into triangles.

- Compute hierarchical representations for all objects being simulated.
- Compute the cell decomposition of the simulated world.
- Implement a set of numerical integrators, including integrators using implicit and explicit methods.
- Implement the collision-detection framework to process collisions.
- Implement the collision-response framework to repond to collisions. This includes separating contacts from collisons.

The dynamic-simulation engine could also support more advanced features, such as:

- Dealing with the dynamics and interactions of articulated systems, supporting one or more types of joints.
- Providing an iteractive user interface to let users design their own articulated bodies.

For each of the above-mentioned features, there are currently available several commercial and open-source-code[1] software packages. The diversity and robustness of such software packages is constantly improving, as new modules are developed to support new capabilities, or remedy known limitations. In this book, we have decided to support a Web site specially devoted to software developers interested in the hands-on experience of implementing a dynamic-simulation engine for rigid-body and particle systems. The URL of the Web site is:

http://hometown.aol.com/animationengine

There you will find links to other related Web sites, categorized according to the list given at the beginning of this appendix. By so doing, the interested reader can obtain entire simulation systems, or separate modules that can be assembled to form a custom-tailored system, or build a simulation engine from scratch using the techniques covered in detail in this book.

---

[1] Usually free for use on non-commercial applications.

References

[AF67] Marcelo Alonso and Edward J. Finn. *Fundamental University Physics*. Addison-Wesley, 1967.

[Arv90] James Arvo. A simple method for box-sphere intersection testing. *Graphics Gems*, 1:335–339, 1990.

[Bar89] David Baraff. Analytical methods for dynamic simulation of non-penetrating rigid bodies. *Computer Graphics (Proceedings SIGGRAPH)*, 23:223–232, 1989.

[Bar90] David Baraff. Curved surfaces and coherence for non-penetrating rigid body simulations. *Computer Graphics (Proceedings SIGGRAPH)*, 24:19–28, 1990.

[Bar91] David Baraff. Coping with friction for non-penetrating rigid body simulation. *Computer Graphics (Proceedings SIGGRAPH)*, 25:31–40, 1991.

[Bar92] David Baraff. *Dynamic Simulation of Non-Penetrating Rigid Bodies*. PhD thesis, Cornell University, 1992.

[Bar93] David Baraff. Issues in computing contact forces for non-penetrating rigid bodies. *Algorithmica*, 10:292–352, 1993.

[Bar94] David Baraff. Fast contact force computation for non-penetrating rigid bodies. *Computer Graphics (Proceedings SIGGRAPH)*, 28:24–29, 1994.

[Bar96] David Baraff. Linear-time dynamics using lagrange multipliers. *Computer Graphics (Proceedings SIGGRAPH)*, 30:137–146, 1996.

[Bau72] J. Baumgarte. Stabilization of constraints and integrals of motion in dynamical systems. *Computer Methods in Applied Mechanics*, pages 1–36, 1972.

[BB88] R. Barzel and A. H. Barr. A modeling system based on dynamic constraints. *Computer Graphics (Proceedings SIGGRAPH)*, 22:179–188, 1988.

[BD92] Chanderjit L. Bajaj and Tamal K. Dey. Convex decomposition of polyhedra and robustness. *SIAM Journal on Computing*, 21(2):339–364, April 1992.

[BHW94] David Breen, Donald H. House, and Michael J. Wonzny. Predicting the drape of woven cloth using interacting particles. *Computer Graphics (Proceedings SIGGRAPH)*, 28:24–29, 1994.

[BJ77a] Ferdinand P. Beer and E. Russell Johnston. *Vector Mechanics for Engineers: Volume 1 - Statics*. McGraw-Hill, 1977.

[BJ77b] Ferdinand P. Beer and E. Russell Johnston. *Vector Mechanics for Engineers: Volume 2 - Dynamics*. McGraw-Hill, 1977.

[Bra91] Raymond M. Brach, editor. *Mechanical Impact Dynamics: Rigid Body Collisions*. John Wiley & Sons, 1991.

[BW97] David Baraff and Andrew Witkin. Partitioned dynamics. Technical Report CMU-RI-TR-97-33, The Robotics Institute at Carnegie Mellon University, 1997.

[BW98a] David Baraff and Andrew Witkin. Physically based modeling. *SIGGRAPH Course Notes*, 13, 1998.

[BW98b] David Baraff and Andrew Witkin. Physically based modeling. *SIGGRAPH Course Notes*, 13, 1998.

[BW99] David Baraff and Andrew Witkin. Large steps in cloth simulation. *Computer Graphics (Proceedings SIGGRAPH)*, 33:43–54, 1999.

[Cam97] S. Cameron. Enhancing GJK: Computing minimum and penetration distances between convex polyhedra. *Proceedings IEEE International Conference on Robotics and Automation*, pages 3112–3117, 1997.

[CD68]   Richard W. Cottle and George B. Dantzig. Complementary pivot theory of mathematical programming. *Linear Algebra and Its Applications*, 1:103–125, 1968.

[Cha84]   Bernard Chazelle. Convex partitions of polyhedra: A lower bound and worst-case optimal algorithm. *SIAM Journal on Computing*, 13(3):488–507, August 1984.

[CKS98]   Swen Campagna, Leif Kobbelt, and Hans-Peter Seidel. Directed edges: A scalable representation for triangle meshes. *Journal of Graphics Tools*, 3(4):1–11, 1998.

[CPS92]   Richard W. Cottle, Jong-Shi Pang, and Richard E. Stone. *The Linear Complementarity Problem*. Academic Press, 1992.

[Cra89]   John J. Craig. *Introduction to Robotics, Mechanics and Control*. Addison-Wesley, 1989.

[CYND92]   Michel Carignan, Yang Ying, Magnenat T. Nadia, and Thalmann Daniel. Dressing animated synthetic actors with complex deformable clothes. *Computer Graphics (Proceedings SIGGRAPH)*, 26:99–104, 1992.

[dBvKOS97] Mark de Berg, Marc van Kreveld, Mark Overmars, and O. Schwartskopf. *Computational Geometry: Algorithms and Applications*. Springer-Verlag, 1997.

[DER86]   I. S. Duff, A. M. Erisman, and J. K. Reid. *Direct Methods for Sparse Matrices*. Oxford University Press, 1986.

[DJAK87]   Terzopoulos Demetri, Platt John, Barr Alan, and Fleischer Kurt. Elastically deformable models. *Computer Graphics (Proceedings SIGGRAPH)*, 21:205–214, 1987.

[Ede87]   Herbert Edelsbrunner. *Algorithms in Combinatorial Geometry*. Springer-Verlag, 1987.

[Fea83]   R. Featherstone. The calculation of robot dynamics using articulated-body inertias. *International Journal of Robotics Research*, 2:13–30, 1983.

[Fea87]   R. Featherstone, editor. *Robot Dynamics Algorithms*. Kluwer, 1987.

[FM96]   Nick Foster and Dimitris Metaxas. Realistic animation of liquids. *Proceedings Graphics Interface*, pages 204–212, 1996.

[FM97]   Nick Foster and Dimitris Metaxas. Modeling the motion of a hot, turbulent gas. *Computer Graphics (Proceedings SIGGRAPH)*, 31:181–188, 1997.

[FS96]   Daan Frenkel and Berend Smit. *Understanding Molecular Simulation from Algorithms to Applications*. Academic Press, 1996.

[FvDFH96]   James D. Foley, Andries van Dam, Steven K. Feiner, and John F. Hughes. *Computer Graphics Principles and Practice*. Addison-Wesley, 1996.

[GJK88]   Elmer G. Gilbert, Daniel W. Johnson, and S. Sathiya Keerthi. A fast procedure for computing the distance between complex objects in three-dimensional space. *IEEE Journal of Robotics and Automation*, 4(2):193–203, 1988.

[GL96]   Gene H. Golub and Charles F. Van Loan. *Matrix Computations*. The Johns Hopkins University Press, Baltimore, 1996.

[Gla90]   Andrew Glassner. Useful 3D geometry. *Graphics Gems*, 1:297–300, 1990.

[Gla94]   Georg Glaeser. *Fast Algorithms for 3D-Graphics*. Springer-Verlag, 1994.

[GLM96]   S. Gottschalk, M. C. Lin, and D. Manocha. Obbtree: A hierarchical structure for rapid interference detection. *Computer Graphics (Proceedings SIGGRAPH)*, 30:171–180, 1996.

[GM78]   Philip E. Gill and Walter Murray. Numerically stable methods for quadratic programming. *Journal of Mathematical Programming*, 14:349–372, 1978.

[GMSW87]   Philip E. Gill, Walter Murray, Michael A. Saunders, and Margaret H. Wright. Maintaining lu factors of a general sparse matrix. *Linear Algebra and Its Applications*, 88/89:239–270, 1987.

[Gol50]   Herbert Goldstein. *Classical Mechanics*. Addison-Wesley, 1950.

[Got96]   S. Gottschalk. The separating axis test. Technical Report TR-96-24, University of North Carolina, Chapel Hill, 1996.

[GR80]   I. S. Gradshteyn and I. M. Ryzhik. *Table of Integrals, Series and Products*. AP Academic Press, 1980.

[Hah88]   James K. Hahn. Realistic animation of rigid bodies. *Computer Graphics (Proceedings SIGGRAPH)*, pages 299–308, 1988.

[Hec00a]   Chris Hecker. How to simulate a ponytail, part 1. *Game Developer Magazine*, pages 34–42, March 2000.

[Hec00b]   Chris Hecker. How to simulate a ponytail, part 2. *Game Developer Magazine*, pages 42–53, April 2000.

[Hel97]   Martin Held. ERIT: A collection of efficient and reliable intersection tests. *Journal of Graphics Tools*, 2(4):25–44, 1997.

[HFK94]   John C. Hart, George K. Francis, and Louis H. Kauffman. Visualizing quaternion rotation. *ACM Transactions on Graphics*, 13(3):256–276, July 1994.

[HJ91]   Roger A. Horn and Charles R. Johnson. *Matrix Analysis*. Cambridge University Press, 1991.

[HM99]   John F. Hughes and Tomas Möller. Building an orthonormal basis from a unit vector. *Journal of Graphics Tools*, 4(4):33–35, 1999.

[Hub96]   Philip M. Hubbard. Approximating polyhedra with spheres for time-critical collision detection. *ACM Transaction on Graphics*, 15(3):179–210, July 1996.

[Joe86]   Barry Joe. Delaunay triangular meshes in convex polygons. *SIAM Journal of Scientific and Statistic Computing*, 7:514–539, 1986.

[Joe91]   Barry Joe. Delaunay versus max-min solid angle triangulations for three-dimensional mesh generation. *International Journal for Numerical Methods in Engineering*, 31:987–997, 1991.

[Joe94]   Barry Joe. Tetrahedral mesh generation in polyhedral regions based on convex polyhedron decompositions. *International Journal for Numerical Methods in Engineering*, 37:693–713, 1994.

[JS86]   Barry Joe and R. B. Simpson. Triangular meshes for regions of complicated shape. *International Journal for Numerical Methods in Engineering*, 23:751–778, 1986.

[Kei96]   Mark Keil. *Handbook of Computational Geometry*, chapter entitled Polygon Decomposition. Elsevier Science Publishing, 1996.

[Kel86] J. B. Keller. Impact with friction. *Transactions of the ASME Journal of Applied Mechanics*, 53:1–4, 1986.

[KPTB99] Evaggelia-Aggeliki Karabassi, Georgios Papaioannou, Theoharis Theoharis, and Alexander Boehm. Intersection test for collision detection in particle systems. *Journal of Graphics Tools*, 4(1):25–37, 1999.

[KSK97] Katsuaki Kawachi, Hiromasa Suzuki, and Fumihiko Kimura. Simulation of rigid body motion with impulsive friction force. *Proceedings IEEE International Symposium on Assembly and Task Planning*, pages 182–187, 1997.

[LH74] C.L. Lawson and R. J. Hanson. *Solving Least Squares Problems*. Prentice-Hall, 1974.

[LK84] Sheue-Ling Lien and James T. Kajiya. A symbolic method for calculating the integral properties of arbitrary nonconvex plyhedra. *IEEE Computer Graphics and Applications*, 4(10):34–41, October 1984.

[Löt84] Per Lötstedt. Numerical simulation of time-dependent contact friction problems in rigid-body mechanics. *SIAM Journal of Scientific Statistical Computing*, 5(2):370–393, June 1984.

[LR82a] Yong Tsui Lee and Aristides A. G. Requicha. Algorithms for computing the volume and other integral properties of solids: Known methods and open issues. *Communications of the ACM*, 25(9):635–641, September 1982.

[LR82b] Yong Tsui Lee and Aristides A. G. Requicha. Algorithms for computing the volume and other integral properties of solids: A family of algorithms based on representation conversion and cellular approximation. *Communications of the ACM*, 25(9):642–650, September 1982.

[Mir96a] Brian V. Mirtich. Fast and accurate computation of polyhedral mass properties. *Journal of Graphics Tools*, 1(2):31–50, 1996.

[Mir96b] Brian V. Mirtich. *Impulse-based Dynamic Simulation of Rigid Body Systems*. PhD thesis, University of California, Berkeley, 1996.

[Mir97] Brian Mirtich. V-clip: Fast and robust polyhedral collision detection. Technical Report TR-97-05, MERL: A Mitsubishi Electric Research Laboratory, 1997.

[Mir98]   Brian Mirtich. Rigid body contact: Collision detection to force computation. Technical Report TR-98-01, MERL: A Mitsubishi Electric Research Laboratory, 1998.

[Möl97]   Tomas Möller. A fast triangle-triangle intersection test. *Journal of Graphics Tools*, 2(2):25–30, 1997.

[OG97]   C. J. Ong and Elmer G. Gilbert. The gilbert-johnson-keerthi distance algorithm: A fast version for incremental motions. *Proceedings IEEE International Conference on Robotics and Automation*, pages 1183–1189, 1997.

[O'R98]   Joseph O'Rourke. *Computational Geometry in C*. Cambridge University Press, 1998.

[Piq90]   Michael E. Pique. Rotation tools. *Graphics Gems*, 1:465–469, 1990.

[Pro95]   Xavier Provot. Deformation constraints in a mass-spring model to describe rigid cloth behavior. *Proceeedings Graphics Interface*, pages 147–155, 1995.

[Pro97]   Xavier Provot. Collision and self-collision handling in cloth model dedicated to design garments. *Proceedings Graphics Interface*, pages 177–189, 1997.

[PS85]   Franco P. Preparata and Michael Ian Shamos. *Computational Geometry: An Introduction*. Springer-Verlag, 1985.

[PTVF96]   William H. Press, Saul A. Teukolsky, William T. Vetterling, and Brian P. Flannery. *Numerical Recipes in C: The Art of Scientific Computing*. Cambridge University Press, 1996.

[Ree83]   William T. Reeves. Particle systems: A technique for modeling a class of fuzzy objects. *Computer Graphics (Proceedings SIGGRAPH)*, 17:359–376, 1983.

[Rit90]   Jack Ritter. An efficient bounding sphere. *Graphics Gems*, 1:301–303, 1990.

[RJKD92]   G. Rodriguez, A. Jain, and K. Kreutz-Delgado. Spatial operator algebra for multibody system dynamics. *Journal of the Astronautical Sciences*, 40:27–50, 1992.

[Sam89]   H. Samet. *Spatial Data Structures: Quadtree, Octrees and Other Hierarchical Methods*. Addison-Wesley, 1989.

[SF95]   Jos Stam and Eugene Fiume. Depicting fire and other gaseous phenomena using diffusion processes. *Computer Graphics (Proceedings SIGGRAPH)*, 29:129–136, 1995.

[Sha94]   Ahmed A. Shabana. *Computational Dynamics.* John Wiley & Sons, 1994.

[Sha98]   Ahmed A. Shabana. *Dynamics of Multibody Systems.* Cambridge University Press, 1998.

[Sho85]   K. Shoemake. Animating rotation with quaternion curves. *Computer Graphics (Proceedings SIGGRAPH)*, 19:245–254, 1985.

[Ski97]   Steven Skiena. *The Algorithm Design Manual.* Springer-Verlag, 1997.

[Sta99]   Joe Stam. Stable fluids. *Computer Graphics (Proceedings SIGGRAPH)*, 33:121–127, 1999.

[Str91]   Gilbert Strang. *Linear Algebra and its Applications.* Academic Press, 1991.

[SV94]   P. W. Sharp and J. H. Verner. Completely imbedded runge-kutta pairs. *SIAM Journal on Numerical Analysis*, 31:1169–1190, 1994.

[Ter87]   Demetri Terzopoulos. Elastically deformable models. *Computer Graphics (Proceedings SIGGRAPH)*, 21:205–213, 1987.

[TW98]   Grit Thürmer and Charles A. Wüthrich. Computing vertex normals from polygonal facets. *Journal of Graphics Tools*, 3(1):43–46, 1998.

[vdB97]   Gino van den Bergen. Efficient collision detection of complex deformable models using aabb trees. *Journal of Graphics Tools*, 2(4):1–13, 1997.

[vdB99]   Gino van den Bergen. A fast robust gjk implementation for collision detection of convex bodies. *Journal of Graphics Tools*, 4(2):7–25, 1999.

[Wil99]   Mark L. Wilkins. *Computer Simulation of Dynamic Phenomena.* Springer Verlag, 1999.

Index

Articulated rigid-body systems
  dynamic equations, 247
  introduction, 245
  single or multiple external collisons, 258
  single or multiple external contacts, 260
  state vector representation, 247
Axis-Aligned Bounding Boxes (AABB), 20–21

Bisection method, 78, 169
Bounding Spheres, 25–27

Closest point between a line and a line segment, 268
Cloth simulation
  overview, 146
Collision and contact-frame computation, 270
Collision detection
  between convex rigid bodies, 174
  between non-convex rigid bodies, 172
  computing the collision and contact frame, 270
  for articulated rigid bodies, 253
  for particle-particle collisions, 78
  for particle-rigid body collisions, 81
  Gilbert-Johnson-Keerthi (GJK) algorithm, 195
  overview, 11
  using hierarchical representations, 17
  Voronoi Clip algorithm, 174
Collision response
  articulated rigid bodies, 256
  for a particle-rigid body collision, 136
  for a particle-rigid body contact, 137, 238
  for a single particle-particle collision, 107
  for a single particle-particle contact, 120

for a single rigid body-rigid
body collision, 205
for a single rigid body-rigid
body contact, 225
for articulated rigid-body collisions, 258
for articulated rigid-body contacts, 260
for multiple particle-particle
collisions, 115
for multiple particle-particle
contacts, 129
for multiple rigid body-rigid
body collisions, 218
for multiple rigid body-rigid
body contacts, 234
for particle-particle collisions,
106
overview, 12
partitioned-matrix representation, 114, 118, 217, 222
Constrained force fields
applied to particles, 71
Contact force computation
contact with friction, 348
for a particle-rigid body contact, 137, 238
for a single particle-particle
contact, 120
for a single rigid body-rigid
body contact, 225
for articulated rigid-body contacts, 260
for multiple particle-particle
contacts, 129
for multiple rigid body-rigid
body contacts, 234
frictionless contact, 339
Convex decomposition
finding cut planes, 329
introduction, 325
Joe's algorithm, 327
termination conditions, 334
tracing cut faces, 331
Convex hull, 27–31

Covariance matrix, 23, 25
Critical coefficient of friction, 113,
212
Cross-product
matrix-vector representation,
271

Damped springs
connecting particles, 70
connecting rigid bodies, 167
Divergence theorem, 303
Dynamic simulation
overview, 3

Euler method
explicit, 276
implicit, 278

Force fields
constrained, 71
unconstrained, 73

Gilbert-Johnson-Keerthi (GJK) algorithm, 195
Gravitational force
applied to particles, 69
applied to rigid bodies, 165
Green's theorem, 308

Hierarchical representations
Axis-Aligned Bounding Boxes
(AABB), 20
Bounding Spheres (BS), 25
convex hull, 27
detecting collisions between
different types of, 42
different types of, 19
multi-level grid, 37
of the simulated world, 32
Oriented Bounding Boxes (OBB),
22
overview, 17
uniform grid, 33

Impulsive-force computation

for a particle-rigid body collision, 136
for a single particle-particle collision, 107
for articulated rigid-body collisions, 258
for multiple particle-particle collisions, 115
for multiple rigid body-rigid body collisions, 218
partitioned-matrix representation, 114, 118, 217, 222
single rigid body-rigid body collision, 205

Intersection
box-box, 42
box-sphere, 51
box-triangle, 52
cylinder-box, 90
cylinder-cylinder, 87
cylinder-sphere, 94
cylinder-triangle, 97
plane-line segment, 267
point-in-cylinder test, 105
point-in-triangle test, 49
sphere-line segment, 54
sphere-sphere, 45
sphere-triangle, 53
triangle-line segment, 55
triangle-triangle, 46

Linear complementarity, 125, 230
Baraff's algorithm, 348
Dantzig's algorithm, 339
introduction, 337

Mass properties
computing line integrals, 309
Mirtich's algorithm, 300
of rigid bodies, 299
projected surface to line integral reduction, 308
surface to projected surface integral reduction, 304
volume to surface integral reduction, 303

Matrix representation
of a single particle-particle collision, 114
of a single rigid body-rigid body collision, 217
of multiple particle-particle collisions, 118
of multiple rigid body-rigid body collisions, 222

Mean vector, 23, 24
Multilevel-grid decomposition, 37–41

Numerical integration
overview, 10

Ordinary differential equations
Euler method, 276
introduction, 273
Runge-Kutta method, 280
step-doubling technique, 286
using adaptive time steps, 285

Oriented Bounding Boxes (OBB), 22–25

Parallel-axis theorem, 302
Particle systems
collision detection overview, 76
detecting particle-particle collisions, 78
detecting particle-rigid body collisions, 81
dynamic equations, 63
interaction forces, 68
introduction, 61
multiple particle-particle collisions, 115
multiple particle-particle contacts, 129
overview of cloth simulation, 146
particle emitter, 138

particle-particle collision response, 106
particle-rigid body collision, 136
particle-rigid body contact, 137
simulation overview, 65–68
single particle-particle collision, 107
single particle-particle contact, 120
state-vector representation, 64
user-adjustable parameters, 143
user-definable parameters, 140
user-interaction force, 76
using damped springs with, 70
using gravity with, 69
using spatially dependent forces with, 71
using viscous drag with, 69
Projections
point on line, 266
point on plane, 266

Quaternions
addition, 290
advantages over rotation matrices, 295
conjugate, 291
dot product, 290
introduction, 289
inverse, 292
module, 292
multiplication, 290
representing rotations, 293
time derivative, 320
unit quaternions, 292

Rigid-body systems
collision detection, 168, 253
collision response, 256
collison between convex bodies, 174
collison between non-convex bodies, 172

dynamic equations, 156
interaction forces, 165
introduction, 155
mass properties, 299
multiple rigid body-rigid body collisions, 218
multiple rigid body-rigid body contacts, 234
particle-rigid body contact, 238
simulation overview, 161–165
single rigid body-rigid body collision, 205
single rigid body-rigid body contact, 225
state-vector representation, 160
user-interaction force, 167
using damped springs with, 167
using gravity with, 165
using viscous drag with, 166
Rotation-matrix representation, 293
time derivative, 318
Runge-Kutta method
fourth order, 283
second order, 280

Separating axis
computing for AABB, 43
computing for OBB, 44
Separating-axis theorem, 42
Simulation engines
collision detection, 11
collision response, 12
design principles, 7
interfacing with renderer, 7
introduction, 3
numerical integration, 10
software implementation, 16, 365, 366
Software packages, 365
Spherical joints, 246
Step doubling, 286

Time derivative
of a rotation matrix, 318

of a unit quaternion, 320
of a vector attached to a rigid
    body, 313
of contact normal vector of
    particle-particle contact,
    314
of contact normal vector of
    rigid body-rigid body contact, 315
of tangent plane of a collision
    or contact, 317

Unconstrained force fields
    applied to particles, 73
Uniform-grid decomposition, 33–36
User interaction
    with particles, 76
    with rigid bodies, 167

Viscous drag
    applied to particles, 69
    applied to rigid bodies, 166
Voronoi Clip algorithm, 174